Lecture Notes in Networks and Systems

Volume 388

AF870075

The series "Lecture Notes in Networks and Systems" publishes the latest developments in Networks and Systems—quickly, informally and with high quality. Original research reported in proceedings and post-proceedings represents the core of LNNS.

Volumes published in LNNS embrace all aspects and subfields of, as well as new challenges in, Networks and Systems.

The series contains proceedings and edited volumes in systems and networks, spanning the areas of Cyber-Physical Systems, Autonomous Systems, Sensor Networks, Control Systems, Energy Systems, Automotive Systems, Biological Systems, Vehicular Networking and Connected Vehicles, Aerospace Systems, Automation, Manufacturing, Smart Grids, Nonlinear Systems, Power Systems, Robotics, Social Systems, Economic Systems and other. Of particular value to both the contributors and the readership are the short publication timeframe and the world-wide distribution and exposure which enable both a wide and rapid dissemination of research output.

The series covers the theory, applications, and perspectives on the state of the art and future developments relevant to systems and networks, decision making, control, complex processes and related areas, as embedded in the fields of interdisciplinary and applied sciences, engineering, computer science, physics, economics, social, and life sciences, as well as the paradigms and methodologies behind them.

Indexed by SCOPUS, INSPEC, WTI Frankfurt eG, zbMATH, SCImago.

All books published in the series are submitted for consideration in Web of Science.

More information about this series at https://link.springer.com/bookseries/15179

Anna Maria Gil-Lafuente ·
Josefa Boria · Agustín Torres ·
José M. Merigó · Janusz Kacprzyk
Editors

Computational and Decision Methods in Economics and Business

Proceedings of the International Workshop "Innovation, Complexity and Uncertainty in Economics and Business"

Springer

Editors
Anna Maria Gil-Lafuente
Department of Business Administration
University of Barcelona
Barcelona, Spain

Josefa Boria
Department of Business Administration
University of Barcelona
Barcelona, Spain

Agustín Torres
Department of Business Administration
University of Barcelona
Barcelona, Spain

José M. Merigó
School of Information, Systems and Modelling
University of Technology
Sydney, NSW, Australia

Janusz Kacprzyk
Systems Research Institute
Polish Academy of Sciences
Warsaw, Poland

ISSN 2367-3370 ISSN 2367-3389 (electronic)
Lecture Notes in Networks and Systems
ISBN 978-3-030-93786-7 ISBN 978-3-030-93787-4 (eBook)
https://doi.org/10.1007/978-3-030-93787-4

This Springer imprint is published by the registered company Springer Nature Switzerland AG
The registered company address is: Gewerbestrasse 11, 6330 Cham, Switzerland

Preface

The Ibero-American Network for Competitiveness, Innovation and Development (REDCID in Spanish) and the Royal Academy of Economic and Financial Sciences (RACEF) are pleased to present the main actualized results of the International Workshop "Innovation, Complexity and Uncertainty in Economics and Business," held in Barcelona, November 14–15, 2019, during XV international academic meeting of RACEF, through this Book of Proceedings published in the Springer Series "Studies in Fuzziness and Soft Computing."

REDCID is part of the Ibero-American Science and Technology for Development Program (CYTED in Spanish), created by the governments of Ibero-American countries to promote cooperation in science, technology and innovation issues for the harmonious development of Ibero-America, through cooperation initiatives that facilitate obtaining scientific and technological results, transferable to productive systems and social policies. The workshop is interdisciplinary and aims to present the latest scientific research on models and applications in innovation, complexity and treatment of uncertainty in economics and business.

The International Workshop "Innovation, Complexity and Uncertainty in Economics and Business" is a meeting to present the results of recent research on different topics related to innovation, complexity, uncertainty, modeling and simulation, fuzzy logic, decision-making, aggregation operators, business and economic applications, among others. Thus, the title of this book is "Computational and Decision Methods in Economics and Business."

We would like to thank all the contributors, reviewers and the scientific and honorary committees for their kind cooperation with REDCID and participation in this meeting; to Jaime Gil-Aluja as President of RACEF for his support and leadership; and to the whole team of the organizing committee, including Luciano Barcellos de Paula, Tulia Carolina Guzmán, Federico Gonzalez-Santoyo, Montserrat Guillen and Aline Rezende. Finally, we would like to express our gratitude to Springer and in particular to Janusz Kacprzyk (Editor in Chief of the book series “Studies in Fuzziness and Soft Computing”) for his support in the preparation of this book.

November 2020

Anna M. Gil Lafuente
Sefa Boria Reverter
José M. Merigó
Agustín Torres Martínez

Organization

Honorary Committee

Special thanks to all the members of the Honorary Committee for their support in the organization of the International Workshop "Innovation, Complexity and Uncertainty in Economics and Business."

President of the Honorary Committee

Jaime Gil-Aluja
President of the Spanish Royal Academy of Financial and Economic Sciences

Honorary Committee

Mario Aguer Hortal, Spain
Fernando Casado Juan, Spain
Isidre Faine Casas, Spain
Korkmaz Imanov, Azarbaiyán
Jorge Nuñez Jover, Cuba
Jose Antonio Redondo López, Spain

Scientific Committee

Special thanks to all the members of the Scientific Committee for their kind support in the organization of the International Workshop "Innovation, Complexity and Uncertainty in Economics and Business."

President of the Scientific Committee

Finn E. Kydland
Nobel Prize for Economics in 2004

Scientific Committee

Víctor Alfaro, Mexico
Francisco-Javier Arroyo, Spain
Luciano Barcellos de Paula, Peru
Fabio Blanco-Mesa, Colombia
Sefa Boria-Reverter, Spain
Beatriz Flores, Mexico
Anna M. Gil-Lafuente, Spain
Federico González Santoyo, Mexico
Montserrat Guillén, Spain
Janusz Kacprzyk, Poland
Yuriy P. Kondratenko, Ukraine
Ernesto León-Castro, Chile
Vicente Liern, Spain
Jose M. Merigó, Australia
Carolina Nicolás, Chile
Lourdes Souto, Cuba
Agustin Torres-Martínez, Spain
Emili Vizuete-Luciano, Spain

Organizing Committee

Thanks to all the members of the Organizing Committee for their kind support in the organization of the International Workshop "Innovation, Complexity and Uncertainty in Economics and Business."

Organizing committee

Anna M. Gil-Lafuente (Chair), Spain
Luciano Barcellos, Peru
Sefa Boria-Reverter, Spain
Tulia Carolina Guzmán-Pedraza, Spain
Jose M. Merigó, Australia
Agustin Torres-Martínez, Spain

Program Committee

Gerardo Alfaro, Mexico
Víctor Alfaro, Mexico
Lluís Amiguet, Spain
Francisco-Javier Arroyo, Spain
Ezequiel Avilés, Mexico
Hugo Baier-Fuentes, Chile
Fabio Blanco-Mesa, Colombia
Jaime Gil-Lafuente, Spain
Ernesto León-Castro, Chile

Carolina Luis Bassa, Spain
Carolina Nicolás, Chile
Mel Solé Moro, Spain
Lourdes Souto, Cuba
Emilio Vizuete Luciano, Spain

Acknowledgement to Reviewers

Special thanks to all the referees for their advice in the revision process of the papers submitted to the International Workshop "Innovation, Complexity and Uncertainty in Economics and Business."

Gerardo Alfaro, Mexico
Víctor Alfaro, Mexico
Francisco-Javier Arroyo, Spain
Luciano Barcellos, Peru
Sefa Boria-Reverter, Spain
Beatriz Flores, Mexico
Anna M. Gil-Lafuente, Spain
Tulia Carolina Guzmán-Pedraza, Spain
Jose M. Merigó, Australia
Aline Rezende, Portugal
Federico Gonzalez-Santoyo, Mexico
Agustin Torres-Martínez, Spain
Emilio Vizuete Luciano, Spain

The International Workshop "Innovation, Complexity and Uncertainty in Economics and Business", is supported by:

Royal Academy of Economic and Financial Sciences

REDCID
Red Iberoamericana para la Competitividad, Innovación y Desarrollo

PROGRAMA IBEROAMERICANO
CYTED
CIENCIA Y TECNOLOGÍA PARA EL DESARROLLO

GOBIERNO DE ESPAÑA
MINISTERIO DE CIENCIA E INNOVACIÓN

Obra Social "la Caixa"

Contents

Uncertainty Characterizations, Typologies and Related Notions in Times of Change

Cynthia M. Montaudon-Tomas(✉) and Ingrid N. Pinto-López

UPAEP, Universidad, 21 Sur 1103 Barrio de Santiago, Puebla, Mexico
{cynthiamaria.montaudon,ingrid.pinto}@upaep.mx

Abstract. Modern times have been declared as the age of uncertainty. In business and management, as well as other disciplines, the study of uncertainty becomes relevant as it is the inherent state of nature, pervading organizational life in a permanent form. Numerous ways of observing reality are linked to the notion of uncertainty, including the perception of the times we live in, which have been conceptualized in different ways, including VUCA, exponential, liquid, chaotic, turbulent, seismic and explosive times, just to cite a few. These time-constraint conditions have paved the way for different ways of using or interpreting the notion of uncertainty, which is an intrinsically unobservable concept. Through a bricolage process, a wide selection of characterizations and notions related to uncertainty were identified, somehow assembling a *theoretical montage* through which meaning is constructed. The objective is to reduce the anxiety that results from not knowing what we are dealing with when considering uncertainty.

Keywords: Uncertainty · Times · Change · Definitions · Collection · Bricolage

1 Introduction

Uncertainty refers to the presence of ignorance in some form, or alternatively, the absence of certainty. The saying that we are living in an uncertain world has become a cliché. Today the world is seen changing in front of our eyes, and the phenomenon of uncertainty is quite common [1]. Sociologists agree that uncertainty, insecurity, and vulnerability are essential traits of modern societies [2].

Understanding of uncertainty has evolved during the development of modernity. Historically, it was only after the Second World War that an accepted theory of uncertainty began to evolve, and now it seems that it is being applied to all kinds of issues, such as safety, health, income distribution, and return on investments, among others [3]. It has been expressed in the notion of risk, which is embedded in the socioeconomic context and grounded in particular temporal orientations, specifically in notions of determinism and indeterminism [4].

Synonyms for uncertainty abound. They may range from a falling short of certainty to an almost complete lack of conviction or knowledge, especially about an outcome or result [5]. Some of the most relevant synonyms in order of importance are the following: ambiguity, ambivalence, anxiety, concern, confusion, distrust, mistrust, skepticism

A. M. Gil-Lafuente et al. (Eds.): REDCID 2019, LNNS 388, pp. 1–28, 2022.
https://doi.org/10.1007/978-3-030-93787-4_1

suspicion, trouble, uneasiness, unpredictability, worry, bewilderment, conjecture, contingency, dilemma, disquiet, doubtfulness, dubiety, guesswork, hesitancy, hesitation, incertitude, inconclusiveness, indecision, irresolution, misgiving, mystification, oscillation, perplexity, puzzle, puzzlement, qualm, quandary, query reserve, scruple, vagueness, wonder, lack of confidence [6].

The notion can be analyzed from multiple perspectives. There are different types, sources, degrees, and layers of uncertainty [7], and the notion has been studied in a variety of disciplines including psychology, sociology, philosophy, communication, finances, health professions, management, probability, and statistics. Because of its multifaceted characteristics and the amplitude of scope for uncertainty, a study was conducted in order to collect different variants of uncertainty as it is used in diverse areas of knowledge. The article was created through a process of bricolage [8] to create awareness of alternative conceptualizations [9] of uncertainty as perceived in organizations in an ever changing environment. Bricolage is associated with resource availability. It reflects the creative bundling of scarce resources [10] to pursue new opportunities [11].

This is a theoretical research based on an analysis of different forms of uncertainty that have been documented in business, as a form of bricolage. Within the last decade, bricolage, as an approach to qualitative inquiry, has gained popularity in academic circles [12]. Bricolage helps bring in concepts and ideas from different areas make-do with anything at hand to concoct whatever needs to be accomplished, solving problems as they emerge. The "intellectual bricolage" that was performed allows piecing together social practices in a given cultural context to construct meaning [13].

In bricolage, everything can be combined and recombined through purposeful manipulation of existing resources [14] without considering rules, regulations or constraints [15], providing elements for innovating on already existing ideas [16] while enacting alternative practices and routines, and incorporating flexibility and innovative adaptation [17].

In the first section of the study, the reality that prevails is analyzed by describing the different conceptualizations that have been made of modern times. These ways of looking at the present and the future have led to multiple characterizations of uncertainty, which are presented in the second section. Finally, conclusions are drawn, and perspectives for future research are presented.

2 Uncertain Times and Times of Change

Uncertain times are analyzed through different perspectives, different cultures, beliefs, and mindsets. The new era has been deeply marked by the politics of paranoia, which started in the Cold War and has since been exacerbated, resulting in a philosophy of suspicion [18] and a culture of terror. This has led to the disruption of critical power resulting in public delusion, cynicism, fear, insecurity, and despair [19] as if humans were living in a society of risk. These mindsets are, in part, the result of available information and how we deal with such information. The new times are marked by information saturation and super saturation; more information equals more uncertainty because complete information is rare.

Moreover, access to information is faster than ever before, exceeding the capacity of processing it. Every situation has multiple options, and knowing about more possible outcomes increases uncertainty because the more one knows, the more one realizes what is unknown. In this sense, and from the perspective of uncertainty, knowledge can be described as an iceberg since two-thirds of the required knowledge cannot be observed [20], hence the need and value of new technologies [21].

Uncertainty is an inevitable part of growth and development [22]. In permanent evolution, uncertainty presents itself as a series of challenges that need to be continuously addressed. The same social structures cannot withstand the forces of change, and some are disappearing. It is a continuous succession of new beginnings, transformations, and reconstructions [23]. Life has become a series of relatively connected phases in which certainties have become blurred, fragile, and less reliable.

The increased speed of developments makes it practically impossible to determine what the future might hold; every event and circumstance that the world faces has a heavy load of uncertainty [24]. Under these conditions, people want the world to make sense, be predictable, and follow precise rules of cause and effect [25]. Nevertheless, the future is not foreseeable; globalization and technology are sweeping away market structures that have historically defined business and the nature of competition, making it impossible to predict, with any confidence, what the future will bring [26]. The new millennium has brought about the most profoundly uncertain business environment ever faced [27]. Decisions today are probably more complex and challenging than at any time in the past [28]. Businesses are operating amid unprecedented uncertainty as economies are becoming more sophisticated, exacerbating challenges. The future is increasingly uncertain, even with some understanding of existing natural, economic, and social systems [29].

Times are changing in unprecedented ways, and they are becoming more fragile and less trustworthy [23], uncovering a reality in which previous social structures are no longer valid [21]. The age of uncertainty has been declared [30], and so has the era of the considerable uncertainty [25], as a contrast between the certainty of economic and social ideas that prevailed in the XIX century, the beginning of the XXI century, and the way in which problems are faced.

When analyzing uncertainty from a historical perspective, six distinct eras or ages have been identified [32] (Table 1):

The six ages show the evolution of how uncertainty is perceived. How people have dealt with uncertainty in each age is substantially different. At the beginning of times, it was considered just as the will of the gods; it evolved to be considered as ignorance, then statistical and probabilistic explanations came about, only to have other disciplines express the importance of randomness and unpredictability, something that has led to the understanding that uncertainty is a part of life that is comprehensible to some extent.

Understanding uncertainty is critical in the current business environment. The notion has been explained by different theories in response to the times we are living in. In the new millennium, it seems there is an array of different perceptions about living under uncertainty. Apparently, the best-known conceptualization is that of the VUCA times. The acronym refers to times that are volatile, uncertain, complex, and ambiguous [33, 34]; it conveys a description of the current business environment [35]. The VUCA times were initially conceived by the American militia to describe the new world order after the Cold War, becoming a guideline for understanding a changing global world [35].

Table 1. Different ages of uncertainty

Age	Description
Age of belief	Uncertainty was rationalized as the will of the gods
Age of Sciences	Showed that perhaps uncertainty was merely temporary ignorance. Nature was less specific than initially thought
Age of probability	The study of uncertainty became a new branch of mathematics. Probability and statistics dominated
Age of quantum revolution	Physicists undermined the notion of determinism The universe seemed to behave in a truly random way
Age of chaos theory	Mathematicians and scientists realized that even an entirely deterministic system could be unpredictable
Age of understanding uncertainty	The realization that uncertainty comes in many forms, each comprehensible to some extent, numerous tools are available, including modeling

Beyond the Cold War scenario, the US Army faced an environment characterized by diverse problems in different geographies that resulted in the creation of special forces to address specific issues, rather than having a single general army [36] as a response to the changes they were facing. In VUCA environments, disruption is a permanent change: all industries face some form of disruption and the new risks that emerge due to globalization, such as the recession. Volatile, uncertain, ambiguous, and complex times are explained in further detail.

2.1 V: Volatility

Volatility is commonly used to explain financial issues, particularly statistical measures of values that present significant changes over time [36]. It also involves environmental conditions that create an out of balance environment due to an increased speed of change. It is important to note, that volatility is not only influenced by the historical data, but can also be influenced by some macroeconomics variables [37]. Change becomes unpredictable and, in an extreme way, particularly negatively affecting strategies that have already been put in place. Businesses need to cope with the volatility of raw materials, changes in regulated prices, and the shortage of skilled workers in developed markets, so they require new strategies that allow them to adapt [38].

Challenges are unexpected and unstable, and their duration cannot be estimated. However, volatility is not necessarily about difficult situations because some knowledge is often available. The problem of volatility lies in the difficulty of recognizing it because there are no repetitive patterns; everything changes rapidly, and stability cannot be maintained [39]. In highly volatile environments, individuals react uncharacteristically. Conversations are loaded with emotionality, and people tend to act impulsively [36]. Coping with volatility requires preparation, and allocating resources, such as capital, material, or talent. This will require significant expenses, but the investments must be balanced with the risks [40].

2.2 U: Uncertainty

Uncertainty as a lack of certainty and security implies not knowing what or who can be trusted, and making predictions becomes difficult [39]; therefore, limiting clarity to assess situations accurately and identify challenges and opportunities is necessary. In the past, linear regression models allowed predictions to be made; however, in VUCA contexts, it is no longer possible to extrapolate future developments [36] to make forecasts. In business, uncertainty refers to those strategies that have not yet been implemented, including the use of imperfect or incomplete information that might have a negative impact. The main problem is the inability to predict the operating environment. Disruptive changes occur frequently, and the past cannot be used as an adequate predictor [39]. Relevant information is not known or definitive; there may be doubts or a lack of clarity regarding a present situation and its future consequences [41].

Uncertainty seems to create bipolar situations between pessimism and optimism. It is common to think of catastrophic scenarios considering everything that could go wrong or take a diametrically opposite and optimistic stance [36]. Economic forecasts are notoriously difficult to make [42]. In order to deal with uncertainty, it is necessary to invest in information, collect, interpret, and share information, as well as create collaborative networks [40].

2.3 C: Complexity

Complexity is a term that does not imply a complicated situation; it indicates that a different treatment is required for its solution. When something has a problematic or entirely diverse situation, it is said to be complex [22]. Complexity has two fundamental aspects: the complex structure of a system and the complex behavior of the system itself [36]. When a problem arises, even with identical external conditions to a previous problem, it is possible to obtain completely different results due to the combination of variables and factors. Some authors suggest that nothing has changed and that the world has always been complex, but that there was no clear perception to capture this situation due to the previous speed of the changes [36] and the adoption of old habits.

Besides uncertainty in decision making, complexity creates different problems as it often confuses immediate with future decisions. Complexity can cause the decision-maker to focus on the wrong problem or decision [28]. Dealing with complexity requires restructuring, development of specialists, and the construction of the necessary resources. It denotes a diversity of interconnected parts and multiple factors for decision making, interactions between the various actors or agents, co-evolution, the need for adaptation, and the existence of weak signals from the environment [41]. In complex scenarios, it is possible that certain information is available or can be predicted, but the volume or its nature can become an overwhelming process [40], making it difficult to understand the causes and mitigating factors involved in the problem [39].

2.4 A: Ambiguity

Ambiguity stands for the lack of a proven recipe or the series of consecutive steps that could reach the same solution. There are multiple possible outcomes to a problem, and there is no single analytical process that can be used for proper decision making [36]. Ambiguity leads to two primary forms of confusion: latent, which is not immediately apparent but eventually arises, and patent, which manifests itself right from the beginning, including the lack of clear goals and objectives. It is present in numerous decision situations in which several decision-makers are involved, and/or, there is a weak decision-maker [28]. Ambiguity has the most similar meaning to uncertainty, as it refers to possible outcomes that have unclear meaning and interpretation.

In ambiguous environments, there is the possibility of more than one interpretation, and the meaning of an event can be understood in different ways. When ambiguous situations arise, there are no previous ones; various interpretations can be generated because the explanation or interpretation depends on the context [39]. In order to deal with this situation, it is necessary to understand the cause and the effect by generating hypotheses and putting them to the test [40]. A problem that arises frequently is that an ambiguous context can become comfortable because it avoids the need to make changes.

3 Other Conceptualizations of Uncertain Times

Modern times have also been described as exponential [43–45] turbulent [46, 47], liquid [21], explosive [48], chaotic, and prone to seismic change. The influence of continuously-changing technology [49] has united the globe in a way that has not been possible until now, and communication has become instant [28]. It is an era that has been dubbed *trumpian uncertainty* [42], a way to explain a new reality that is destructive, impulsive, and unpredictable [50], one in which ubiquitous, continuous and dynamic changes are all-pervasive in organizations [51].

Influence of these conditions is visible in all aspects of business, including knowledge management and networks [52], strategy and decision making [53], lean enterprises and agility [54], exploration and innovation [55], institutional complexity [56], leadership development [46], leadership resilience [57], and employee involvement and participation [58], among others.

Exponential times are described as the complexity of the permanent [59]. They denote the multiple global forces that affect both the present and the future. It is a change in mindset and processing; the new imperative is to understand the evolution in IT capabilities that are increasingly faster and create multiple effects. The time between innovations seems to be reducing [43]; there is an exponential increase in the power of computers that resulted in the digital revolution and the vast amounts of data available for use [44]. New forms of work and business models arise; however, they can generate inequality, mainly because of their disruptive potential [60].

Technological advances are increasingly rapid, as is their adoption or abandonment. The most important thing is not the speed of the developments, but the unexpected convergent consequences [61], such as the incorporation of robots in all the environments of life, the displacement of some human activities, and the creation of new ways of working. Constant risks rise due to the number of factors that can affect businesses. Just a decade ago, companies took up to 20 years of sustainable growth to reach billionaire valuations. Today digital startups can achieve it even in less than four years [62].

Turbulent times are downturn periods characterized by volatility, uncertainty and change, structural brakes, and shift phases. Changes are fast, and directions of economic change are not very clear [46, 47]. The idea of turbulence comes from air travel, in which even on a clear day, major invisible forces such as the wind can strike with no previous notification. Turbulence can come about as a result of both external and internal factors. Some are cases of force *majeure*, such as natural disasters or acts of war or terrorism, while others are more commonplace, such as activity in the international commodity or financial markets, or changes in the geopolitical landscape [63].

Liquid times reflect the transition from solid modernity to a liquid one; a series of challenges never before encountered are appearing; cultural forms, structures, and institutions can no longer keep their shape; there is a collapse of long term thinking, planning, and actions [21]; there are several negative consequences of globalization, such as the generation of surplus people who have nowhere to go in a world that is full; there are increasingly visible inequalities as the rich and the poor come to live closer together; and, there is a world in which it is increasingly difficult for communities and nations to provide collective security [64].

In explosive times there is a high discontinuity between past and future, and it has even been suggested that the past and the present are dissolved in the creation of the future [48]. Traditions no longer hold. Decisions need to be taken urgently, as everything might disappear in a blast. These times show the disruptive events that are remaking the world.

Chaotic times are based on decision-making problems, corporate profitability, loss, economic situations, and others. These times are made of multiple and simultaneous frameworks that demand instantaneity. Many complex problems are occurring at once, which, by definition, cannot be fully understood. Solutions cannot be reduced to a formula or easy answers [50].

Seismic times have been used to describe the financial world's problems, particularly the worldwide financial crunch of 2008. There are seismic changes in the sense of distrust for everything that happens around, particularly scandals [65]. Tectonic market shifts are transforming the global business landscape. Economic realignment, advances in technology, the globalization of markets, changing demographic trends, new customer needs, and increased competition are radically altering how companies operate in virtually every industry and region of the world [66].

4 Defining Uncertainty from a Business Perspective

Uncertainty has been a common buzzword in strategic planning and decision making and has been used in numerous scientific fields, often encompassing a multiplicity of related concepts, although most uncertainties deal with mathematical treatments. It has even been suggested that different financial crisis have been responsible for an unprecedented surge in the use of the term 'uncertainty' among academics and the general public, although there is evidence that it was brought forward very early by Knight and Keynes in the 1920s and 1930s, respectively [67]. Uncertainty acknowledges the existence of an underlying truth: it is not possible to be uncertain of nothing; nothing is the complete absence of anything [68].

Uncertainty may exist due to a lack of information, the complexity of the information, questions about the quality of information, inability to structure information in a meaningful way, or the inability to associate information with other information [69]. "The degree of uncertainty represents the availability and reliability of the information that is relevant in a given situation" [70]. In terms of degrees, uncertainty can range from very low to very high, and it is linked to the amount of relevant knowledge that has been acquired [71]. There can be a complete lack of knowledge: facts are not known or are known only imprecisely, also lack of knowledge of available facts, or the state of limited knowledge.

Nevertheless, uncertainty is not only the absence of knowledge, but it can also prevail in situations where there is a lot of information available, and new information can either increase or decrease uncertainty [29]. Uncertainties are things that are not known [28] or known only imprecisely. Uncertainties are factual and can be measured as things can be known, not known, or known to a quantifiable degree within quantifiable bounds [72]. In this sense, uncertainty is a situation of inadequate information that can be of three sorts: inexactness, unreliability, and bordering with ignorance [73], depending on the limited knowledge about future, past or current events [29].

In decision-making, uncertainty is relevant [25] as it refers to the gap between available knowledge and the knowledge needed to make the best choice. Information exchange is increasing the uncertainty organizations face when making decisions [28]. Problems emerge form not being able to see the full picture and having missing information, something that results in an impossibility to make judgments and decisions about the appropriate course of action and its outcome [74].

The fact that economic phenomena are characterized by their constant mutability does not allow, in most cases, to consider data from the past to establish a sufficiently valid prevision of the future [24]. Traditionally, people deal with uncertainty by ignoring it, worrying about it, or trying to hedge against it depending upon the significance and the consequences [28], in part due to their inability to explain or predict [75]. In business, uncertainty refers to management's difficulty in predicting environmental or organizational variables that may have an impact on a company, particularly strategic choices [76].

Uncertainty is a problem that complicates decision making, and it needs to be handled. There are numerous things that cannot be controlled, including the fate of the economy, political events, disasters, and the actions of others [77]. Most people tend to ignore uncertainty rather than to deal with it. Underestimating the range of uncertainty in a problem can have catastrophic consequences [28] since it surrounds the very future of global economic order. Business leaders must overcome the paralysis that dooms organizations when dealing with the unknown because no matter what is being done, uncertainty will remain [27].

Uncertainty has also been analyzed from a cultural standpoint, precisely uncertainty avoidance, expressed as the extent to which a society or group relies on social norms, rules, and procedures to alleviate unpredictability or unforeseen events [78]. Uncertainty avoidance affects the implementation of deadlines, organizational structures, and stress levels, among others [79]. It is correlated to country conditions and individual and collective behavior [80].

Uncertainty may have different types of sources, from quantifiable errors in the data to ambiguously defined terminology or uncertain projections of human behavior. Uncertainty measurements can, therefore, be represented by quantitative metrics (e.g., a range of values calculated by various models) or by qualitative statements (e.g., reflecting the judgment of a team of experts) [81].

Uncertainty is at the heart of science and decision making [20]. Probability is the language of uncertainty [68], and uncertainty is the soul of modeling, probability, and statistics. Uncertainty can be associated with all aspects of a problem of interest, such as the system comprising the decision domain, the world outside the system, the outcomes of the system, and the importance of the stakeholders' place on the various outcomes from the system [82].

An unknown cannot be known until the event happens, nearly all important decisions (business or personal) are made under conditions of uncertainty; the difficulty is in understanding the consequences or outcomes of the different causes of actions [28]. An essential aspect of analyzing is that uncertainties are neutral. Although it is considered as unfavorable, uncertainty is not necessarily bad [72].

Change and uncertainty are the basic principles of the universe. There are multiple perspectives from which to analyze them: technical, organizational, and personal. The technical perspective analyzes uncertainty as a problem that complicates decision-making and needs to be handled; the organizational perspective analyses uncertainty as something diffuse and non-abstract, something that arises in contexts where the problems are forced upon people, not chosen. There is a focus on processes, including planning and decision making; uncertainty is viewed as a kind of information that is needed recurrently but is not adequately available. Finally, from a personal perspective, the experience of uncertainty includes strong emotions such as excitement and fear, which can be associated with stress and complications due to variations between individuals [83].

The study of uncertainty has become relevant as it helps understand what might be expected and develop actions that would eventually mitigate its previewed consequences [84]. Traditionally, uncertainty has been analyzed from a binary perspective, in which the world is either certain and, therefore, open to predictions, or uncertain, and therefore completely unpredictable [85]. Underestimating uncertainty can lead to strategies that neither defend against threats nor take advantage of the opportunities at hand [85].

Decision making is linked to uncertainty. Decision-making skills under uncertainty originate as a result of previous uncertainties, which arrive at a resolution point, or as anticipated actions to uncertainty that are either present or that are foreseen [84].

There are different levels of uncertainty, as there does not seem to be a generally accepted theory. The level of uncertainty refers to the degree to which the object of study is uncertain from the decision maker's point of view. Some authors consider more or fewer levels.

Three levels of uncertainty:

- Clear future. Forecasts with small error margins and uncertainty become a determining variable in decision making.

- People in decision positions in businesses have limited experience and pattern recognition abilities; they acknowledge the issue of uncertainty without really dealing with its implications. Uncertainty is frequently ignored because of the incapability of humans to handle multiple realities. Uncertainty is sometimes treated as an afterthought as potential risk issues are not clearly addressed. In this sense, people just pretend to deal with uncertainty [25].
- Alternate scenarios: It commonly presents a few probable futures that are mutually exclusive, and if they become real or not, they will completely transform strategy [86].

Three sorts of uncertainty have been identified [73].

- Inexactness, i.e., a technical level of uncertainty involving random and systematic errors in empirical quantities.
- Unreliability is based on methodological uncertainties arising, including incomplete understanding and the approximations made when describing the structural and functional characteristics of a system under study.
- Border with ignorance or the epistemological level of uncertainty, including omissions of processes and parameters due to ignorance (Table 2).

Table 2. Four levels of uncertainty [85]

L	DG	What can be known	Analytical tools
1	A clear enough future	Single forecasts are enough for strategy development	Traditional strategy toolkit
2	Alternate futures	Few discrete outcomes are required. Scenarios. It cannot identify which outcomes will occur but may help establish probabilities	Decision analysis Option valuation models Game theory
3	A range of futures	Range of possible outcomes but not natural scenarios, a limited number of crucial variables. The actual outcome may lie somewhere along the continuum grounded by the range	Demand research Technology forecasting Scenario planning
4	True ambiguity	Multiple dimensions of uncertainty interact to create an environment that is virtually impossible to predict. There is no basis for forecasting the future	Analogies Pattern recognition

L: Level, DG: Degree of uncertainty

The table two presents four levels of uncertainty based on the degree in which it occurs and the essential tools that can be used to analyze uncertainty, which range from traditional strategy tools to pattern recognition, which nowadays can be performed using algorithms.

Five levels have also been proposed [87] (Table 3).

Table 3. Five levels of uncertainty

Level	Description
Level 1 (recognized uncertainty)	There is no absolute certainty without being able or willing to measure the uncertainty explicitly
Level 2 (shallow uncertainty)	There are multiple alternatives, providing probabilities and multiple possible futures or alternative model structures
Level 3 (medium uncertainty)	There are multiple possibilities, being able to rank order the possibilities in terms of perceived likelihood and multiple possible futures or alternative model structures that can be judged in terms of perceived likelihood
Level 4 (deep uncertainty)	There are multiple possibilities without being able to rank order the possibilities in terms of how likely or plausible they are judged to be, and multiple possible futures or multiple alternative model structures, without being able to specify their likelihood
Level 5 (recognized ignorance)	It is not possible to enumerate multiple possibilities while admitting the possibility of being surprised keeping open the possibility of being

Uncertainty has also been described as a three-dimensional concept. The dimensions involved are the location of uncertainty, which refers to where the uncertainty manifests itself; the level or degree of uncertainty, explaining where the uncertainty manifests along the spectrum between deterministic knowledge and total ignorance, and the nature of uncertainty, which explains whether the uncertainty is due to the imperfection of our knowledge or the inherent variability of the phenomenon being described [29].

5 Uncertainty Characterizations, Typologies, and Related Notions

The environment of ever-growing change and the responses it brings with it have resulted in the consideration of uncertainty from multiple perspectives. Nearly 100 characterizations, typologies, and related notions were identified in literature coming from multiple disciplines. The notions are presented in alphabetical order, and a short definition is included (Table 4).

Table 4. Uncertainty characterizations, typologies, and related notions

Notion	Definition
Aleatory uncertainty	A type of irreducible uncertainty important to natural resources decision making. Represents natural unpredictable variation [88], also known as stochastic uncertainty [89]
Approximation uncertainty	Refers to limits to precision due to unavoidable approximations in the measuring process [89, 90]
Authority uncertainty	Uncertainty about who will select the alternatives, approve or disapprove the selection, evaluate options, competition for authority that affects decision making, and formal and informal authority in the process [82, 83]
Behavioral uncertainty	Reflects ambiguity about how an individual should act and how a partner will act. Behavioral uncertainty is evident in questions about the norms for appropriate behavior in a relationship [91]
Blind uncertainty	Epistemic uncertainty for which it is not recognized that the knowledge is incomplete and that the knowledge is relevant. It can be caused by inadequate communication [92]
Collaborative uncertainty	Addresses the uncertainty that arises due to (a lack of knowledge about) the individuals involved and their different expectations, preferences, abilities, and actions [93]
Combined uncertainty	The result of the combination of two or more independent sources of uncertainty [94]
Contingency/agency uncertainty	Uncertainty which arises from human intentionality, whereby the right policy decision made in a particular study contributes to uncertainty itself [95]
Cyclical uncertainty	Uncertainty of, or, relating to occurring in an interval of time during which a sequence of a recurring succession of events or phenomena is completed [97]
Data uncertainty	Not being sure where, within a range, data falls, and instead of having value and confidence in that value, there is a range of possible values [98]
Decision uncertainty	Presence of ambiguity or controversy about how to quantify or compare social objectives [99]

(*continued*)

Table 4. (*continued*)

Notion	Definition
Deep uncertainty	Experts do not know other parties to a decision and cannot agree upon the system's external context or its relative importance [100]. It arises from actions taken over time in response to unpredictable evolving situations [82, 101]
Economics of uncertainty	Each person adapts to his/her given state of limited information by choosing the best terminal action available. The individual is pressured to act based on current field beliefs [3]
Effect uncertainty	It is difficult to assess the possible impact of these changes on a particular company, decision, or situation [102]
Endemic uncertainty	The uncertainty that is native or restricted to a particular place [21]
Endogenous uncertainty	Doubt surrounding a company's financial situation could be resolved if the company announced the uncertainty [103]
Environmental uncertainty	Reflects the sense of doubt experienced by managers when facing the problem of predicting future competitive conditions [102]; describes the inability to predict how changes in the environment will influence something [104]
Epistemic uncertainty	Related to lack of knowledge or precision of a model parameter, also considered knowable uncertainty [105]
Epistemological uncertainty	Knowledge guided decisions; concerns the conception of a phenomenon [105, 106]
Essential uncertainties	Those in which no degree of observation or information processing abilities would reduce the alternatives to reach a specific result before the occurrence of the event [83, 84]
Event uncertainty	Each individual would be certain only about exogenous concurrencies as they affect his/her circumstances of the market as a whole (exogenous variables) [3]
Exogenous uncertainty	Anything outside a company's control. Similar to external uncertainty [107]

(*continued*)

Table 4. (*continued*)

Notion	Definition
Expanded uncertainty	Uncertainty calculations that have a higher confidence level. Since most uncertainty formulas are based on standard deviation, an underlying uncertainty value would represent the potential range of error plus or minus one standard deviation [108]
Experimental uncertainty	Also called error analysis. It is the study and evaluation of uncertainty in measurement [109]
Explicit uncertainty	Uncertainties that are verbalized [110]
External uncertainty	Uncertainty that is attributed to the dispositions of causal systems in the outside world [111]
False uncertainty	Pretending not to be sure when, in fact, the decision-maker is [112]
Fuzzy uncertainty	Deals with situations where set boundaries are not sharply defined [113] Part of a more general field known as soft computing [114]
Grey uncertainty	Deals with problems of uncertainty with few data points and inadequate information, which have incomplete information known, partial information unknown [115]
Individual uncertainty	Concerns the individual experience of task, collaborative, and process uncertainty and the reaction to, or 'management' of, that experience [93]
Inherited uncertainty	Uncertainty that has been received from previous processes, situations, operations, data [116]
Internal uncertainty	Uncertainty attributed to a personal state of knowledge [111]
Institutional uncertainty	Refers to the roles and actions of institutions and their members. High institutional uncertainty can hinder collaboration or understate it, making predictions difficult [72]
Irreducible uncertainty	Uncertainty in the physical world that no amount of measurement or knowledge can eliminate [117]; generated by causal processes underlying ecologically significant events that are fundamentally untraceable at the spatial and temporal scales at which an organism operates [118]; events which remain unpredictable whatever the amount of data available [88, 119]

(*continued*)

Table 4. (*continued*)

Notion	Definition
High order uncertainty	Measures the uncertainty that people have about what others believe, which usually arises when forecasts differ [120]
Knightian uncertainty	Based on the works of Frank Knight from the University of Chicago. It refers to a lack of quantifiable knowledge about some possible occurrence, as opposed to the presence of quantifiable risk (The concept acknowledges some fundamental degree of ignorance, a limit to knowledge, and an essential unpredictability of future events, what Knight called 'true uncertainty' and is now known as Knightian uncertainty) [121]
Language uncertainty	Uncertainty that emerges due to imprecision of language [88]
Legal uncertainty	It is relevant when future contingencies need to be assessed. High legal uncertainty may result in defensive responses [122]
Likelihood uncertainty	Knowing the set of possible states of the world but not knowing the probability of their occurrence [121, 123]
Liquid uncertainty	A flowing state of continuous uncertainty which is unformed and unstable in nature [21]
Macro uncertainty	Measures of uncertainty about macroeconomic outcomes [120]
Market uncertainty	Being unsure about an offering of other economic agents (endogenous variables) [3]
Measurable uncertainty	Has been called risk [124]
Measurement uncertainty	It is the expression of the statistical dispersion of the values attributed to a measured quantity [125]
Methodological uncertainty	Also called error analysis. It is the study and evaluation of uncertainty in measurement [126]
Micro Uncertainty	Uncertainty that can be measured at industry, firm, plant, or even individual product level [127]
Model uncertainty	Lack of confidence that the mathematical model is a "correct" formulation of the problem [128]
Moral uncertainty	Not being certain about what should be done morally [129, 130]. A person is fundamentally morally uncertain when his/her uncertainty directly concerns whether the act's properties are wrong-making [131]. It will be high when moral and ethical concerns become critical [72]

(*continued*)

Table 4. (*continued*)

Notion	Definition
Objective uncertainty	Objective uncertainty also emerges as a result of individual perception, and, consequently, involves an element of subjectivity [102, 132]
Ontological uncertainty	Quasi rational decision [133, 134]
Option uncertainty	Feasibility in the options. Not knowing additional options and/or the outcomes of additional outcomes [135]
Positive uncertainty	It helps deal with change and ambiguity, accepting uncertainty and inconsistency, and utilizes the non-rational and intuitive side of thinking and choosing. It is a strategy that promotes positive attitudes and paradoxical methods in the presence of increasing uncertainty [20]
Pragmatic uncertainty	It is derived from insufficient and inadequate means to deal with the complexity of the problems at hand [136]
Process Uncertainty	Arises at different times during the development process, namely in the beginning, middle, and when deliverables have to be handed in along the process or at the end [93]
Parameter uncertainty	Insufficient scientific knowledge to determine the parameter accurately. In general, it is reducible with sufficient investment [137]
Quality uncertainty	Uncertainty about product quality due to information asymmetry [138]
Radical uncertainty	The kind of uncertainty that statistical analysis cannot deal with. And the outcome of standard market activity [66]
Recognized uncertainty	When even the likely consequences are not clear; there is a suspected hazard, but an unknown outcome [139]
Reducible uncertainty	Uncertainty about the external world and can be reduced through exploration and learning [88, 117]
Relative uncertainty	It is the ratio of the absolute uncertainty of a measurement to the best estimate. It expresses the relative size or precision of the uncertainty of a measurement [140]

(*continued*)

Table 4. (*continued*)

Notion	Definition
Response uncertainty	Difficulty knowing which type of response may prove successful [102]. Response uncertainty is defined as a lack of knowledge of response options and/or an inability to predict the likely consequences of a response choice [104]
Rule uncertainty	It is usually used to refer to any sort of uncertainty, from the standpoint of the unfavorable contingency [141]
Scenario uncertainty	Where the likely consequences are known, but their probabilities cannot be quantified. Hazards and outcomes are known, but probabilities are unknown. It is used to analyze uncertainty related to the external environment [29, 82]
Scientific uncertainty	Rises from the scientific and technical dimensions of a problem and is intrinsic to the processes of risk assessment and forecasting [142]
Shallow uncertainty	The uncertainty resulting from undefined terms, careless formulation, lack of operational definitions, and overall intellectual sloth [143]
Situational uncertainty	It refers to individual behaviors or personal interventions in crisis situations [122]. High situational uncertainty would be characterized by situations where individual decisions play a substantial role, and there is uncertainty about the nature of those decisions
Societal uncertainty	Arises when different communities have diverse approaches to decision-making and assessment. Societal uncertainty would typically be high when decisions involve strong collaboration among groups characterized by divergent decision-making styles [122]
State uncertainty	Refers to the difficulty in knowing in which direction the environment may change [102]. "Perception by an individual that a particular component of the environment is unpredictable, more specifically, that one does not understand how the components of the environment are changing" [104]
Statistical uncertainty	Where likely consequences are known, and their probabilities can be quantified. Hazard, outcomes, and probabilities are known [144]
Stochastic uncertainty	Aleatory uncertainty related to the real variability of a population or a physical property cannot be reduced. Includes temporary and spatial variability [145]

(*continued*)

Table 4. (*continued*)

Notion	Definition
Structural uncertainty	Not knowing how a system would respond to external developments [82]
Subjective uncertainty	The confidence observers report [105, 145, 146]
Systematic uncertainty	Measurement uncertainty is due to uncertainties on external input or uncertainties not related to the statistics of data [147]
Task uncertainty	Addresses the uncertainty that arises due to a lack of understanding of or doubts about a task ahead. It is not only concerned with a lack of knowledge or overview over what, how, and when; it also has to do with doubts about one's abilities and the time frame, as well as the quality of the result [93]
Technical uncertainty	Arises from the equality or appropriateness of the data as well as from the lack of data and approximation [106]
Temporal uncertainty	The well-known openness of the future [148]. Uncertainty that impairs coordination by weakening the temporal structure of an organization [149]
Theoretical uncertainty	It occurs when there is no theoretical framework that unifies a scientific field or a specific topic or problem [136]
Total uncertainty	The sum of fuzzy and stochastic uncertainty [150]
True uncertainty	The belief that one does not know and guesses truly not knowing [83]
Ubiquitous uncertainty	Omnipresent uncertainty. The type of uncertainty that exist in different places at the same time and is encountered continuously [7, 151]
Uncertaintism	The view that this line of uncertainty is correct [152]
Uncertainty analysis	Used to identify and analyze the uncertainty of variables that are used in decision-making problems in which observations and models represent the knowledge base, aiming to make a technical contribution to decision-making through the quantification of uncertainties in the relevant variables [119, 153]
Uncertainty avoidance	The tolerance for uncertainty. It reflects the extent to which members of a society feel threatened by uncertainty and/or prefer having clear rules [78, 154]

(*continued*)

Table 4. (*continued*)

Notion	Definition
Uncertainty conditions	Conditions in which the future environment is unpredictable and everything is in a state of flux. Very little information is available, and decision-makers are unsure about the reliability of such information [155]
Uncertainty cone/cone of uncertainty	A graphic depiction of the increasing accuracy that is possible for estimates as the details of a project become more known over time [154, 156]
Uncertainty interval	It is used in place of the confidence interval. The difference between the two is philosophical rather than mathematical [157]
Uncertainty machine	A Web-based software application to evaluate the measurement uncertainty associated with an output quantity defined by a measurement model of the form: y = f(x0, …, xn), by the National Institute of Standards and Technology in the US [158]
Uncertainty management	Focuses on uncertainty and information seeking, and on how people respond to uncertainty. It includes considering uncertainty as an aversive state and providing conditions under which uncertainty might be a desired state [159]
Uncertainty map	A graphical representation developed by Pearson & BlockHoff [160], which plots process uncertainty versus output uncertainty according to high and low levels. It is divided into four distinct quadrants: exploratory research, development engineering, applications engineering, and combining market opportunities with technical capabilities
Uncertainty matrix	Matrix developed by van der Sluijs et al.; Ekström et al. Kwakkel et al. and Janssen et al. [95, 96] that illustrates the location, level, and nature of the uncertainty associated with models and provides a systematic and graphic overview of the range of essential uncertainties. It represents the most utilized typology relating to the modeling process found in the selected literature
Uncertainty node	A node in a belief or decision network that represents an uncertain state of outcome [88]

(*continued*)

Table 4. (*continued*)

Notion	Definition
Uncertainty principle	Heisenberg's uncertainty principle states that the more precisely a particle's position is measured, the less precisely its momentum can be known, and vice versa. It has long been invoked to describe the way measuring an object disturbs that object. Historically, it has been confused with the observer effect [161]
Uncertainty quantification	The science of quantitative characterization and reduction of uncertainties in both computational and real-world applications. It tries to determine how likely specific outcomes are if some aspects of the system are not exactly known [29]
Uncertainty shocks	Countercyclical uncertainty [120]. It is used to explain business cycle fluctuations [162]
Uncertainty tree	A graphical, hierarchical structure to assess the flow of uncertainty in experimental measurements [163]

6 Conclusions

Uncertainty means working with imperfect or incomplete information. It has become one of the most frequently heard words in politics and businesses. It is, in part, the result of different ways of conceptualizing and analyzing the times we are living in. Uncertainty is a crucial part of business decisions, and its study has become essential for leaders and managers, primarily due to increased complexity in the business ecosystem. In the new era of business, it becomes even more relevant as machine learning becomes central because applied machine learning requires managing uncertainty.

Different perspectives for analyzing uncertainty have been presented, describing a variety of levels, degrees, stages, and even ages of uncertainty. Although there is no universal agreement on how uncertainty can be defined, general elements and considerations about this notion have been identified.

In the new millennium uncertainty has become one of the most frequently used words for the future. One hundred notions related to the idea of uncertainty were collected to provide a general framework of what uncertainty means and how it is analyzed and measured in different contexts. The list is far from complete; many other terms and associated notions are still to be identified. It is relevant to state that there has been limited analysis that could allow classifying the notions identified in a better way. The collection of terms and notions associated with uncertainty was done to help describe uncertainty when analyzed from different perspectives.

As a final remark, there is no doubt that we live in a time of uncertainty –because that has always been the case [164]. Uncertainty is nothing but a fact of life.

7 Future Research Directions

As mentioned above, the list of uncertainty characterizations, typologies, and related notions is not complete. Additional areas of knowledge need to be explored, and additional research on classical theories on uncertainty can yield additional characterizations. A theoretical analysis contrasting different perspectives of uncertainty might also be relevant. Once the list of notion is reviewed and completed, it will be possible to develop additional studies by modeling the notions and weighting them in a way in which quantitative analysis can be performed, and even fuzzy logic can be applied to help identify specific patterns about the way how uncertainty is understood in various areas of business.

References

1. Lindley, D.V.: Understanding Uncertainty. Wiley, New York (2013)
2. Paniker, S.: Asimetrías: apuntes para sobrevivir en la era de la incertidumbre. Debate Editorial, Barcelona (2008)
3. Hirshleifer, J., Riley, J.G.: The Analytics of Uncertainty and Information. Cambridge University Press, New York (1992)
4. Reith, G.: Uncertain times: the notion of 'risk' and the development of modernity. Time Soc. **13**(2–3), 383–402 (2012)
5. Merriam-Webster Thesaurus: Uncertainty, synonyms (2019). https://www.merriam-webster.com/thesaurus/uncertainty
6. Thesaurus.com: Uncertainty; synonyms and antonyms (2019). https://www.thesaurus.com/browse/uncertainty
7. Schedler, A.: The Politics of Uncertainty: Sustaining and Subverting Electoral Authoritarianism. OUP, Oxford (2013)
8. Boxenbaum, E., Rouleau, L.: New knowledge products as bricolage: metaphors and scripts in organizational theory. Acad. Manag. Rev. **36**(2), 272–296 (2011)
9. Alvesson, M., Sandberg, J.: Generating research questions through problematization. Acad. Manag. Rev. **36**(2), 247–271 (2011)
10. Halme, M., Lindeman, S., Linna, P.: Innovation for inclusive business: Intrapreneurial bricolage in multinational corporations. J. Manage. Stud. **49**(4), 743–784 (2012)
11. Chinyoka, S.V.: Bricolage: a state of entrepreneurial excellence. Asian Bus. Res. **2**(3), 46 (2017)
12. Rogers, M.: Contextualizing theories and practices of bricolage research. Qual. Rep. **17**(48), 1–17 (2012)
13. Levi-Strauss, C.: The Savage Mind. University of Chicago Press, Chicago (1966)
14. Stenholm, P., Renko, M.: Passionate bricoleurs and new venture survival. J. Bus. Ventur. **31**(5), 595–611 (2016)
15. Fisher, G.: Effectuation, causation, and bricolage: a behavioral comparison of emerging theories in entrepreneurship research. Entrep. Theory Pract. **36**(5), 1019–1051 (2012)
16. Bansler, J.P., Havn, E.: Technology-use mediation: making sense of electronic communication in an organizational context. In: Proceedings of the 2003 international ACM SIGGROUP Conference on Supporting Group Work, pp. 135–143. ACM, Sanibel (2003)
17. Baker, T., Nelson, R.E.: Creating something from nothing: resource construction through entrepreneurial bricolage. Adm. Sci. Q. **50**(3), 329–366 (2005)
18. Vignale, S.: Pedagogía de la incertidumbre. Rev. Iberoamericana Educ. **48**(2), 1–7 (2009)

19. Giroux, H.A.: Public pedagogy and the politics of neoliberalism: making the political more pedagogical. Policy Futures Educ. **2**(3–4), 494–503 (2019)
20. Gelatt, H.B.: Positive uncertainty: a new decision-making framework for counseling. J. Couns. Psychol. **36**(2), 252 (1989)
21. Bauman, Z.: Education challenges in the liquid modern era. Sage J. **50**(1), 15–26 (2013). https://doi.org/10.1177/039219210305000103
22. Nieto, J., Rodriguez-López, R., Franco, D.: Linear first-order fuzzy differential equation. Int. J. Uncertain. Fuzziness Knowl.-Based Syst. **1**(14), 687–709 (2006)
23. Garrison, J., Neubert, S., Reich, K.: John Dewey's Philosophy of Education. An Introduction and Recent Textualization for Our Times. Palgrave, Mc Millan, New York (2012)
24. Gil Lafuente, A.M.: El análisis financiero en la incertidumbre. 2nd edn. Ariel Economía, Barcelona (1993)
25. Rosenzweig, P.: The halo effect and other managerial delusions. McKinsey Q. **1**(76), 77–85 (2007)
26. Bryan, L.L.: Just-in-time strategy for a turbulent world. McKinsey Q. 17–28 (2002)
27. Bryan, L. L., Farrell, D.: Leading through uncertainty. McKinsey Global Institute, Mc Kinsey Quarterly Online (2008). http://www.mckinseyquarterly.com/Leading_through_uncertainty_2263
28. Skinner, D.: Introduction to Decision Analysis A Practitioner's Guide to Improving Decision Quality, 2nd edn. Probabilistic Publishing, Gainsville (2001)
29. Walker, W.E., et al.: Defining uncertainty: a conceptual basis for uncertainty management in model-based decision support. Integr. Assess. **4**(1), 5–17 (2003)
30. Ferrari, C.: Tiempos de incertidumbre. Causas y consecuencias de la crisis mundial. Rev. Econ. Inst. **10**(19), 55–78 (2008)
31. Bird, M.: Measurement uncertainty from a historical perspective. In: The 230th ACS National Meeting, in Washington, DC, vol. 1, no. 230 (2005). http://oasys2.confex.com/acs/230nm/techprogram/S18850.HTM
32. Stewart, I.: Do Dice Play God?: The Mathematics of Uncertainty. Profile Books, London (2019)
33. Kaivo-Oja, J.R.L., Lauraeus, I.T.: The VUCA approach as a solution concept to corporate foresight challenges and global technological disruption. Foresight **20**(1), 27–49 (2018)
34. Mishra, V., Joshi, G.: The VUCA world in IT industry. Paripex Indian J. Res. **5**(8), 210–212 (2016)
35. Miles, T.: VUCA, a useful acronym for today's supply chain. Kinaxis, weblog post (2001). http://blog.kinaxis.com/2011/06/vuca-a-useful-acronym-for-today's-supply-chain/
36. Halek, P., Strobl, G.: Keeping the flow: creating opportunities based on well-structured collaboration. In: Mack, O., Khare, A., Krämer, A., Burgartz, T. (eds.) Managing in a VUCA World, pp. 59–76. Springer, London (2016). https://doi.org/10.1007/978-3-319-16889-0_4
37. Leon-Castro, E., Espinoza-Audelo, L.F., Merigó, J.M., Gil-Lafuente, A.M., Kacprzyk, J.: A new measure of volatility using induced heavy moving averages. Technol. Econ. Dev. Econ. **25**(4), 576–599 (2019)
38. EY Ernst & Young: Beacon Executive roundtable. Davos Switzerland (2017). https://www.ey.com/gl/en/issues/ey-beacon-institute-do-uncertain-times-need-certainty-of-purpose
39. Brynjolfsson, E., McAfee, A.: The Second Machine Age: Work, Progress, and Prosperity in a Time of Brilliant Technologies. WW Norton & Co., New York (2014)
40. Bennett, N., Lemoine, G.J.: What VUCA really means to you. Harv. Bus. Rev. **92**(1/2), 27–28 (2014)
41. Bawany, S.: Leading in a VUCA business environment. Leadersh. Excellence Essent. **3**(7), 39–40 (2016)
42. Stiglitz, J.E.: Incertidumbre trumpiana. El País, España (2017). https://elpais.com/economia/2017/01/12/actualidad/1484238041_245267.html

43. Demirdjian, A.: Challenges and Opportunities in Exponential Times. Xlibris, Bloomington (2016)
44. Kaggermann, H.: Change through digitalization-value creation in the age of Industry 4.0. In: Albach, H., Meffert, H., Pinkwart, A., Reichwald, R. (eds.) Management of Permanent Change, pp. 23–45. Springer, Wiesbaden (2015). https://doi.org/10.1007/978-3-658-05014-6_2
45. Ismail, S.: Exponential Organizations: Why New Organizations are Ten Times Better, Faster, and Cheaper than Yours (and What to Do About It). Diversion Books, New York (2014)
46. Hall, R.D., Rowland, C.A.: Leadership development for managers in turbulent times. J. Manag. Dev. **35**(8), 942–955 (2016)
47. Alexopoulos, M., Cohen, J.: Uncertain times, uncertain measures. University of Toronto Department of Economics Working Paper, p. 352. (2009). https://www.economics.utoronto.ca/public/workingPapers/tecipa-352.pdf
48. Peterson, K., Poli, R.: New Research on the Philosophy of Nicolai Hartmann. De Gruyter, Berlin (2016)
49. Coccia, M.: Technometrics: origins, historical evolution, and new directions. Technol. Forecast. Soc. Chang. **72**(8), 944–979 (2005)
50. Samli, A.C.: Surviving in chaotic modern markets: strategic considerations in turbulent times. J. Mark. Theory Pract. **14**(4), 315–322 (2013)
51. Küpers, W., Weibler, J.: How emotional is transformational leadership, really? Some suggestions for a necessary extension. Leadersh. Org. Dev. J. **27**(5), 368–383 (2006)
52. Parmar, I.: Transnational elite knowledge networks: managing American hegemony in turbulent times. Secur. Stud. **1–33**, 532–564 (2019)
53. Giones, F., Brem, A., Berger, A.: Strategic decisions in turbulent times: lessons from the energy industry. Bus. Horiz. **62**(2), 215–225 (2019)
54. Brown, K.A., Schmitt, T.G., Schonberger, R.J.: ASP, the art, and science of practice: three challenges for a lean enterprise in turbulent times. Interfaces **45**(3), 260–270 (2015)
55. D'agostino, L., Moreno, R.: Exploration during turbulent times: an analysis of the relation between cooperation and innovation activities and radical innovation performance during the economic crisis. Ind. Corp. Chang. **27**(2), 387–412 (2018)
56. Ramus, T., Vaccaro, A., Brusoni, S.: Institutional complexity in turbulent times: formalization, collaboration, and the emergence of blended logics. Acad. Manag. J. **60**(4), 1253–1284 (2012)
57. Holmberg, R., Larsson, M., Bäckström, M.: Developing leadership skills and resilience in turbulent times: a quasi-experimental evaluation study. J. Manag. Dev. **35**(2), 154–169 (2016)
58. Marchington, M., Kynighou, A.: The dynamics of employee involvement and participation during turbulent times. Int. J. Hum. Resour. Manag. **23**(16), 3336–3354 (2012)
59. Dell, H., Marx, D., Husfeldt, T., Taslaman, N., Wahlen, M.: Exponential times. Complexity of the permanent and the tutte polynomial. ACM Trans. Algorithms **10**(4), 1–24 (2012)
60. Schwab, K.: The fourth industrial revolution: what it means and how to respond. World Economic Forum (2016). www.weforum.org/agenda/2016/01/the-fourth-industrialrevolution-what-it-means-and-how-to-respond
61. Diamandis, P.: Top 10 tech trends transforming humanity. XPRIZE Foundation (2019). https://www.diamandis.com/blog/top-10-tech-trends-transforming-humanity
62. World Economic Forum WEF How uncertainty over business conditions affects investment decisions (2016). https://www.weforum.org/agenda/2016/04/how-uncertainty-over-business-conditions-affects-investment-decisions/
63. Jelassi, T., Avagyan, K., Perrinjaquet, M.: Leading in turbulent times. IMD (2017). https://www.imd.org/research-knowledge/articles/leading-in-turbulent-times/

64. Thompson, K.: Zygmunt Bauman's liquid times – a summary (2016). https://revisesociology.com/2016/08/09/zygmunt-bauman-liquid-times-summary/
65. White, C.: Seismic Shifts: Leading in Times of Change. United Church Publishing, Toronto (2006)
66. Oxford Economics. Economic uncertainty (2017). https://www.oxfordeconomics.com/. & http://blog.oxfordeconomics.com/topic/economic-uncertainty
67. Müller, C.: Radical uncertainty: sources, manifestations, and implications. Economics Discussion Papers, 2015-41, Kiel Institute for the World Economy (2015). http://www.economics-ejournal.org/economics/discussionpapers/2015-41
68. Briggs, W.: Uncertainty: The Soul of Modeling, Probability & Statistics. Springer, New York (2016). https://doi.org/10.1007/978-3-319-39756-6
69. Babrow, A.S.: Uncertainty, value, communication, and problematic integration. J. Commun. **51**(3), 553–573 (2001)
70. Mathiassen, L., Stage, J.: The principle of limited reduction in software design. Inf. Technol. People **6**(2), 171–185 (1992)
71. O'Donell, R.: Rethinking uncertainty. In: Rochon, L.P., Rossi, S. (eds.) A Modern Guide to Rethinking Economics. Edward Elgar Publishing, North Hampton (2017)
72. Mc Manus, H., & Hastings, D.: A framework for understanding uncertainty and its mitigation and exploitation in complex systems. In: INCOSE International Symposium, vol. 15, no. 1, pp. 484–503 (2005)
73. Funtowicz, S.O., Ravetz, J.R.: Uncertainty and Quality in Science for Policy 15 Theory and Decision Library. Springer, London (1990). https://doi.org/10.1007/978-94-009-0621-1
74. Streatfield, P.J.: The Paradox of Control in Organizations. Routledge, London (2001)
75. Salem, P., Williams, M.L.: Uncertainty and satisfaction: the importance of information in hospital communication. J. Appl. Commun. Res. **12**(2), 75–89 (1984)
76. Miller, K.D.: Industry and country effects on manager's perceptions of environmental uncertainties. J. Int. Bus. Stud. **24**(3), 693–714 (1993)
77. Bradley, C., Hirt, M., Smit, S.: How to confront uncertainty in your strategy. Mc Kinsey (2018). https://www.McKinsey.com/business-functions/strategy-and-corporate-finance/ourinsights/how-to-confront-uncertainty-in-your-strategy. Accessed 15 May 2018
78. Hofstede, G.: Culture and organizations. Int. Stud. Manag. Organ. **10**(4), 15–41 (1980)
79. Dorfman, P., Javidan, M., Hanges, P., Dastmalchian, A., House, R.: GLOBE: a twenty years' journey into the intriguing world of culture and leadership. J. World Bus. **47**(4), 504–518 (2012)
80. Gatignon, H., Anderson, E.: The multinational corporation's degree of control over foreign subsidiaries: an empirical test of a transaction cost explanation. J. Econ. Organ. **4**, 305 (1988)
81. Millennium Ecosystem Assessment: Ecosystems and Human Well-Being: A Framework for Assessment. Island Press, Washington, DC (2003)
82. Marchau, V.A., Walker, W.E., Bloemen, P.J., Popper, S.W.: Decision Making under Deep Uncertainty. From theory to practice. Springer, Cham (2019). https://doi.org/10.1007/978-3-030-05252-2
83. Lovell, B.E.: A taxonomy of types of uncertainty. Dissertation Doctor in Philosophy, Portland State University. Systems Science (1995). https://doi.org/10.15760/etd.1395
84. Bautista, R.: Incertidumbre y riesgos en decisiones financieras. Ecoe Ediciones, Bogotá (2012)
85. Courtney, H., Kirkland, J., Viguerie, P.: Strategy under uncertainty. Harv. Bus. Rev. **75**(6), 67–79 (1997)
86. Olarte, J.C.: Incertidumbre y evaluación de riesgos financieros. Sci. Tech. **12**(32), 347–350 (2006)

87. Kwakkel, J.H., Mens, M.J.P., de Jong A., Wardekker J.A., Thissen W.A.H., van der Sluijs J.P.: Uncertainty terminology. In: Knowledge for Climate. Utrech University (2011). http://goo.gl/H91ei
88. Conroy, M.J., Peterson, J.T.: Decision Making in Natural Resource Management: A Structured, Adaptive Approach. Wiley, Hoboken (2013)
89. Martorell, S., Soares, C.G., Barnett, J. (eds.): Safety, Reliability, and Risk Analysis: Theory, Methods, and Applications. Taylor & Francis, London (2008)
90. Morgan, M.G., Henrion, M.: Uncertainty: A Guide to Dealing with Uncertainty in Quantitative Risk. Cambridge University Press, Cambridge (1990)
91. Theiss, J.A.: The Experience and Expression of Uncertainty in Close Relationships. Cambridge University Press, Cambridge (2017)
92. Oberkampf, W.L., Roy, C.J.: Verification and Validation in Scientific Computing. Cambridge University Press, Cambridge (2010)
93. Hjort-Madsen, K., Pries-Heje, J.: Enterprise architecture in government: fad or future? In 2009 42nd Hawaii International Conference on System Sciences, pp. 1–10. IEEE (2009)
94. Missouri University of Science and Technology. Experimental Uncertainty (2018). http://web.mst.edu/~dux/repository/me242/Experimental%20Uncertainty%20ppt.pdf
95. Doyle, E.E., Johnston, D.M., Smith, R., Paton, D.: Communicating model uncertainty for natural hazards: a qualitative systematic thematic review. Int. J. Disast. Risk Reduct. **33**, 449–476 (2018)
96. van der Sluijs, J.P., Petersen, A., Funtowicz, S.: Reflective approaches to uncertainty assessment and communication. In: Lentsch, J., Weingart, P. (eds.) Politics of Scientific Advice, pp. 259–269. Cambridge University Press, Cambridge (2011)
97. Smith, R.T.: Cyclical uncertainty, precautionary saving, and economic growth. Economica **63**(251), 477–494 (1996)
98. Vluymans, S., Cornelis, C., Herrera, F., Saeys, Y.: Multi-label classification using a fuzzy rough neighborhood consensus. Inf. Sci. **443**, 96–114 (2018)
99. Gil-Aluna, J.: Elementos para una teoría de la decisión en la incertidumbre. Editorial Milladoiro, Vigo (1999)
100. Lempert, R.J.: Shaping the Next One Hundred Years: New Methods for Quantitative, Long-Term Policy Analysis. Rand Corporation, Pittsburg (2003)
101. Haasnoot, M., Kwakkel, J.H., Walker, W.E., ter Maat, J.: Dynamic adaptive policy pathways: a method for crafting robust decisions for a deeply uncertain world. Glob. Environ. Chang. **23**(2), 485–498 (2013)
102. Santos Álvarez, M., García Merino, M.T.: Environmental uncertainty: the side object of perception. Innovar **18**(32), 65–74 (2013)
103. Business dictionary: Endogenous uncertainty (2019). http://www.businessdictionary.com/definition/endogenous-uncertainty.html
104. Milliken, F.J.: Three types of perceived uncertainty about the environment: state, effect, and response uncertainty. Acad. Manag. Rev. **12**(1), 133–143 (1987)
105. Fox, C.R., Ulkümen, G.: Distinguishing two dimensions of uncertainty. In: Brun, W., Kirkebøen, G., Montgomery, H. (eds.) Essays in Judgment and Decision Making. Universitetsforlaget, Oslo (2011)
106. Van Asselt, M.B.: Perspectives on uncertainty and risk. In: Van Asselt, M.B. (ed.) Perspectives on uncertainty and Risk, pp. 407–417. Springer, Dordrecht (2000). https://doi.org/10.1007/978-94-017-2583-5_10
107. Farley Financial Dictionary Exogenous uncertainty (2012). https://financial-dictionary.thefreedictionary.com/Exogenous+Uncertainty
108. Bell, S.: A beginner's guide to uncertainty of measurement. Meas. Good Pract. Guide Natl. Phys. Lab. **11**(2), 11–21 (2001)

109. Cimbala, J.M.: Experimental uncertainty analysis. Online Edition, pp. 1–4 (2013). https://www.me.psu.edu/cimbala/me345web_Fall_2014/Lectures/Exper_Uncertainty_Analysis.pdf
110. Nicholson, T., Williams, D.M., Grainger, C., Lind, S.E., Carruthers, P.: Relationships between implicit and explicit uncertainty monitoring and mindreading: evidence from autism spectrum disorder. Conscious. Cogn. **70**, 11–24 (2019)
111. Kahneman, D., Tversky, A.: Variants of uncertainty. Cognition **11**(2), 143–157 (1982)
112. Rojas, M.: Equipoise research and the development of neonatal interventions for the management of respiratory distress syndrome: a historical perspective. Am. J. Perinatol. **32**(10), 910–915 (2015)
113. Pal, N.R., Bezdek, J.C.: Measuring fuzzy uncertainty. IEEE Trans. Fuzzy Syst. **2**(2), 107–118 (1994)
114. Merigó, J.M., Gil-Lafuente, A.M., Yager, R.R.: An overview of fuzzy research with bibliometric indicators. Appl. Soft Comput. **27**, 420–433 (2015)
115. Liu, S., Forrest, J., Vallee, R.: Emergence and development of grey systems theory. Kybernetes **38**(7/8), 1246–1256 (2009)
116. Arbani, A., Farahani, R.: Facility location dynamics: an overview of classifications and applications. Comput. Ind. Eng. **62**(1), 408–420 (2012)
117. Night, P.R.: Uncertainty rules. Neuroeconomics seeks to move brain studies beyond the fixed reflex. Books and arts. Nature **424**, 371–372 (2003)
118. Dall, S.R.: Managing risk: the perils of uncertainty. Evol. Behav. Ecol. 194–206 (2010)
119. De Rocquigny, E., Devictor, N., Tarantola, S. (eds.): Uncertainty in Industrial Practice: A Guide to Quantitative Uncertainty Management. Wiley, Cornwall (2008)
120. Kozeniauskas, N., Orlik, A., Veldkamp, L.: What are uncertainty shocks? J. Monet. Econ. **100**, 1–15 (2018)
121. Ben-Haim, Y., Demertzi, M.: Decision making in times of Knightian uncertainty: an info-gap perspective. Economics. **10**(2016–23), 1–29 (2016). The Open-Access, Open-Assessment E-Journal
122. De Marchi, B., Funtowicz, S.O., Ravetz, J.: The management of uncertainty in the communication of major hazards. Commission of the European Communities (1993). http://publications.europa.eu/resource/cellar/cb61fb83-2672-472e-b457-bbfe096a8ed5.0001.02/DOC_1
123. Freer, J., Beven, K., Ambroise, B.: Bayesian estimation of uncertainty in runoff prediction and the value of data. An application of the GLUE approach. Water Resour. Res. **32**(7), 2161–2173 (1996)
124. Maritorena, S., Siegel, D.A.: Consistent merging of satellite ocean color data sets using a bio-optical model. Remote Sens. Environ. **94**(4), 429–440 (2005)
125. Bich, W., Cox, M.G., Harris, P.M.: Evolution of the guide to the expression of uncertainty in measurement. Metrologia **43**(4), 16–166 (2006)
126. Ferson, S., Kreinovich, V., Hajagos, J., Oberkampf, W., Ginzburg, L.: Experimental uncertainty estimation and statistics for data having interval uncertainty. Sandia National Laboratories, Report SAND2007-0939, 162 (2007). https://prod-ng.sandia.gov/techlib-noauth/access-control.cgi/2007/070939.pdf
127. Bloom, N.: Fluctuations in uncertainty. J. Econ. Perspect. **28**(2), 153–176 (2014)
128. Kass, R.E., Raftery, A.E.: Bayes factors. J. Am. Stat. Assoc. **90**(430), 773–795 (1995)
129. Bykvist, K.: Moral uncertainty. Philos. Compass **12**(3), 2408 (2017)
130. Hartman, R.J.: In Defense of Moral Luck: Why Luck Often Affects Praiseworthiness and Blameworthiness. Routledge, New York (2017)
131. Eriksson, J., Olinder, R.F.: Non-cognitivism and the classification account of moral uncertainty. Australas. J. Philos. **94**(4), 719–735 (2010)
132. Jauch, L.R., Kraft, K.L.: Strategic management of uncertainty. Acad. Manag. Rev. **11**(4), 777–790 (1986)

133. Turner, B.S.: Outline of a theory of human-rights. Sociology. J. Br. Sociol. Assoc. **27**(3), 489–512 (1993)
134. Lane, D.A., Maxfield, R.R.: Ontological uncertainty and innovation. J. Evol. Econ. **15**(1), 3–50 (2005)
135. Ruiz-Ramos, M., Minguez, M.I.: Evaluating uncertainty in climate change impacts on crop productivity in the Iberian Peninsula. Clim. Res. **44**(1), 69–82 (2010)
136. Arias, M.: Democracia y sociedad del riesgo: deliberación, complejidad, incertidumbre. Rev. Estud. Polít. **122**, 233–268 (2003)
137. Xie, L.H.: Output feedback H infinity control system with parameter uncertainty. Int. J. Control **63**(4), 741–750 (1996)
138. Wankhade, L., Dabade, B.: Quality uncertainty due to information asymmetry. In: Quality Uncertainty and Perception, pp. 13–25. Physica, Heidelberg (2010)
139. Pei, Z.: Intuitionistic fuzzy variables: concepts and applications in decision making. Expert Syst. Appl. **42**(22), 9033–9045 (2015)
140. Moorthy, S., Ratchford, B.T., Talukdar, D.: Consumer information search revisited: theory and empirical analysis. J. Consum. Res. **23**(4), 263–277 (1997)
141. Knight, F.H.: Risk, Uncertainty, and Profit. Courier Corporation, New York (1982)
142. Tannert, C., Elvers, H.D., Jandrig, B.: The ethics of uncertainty. EMBO Rep. **8**(10), 892–896 (2007)
143. Cooke, R.M.: Deep and shallow uncertainty in messaging climate change. Discussion paper. Resources for the future. (2014). http://citeseerx.ist.psu.edu/viewdoc/download?doi=10.1.1.569.3082&rep=rep1&type=pdf
144. Falush, D., Stephens, M., Pritchard, J.K.: Inference of population structure using multilocus genotype data: linked loci and correlated allele frequencies. Genetics **164**(4), 1567–1587 (2003)
145. Mistrík, I., Bahsoon, R., Kazman, R., Zhang, Y. (eds.): Economics-Driven Software Architecture. Elsevier, Waltham (2014)
146. Barthelmé, S., Mamassian, P.: Evaluation of objective uncertainty in the visual system. PLoS Comput. Biol. **5**(9), e1000504 (2009)
147. Ellinghaus, F.: Systematic uncertainties. University of Mainz, Mainz (2010)
148. Thygesen, N. (ed.): The Illusion of Management Control: A Systems Theoretical Approach to Managerial Technologies. Springer, New York (2011). https://doi.org/10.1057/9780230365391
149. Starbuck, W., Farjoun, M. (eds.): Organization at the Limit: Lessons from the Columbia Disaster. Blackwell, Malden (2005)
150. Dompere, K.K.: Fuzzy rationality, uncertainty, and expectations. In: Dompere, K.K. (ed.) Fuzziness and Approximate Reasoning, vol. 237, pp. 1–26. Springer, Berlin (2009). https://doi.org/10.1007/978-3-540-88087-5_1
151. Vives, M.L., FeldmanHall, O.: Tolerance to ambiguous uncertainty predicts prosocial behavior. Nat. Commun. **9**(1), 2156 (2018)
152. Sepielli, A.: Moral uncertainty and fetishistic motivation. Philos. Stud. **173**(11), 2951–2968 (2016). https://doi.org/10.1007/s11098-016-0645-z
153. Kennedy, M.C., O'Hagan, A.: Bayesian calibration of computer models. J. Roy. Stat. Soc. Ser. B-Stat. Methodol. **63**, 425–450 (2001)
154. Harris, H., Brewster, C., Sparrow, P.: International Human Resource Management. CIPD Publishing, London (2003)
155. Lind, D.A., Marchal, W.G., Wathen, S.: Estadística Aplicada a los Negocios y la Economía. Mc Graw Hill, Mexico (2019)
156. Boehm, B.: Future challenges for software data collection and analysis. In: Proceedings of the 5th International Conference on Predictor Models in Software Engineering. ACM (2009). https://pdfs.semanticscholar.org/2542/d9459dc847cbd0f2b89de4eb1a2ce90953d7.pdf

157. Lauret, P., Fock, E., Randrianarivony, R.: Bayesian neural network approach to short-time load forecasting. Energy Convers. Manage. **49**(5), 1156–1166 (2008)
158. National Institute of Standards and Technology NIST. Uncertainty machine (2018). https://uncertainty.nist.gov/
159. Hogg, M.A., Belavadi, S.: Uncertainty management theories. In: Oxford Research Encyclopedia of Communication (1980)
160. Pearson, A., Brockhoff, K.: The uncertainty map and project management. Proj. Appraisal **9**(3), 211–215 (1994)
161. Sen, D.: The uncertainty relations in quantum mechanics. Curr. Sci. **107**(2), 203–218 (2014)
162. Bloom, N.: The impact of uncertainty shocks. Econometrica **77**, 623–685 (2009)
163. Longtin, J.P.: The uncertainty tree: reducing the uncertainty of uncertainty analysis. Rev. Sci. Instr. **73**(10), 3698–3700 (2002)
164. Matthews, R.: Furore around 'uncertainty' buzzword shows experts that it is really all about chance The National AE (2017). https://www.thenational.ae/uae/furore-around-uncertainty-buzzword-shows-experts-that-it-s-really-all-about-chance1.13319?videoId=5766484581001

Mexico and the Challenges of Achieving the 2030 Sustainable Development Goals

I. N. Pinto-López[1(✉)], C. M. Montaudon-Tomas[1], and Anna Maria Gil-Lafuente[2]

[1] UPAEP, Universidad, 21 Sur 1103 Barrio de Santiago, Puebla, Mexico
{ingrid.pinto,cynthiamaria.montaudon}@upaep.mx
[2] University of Barcelona, Avinguda Diagonal, 690, Barcelona, Spain
amgil@ub.edu

Abstract. The Sustainable Development Goals (SDGs) proposed by the United Nations in the 2030 Agenda, constitute a universal call to adopt measures to end poverty, protect the planet and ensure that all individuals can enjoy peace and prosperity. This article aims to present a multidimensional analysis using Subjective Preferences Fuzzy Logics to identify every state's position in the Mexican Republic when facing the challenge of reaching the SDGs. Results show that the states with the highest sustainable development in the country are México City, Nuevo León, Aguascalientes, Querétaro, and Jalisco and that the states with the lowest level of sustainable development are Chiapas, Guerrero, Oaxaca, Campeche, and Puebla. A relevant fact is that Mexico City presents an improvement in sustainable development that is almost double the results found in Chiapas. This analysis allows us to identify areas of opportunity for the design of public policies directed to the states' need to generate more equality in sustainable development in Mexico.

Keywords: Sustainable development · Subjective preferences methods · Fuzzy logics · Arrangement

1 Introduction

In the Sustainable Development Summit held in September 2015, the UN member states approved the 2030 Agenda for Sustainable Development to end poverty, fight against inequality and injustice, and face climate change [1].

The 2030 Agenda includes a collection of 17 Sustainable Development Goals (SDGs) and 169 targets that create opportunities to improve the quality of life under the framework of sustainable development. This agenda serves as an action launching platform for the international community, governments, civil society, academia, and private sector organizations. The efforts will help face the three interconnected elements of sustainable development: economic growth, inclusion, and environmental sustainability [2].

In the 2030 Agenda meetings, México was an active and engaged participant accepting to adopt the decisions made in a quest to achieve the SDGs. In terms of transparency in the commitments acquired, the Mexican government provides information about the

A. M. Gil-Lafuente et al. (Eds.): REDCID 2019, LNNS 388, pp. 29–45, 2022.
https://doi.org/10.1007/978-3-030-93787-4_2

advances of each target through the Sustainable Development Goals Information System (SDGID) [3].

This article's objective is to develop a multidimensional analysis using fuzzy logic methodologies to identify each state's position in the Mexican Republic when achieving the SDGs. The 44 targets that Mexico reports through the SDGID are analyzed by federal entities.

2 Sustainable Development

In the seventies, the debate about economic growth changed direction. At that time, the first voices against unlimited growth stated to be heard, alerting about the need to consider the biologic limits of the planet in a context in which the pernicious effects that economic crisis was causing over the environment and the quality of life became more palpable [4, 5].

Facing the increasing and alarming magnitude of undesirable effects of economic development, the concept of Sustainable Development (SD) has become one of the most common international, national, and local referents in the last decades. Scientific research associated with this topic has become relevant in the past few years for two main reasons, first, because of the UN invitation to government representatives, civil society, academia and the private sector to appropriate the concept, its targets and its execution [6] and second, because it represents a vital thermometer that measures the importance of SD for society as a whole.

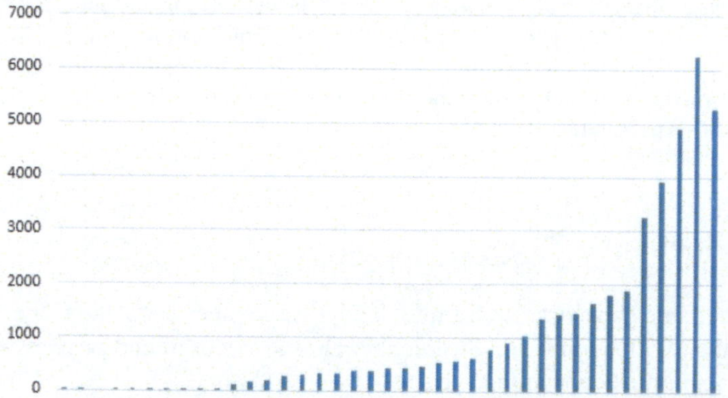

Fig. 1. Number of published articles. Developed by authors with information from the WoS, 2019

According to information from the Web of Science (WoS), scientific research related to SD has shown unremitting growth since 1981, a year in which three articles were published. In 2018, 6244 articles were published, and by September 2019, 5277 articles had been published. Yearly analysis of the evolution in scientific Production is presented in Fig. 1.

From 1981 to 2019, the total number of published articles was 41,449; 12,284 of which are open access. Furthermore, 76.7% of all publications have been made in the last decade. Different universities and organizations are committed to developing research in this area. Some of the most relevant include the following: "Chinese Academy of Sciences", "University of London", "University of Chinese Academy of Science Cas", "Wageningen University Research", among others. The most productive countries in terms of scientific Production on the concept of Sustainable Development are: "Peoples R. China", "USA", "England", "Australia", "Canada", "Germany", "Netherlands", "Spain", "France", "India" and "Brazil" [7].

3 2030 Sustainable Development Goals

Prevalent slow economic growth, social inequalities and environmental degradation present unique challenges for all inhabitants. In order to face these challenges, the 193 member states of the United Nations, along with multiple stakeholders from the civil society, academia, and the private sector, have participated in an open, democratic and participative negotiation process, which resulted in the proclamation of the 2030 Agenda for Sustainable Development [8].

The 2030 Agenda for Sustainable Development is a global plan in favor of dignity, peace, and prosperity for today and tomorrow, both for the people and the planet [9]. The agenda establishes a transformational vision towards the economic, social, and environmental sustainability of the 193 member states that subscribed to it [8]. The SDGs can be used as a planning tool for countries at the national and local levels. It is focused on a long term vision, which will constitute support for every country in their quest for sustained, inclusive, and harmonious development, through public policies and instruments that have been expressly developed for monitoring and evaluation (Ibídem). Since the agenda started to be implemented in 2015, countries have been translating this shared vision into their national development plans and strategies [9].

According to the CEPAL [8], this new plan represents a historic opportunity for Latin America and the Caribbean because it includes high priority topics for the region, which include the reduction of inequalities in all its dimensions, inclusive economic growth with decent employment for all, sustainable cities and climate change, among others.

In this new landscape, a vast amount of scientific research is dedicated to contributing to attaining the SDGs proposed by the 2030 Agenda. Some of the most recent include Bjegovic-Mikanovic, V. [10]; Van Beynen, P., Akiwumi, FA., Van Beynen, K. [11]; Ngunyulu, RN., Peu, MD., Mulaudzi, FM., Mataboge, MLS., Phiri, SS. [12]; Alola, AA. (2019) [13]; Ugwu, CLJ. & Zewotir, TT. [14]; Klinberg-Allvin, M., Atuhairwe, S., Cleeve, A., Byamugisha, JK., Larsson, EC., Makensius, M., Oguttu, M., Gemzell-Danielsson, K. [15]; Lei, Y., Cui, P., Regmi, AD., Murray, V., Pasuto, A., Titti, G., Shafique, M., Priyadarshana, DGT. [16]; Akinyemi, BE., Mushunje, A., Fashogbon, AE. [17]; Majinge, RM. & Mutula, SM. [18], Jasmine, B. [19], and Gil-Lafuente, AM., Torres-Martinez, A., Amiguet-Molina, L., Boria-Reverter, S. [20].

With its 17 objectives and 169 targets, the 2030 Agenda for Sustainable Development presents an ambitious vision and commitment for both developed and developing countries under the framework of a global reinforced alliance which provides means for its implementation and the promotion of change, as well as the mitigation and acceptance of climate change [2, 8, 9].

The 17 objectives are presented in Table 1, along with the targets associated with each objective. Each target is related to a set of indicators through which the advance of each objective, which should be attained in the next 15 years, is linked to a target [9].

Table 1. Sustainable development goals 2030 agenda

Goal	Description	NG	TI
1. End of poverty	Economic growth must be inclusive to provide sustainable jobs and promote equality	7	14
2. Zero hunger	Ending hunger achieve food security and nutrition improvement and promotion of sustainable agriculture	8	13
3. Good health and wellbeing	Ensure a healthy life and promote wellbeing for everyone in all age groups	13	27
4. Quality Education	Guarantee inclusive, equitable, and quality education and promote lifelong learning opportunities for all	10	11
5. Gender Equality	Achieve gender equality and empower all women and girls	9	14
6. Clean Water and Sanitation	Guarantee access to water, its sustainable management, and sanitation for all	8	11
7. Affordable Clean Energy	Guarantee access to accessible, safe, sustainable and modern energy for all	5	6
8. Decent Work and Economic Growth	Promote sustained, inclusive economic growth, full and productive employment, and decent work for all	12	17
9. Industry, Innovation, and Infrastructure	Build resilient infrastructure and promote inclusive and sustainable industrializations and promote innovation	8	12
10. Reduced Inequalities	Reduce inequalities in and among countries	10	11
11. Sustainable Cities and Communities	Make cities and human settlements inclusive, safe, resilient, and sustainable	10	15
12. Responsible consumption and Production	Ensure sustainable forms of consumption and Production	11	13
13. Climate Action	Adopt urgent measures to combat climate change and its effects	5	8

(*continued*)

Table 1. (*continued*)

Goal	Description	NG	TI
14. Life Below Water	Conservation and sustainable use of oceans, seas, and marine resources for sustainable development	10	10
15. Life on Land	Sustainable management of forest, fight against desertification, stop and invest soil degradation and loss of biodiversity	12	14
16. Peace, Justice and Strong Institutions	Promote fair, pacific and inclusive societies	12	23
17. Partnerships for the Goals	Revitalize the Global Alliance for Sustainable Development	19	25
TOTAL		**169**	**244**

NG: Number of Goals, TI: Total indicators

4 2030 Sustainable Development Goals

As a member state of the UN, Mexico is committed to achieving the 2030 Agenda SDGs. This framework for development provides global opportunities to the System of the UN. Locally, in Mexico, prospects are focused on cooperation and planning to continue advocating and promote the topics of inclusion and equality under the framework of human rights, to build better citizenship for Mexicans. As a global actor, Mexico actively participated in the definition of the agenda. The country was one of the most active in different forums, participating, and leading negotiation processes. Not only did the country presented particular proposals to incorporate the principles of equality, social and economic inclusion; it also promoted universality, sustainability, and human rights as the principal axis of the 2030 Agenda. Furthermore, the country advocated the adoption of a multidimensional consideration of poverty, which includes income but also considers sufficient access to other fundamental rights such as food, education, health, social security, and essential living services and conditions [2].

In Mexico, responsibility for the implementation of the 2030 Agenda lies in the Office of the Presidency. This provides a state commitment that permeates through all levels of government. In-state and municipal governments, two spaces for deliberation to promote local development and institutional transformations have been proposed: The National Conference of Governors (CONAGO) and the National Conference of Municipalities in Mexico (CONAMM). Additionally, the Legislative Power created parliamentary mechanisms to strengthen the 2030 Agenda. During the SDG negotiation process and the agenda, it was possible to establish a constructive dialogue with organizations from the civil society who were involved in the process. Participation of the private sector is essential; therefore, the Mexican Agency of International Cooperation for Development (AMEXID) established the alliance for sustainability as a platform for dialogue and action. It includes 80 Mexican and multinational businesses operating in

Mexico to exchange information about the SDGs in their business models and design international cooperation projects on the 2030 Agenda, among other actions [21].

In Mexico, the Information System of the Sustainable Development Goals presents a total of 120 items developed by objective and target, 68 of which correspond to global issues and 52 to national matters.

In this research, 44 out of the 52 targets were considered for each of the 32 states in the Mexican Republic. Specifications on the 44 national items and their objective and corresponding target are presented in Table 2.

Table 2. Goals, national data Mexico

Goal	Target	Indicator	Description
1	1.1	1.2.1	The ratio of population living under the national poverty line, distributed by age and gender
	1.2	1.2.2	The ratio of men, women, and children in all age groups living in poverty in all its dimensions with regards to national dimensions
2	2.1	2.1.2	Ratio of the population with moderate or severe food insecurity (lacking because of access to food)
3	3.1	3.1.1	Maternal mortality rate
		3.1.2	Ratio of births attended by specialized health professionals
	3.2	3.2.1	Mortality rate in children under five years
		3.2.3	Child mortality rate
		3.2.4	Ratio of children age one with the basic immunization plan
		3.2.5	Mortality rate of children under the age of 5 due to diarrheic illnesses (deaths per 100, thousand children under the age of 5)
		3.2.6	Mortality rate of children under the age of 5 due to acute respiratory illnesses (deaths per 100, thousand children under the age of 5)
	3.3	3.3.3	Incident rate associated with malaria (per 100 thousand inhabitants)
		3.3.7	Ratio of children age one who have been vaccinated against measles
		3.3.9	Mortality rate associated with AIDS (per 100 thousand inhabitants)
		3.3.12	Ratio of new cases of pulmonary tuberculosis that are cured at the end of treatment
	3.7	3.7.1	Percentage of women in fertility age (ages 15 to 49) joined by a satisfied demand for modern contraceptive methods
		3.7.2.a	Fecundity rate in girls and teens (ages10 to 14) per 1000 girls and teens in that age group
		3.7.2.b	Fecundity rate in teens (ages 15 to 19) per 1000 women in that age group

(continued)

Table 2. *(continued)*

Goal	Target	Indicator	Description
4	4.1	4.1.2	Net rate of children enrolment in primary education (ages 6 to 11)
		4.1.3	Net rate of secondary education enrolment (ages 12 to 14)
		4.1.4	Completion rate in primary education
		4.1.5	Completion rate in secondary education
		4.1.6	Absorption rate of those who finished primary education
	4.2	4.2.2	Participation rate on organized learning (a year before the official age for entering primary education) divide by gender
		4.2.3	Net rate of enrolment in preschool (ages 3 to 5)
	4.3	4.3.1	Youth and adult participation rate in education and academic and non-academic development in the last 12 months, divided by gender
	4.4	4.4.1	Ratio of young and adults with knowledge information and communication technologies (TIC), divided according to technical knowledge
	4.6	4.6.1	Percentage of population in a specific age group that reaches at least a fixed level of functional competence in a) literacy and b) elementary arithmetic, broken down by gender
	4.c	4.c.1.a	Ratio of preschool teachers who have received at least the minimum organized training before employment or in employment (for example, pedagogical training) required to provide preschool education in Mexico
		4.c.1.b	Ratio of primary school teachers who have received at least the minimum organized training prior to employment or employed (for example, pedagogical training) required to provide primary education in Mexico
		4.c.1.c	Ratio of secondary school teachers who have received at least the minimum organized training before employment or in employment (for example, pedagogical training) required to provide secondary education in Mexico
5	5.2	5.2.1.a	Ratio of women 15 years of age or older, who have suffered physical, sexual, or psychological violence inflicted by an intimate partner, current or former, in the previous 12 months divided by state, according to the type of violence
		5.2.2.a	Ratio of women 15 years of age or older, who have suffered sexual violence at the hands of people who were not their partners in the 12 months, by state and place of fact
	5.6	5.6.3	Prevalence of contraceptive methods in women of childbearing age (15 to 49 years old)

(continued)

Table 2. (*continued*)

Goal	Target	Indicator	Description
		5.6.4	Unsatisfied needs of contraceptive methods
8	8.5	8.5.2	Unemployment rate divided by gender and age
		8.5.5	Correlation between occupation and population at working age
		8.5.6	Ratio of independent workers and non-paid work
		8.5.7	Ratio of women in salaried positions in the agriculture sector
	8.7	8.7.1	Proportion of the population between 5 and 17 years of age who are carrying out a non-permitted child occupation, broken down by sex and age
9	9.1	9.1.3	Ratio of people living in rural areas in a perimeter less than 2 km from a usable road all year
	9.c	9.c.2	Number of fixed telephone subscriptions per 100 inhabitants
		9.c.3	Number of cellphone subscribers per 100 inhabitants
11	11.1	11.1.3	Ratio of urban population living in precarious conditions
17	17.8	17.8.1	Ratio of people that use the internet in states and main cities

Source: Developed by authors with data from the Information System of the Sustainable Development Goals, INEGI, México. 2019

5 Subjective Preferences Method

Throughout the years, different fuzzy models with applications in broad and diverse areas have been developed [22]. These techniques can be divided into diffuse numerical models [23] and non-numerical diffuse models [24–26].

Among these models are Hamming's Distance, developed by Richard Wesley Hamming in 1950 [27–29], the Theory of Expertons introduced by Arnold Kaufmann in 1987 [30, 31], the method of Subjective Preferences proposed by Kaufmann and Gil Aluja in 1987 [31–33], and the OWA Operators, a model presented by R. Yager in 1988 [34–37].

The subjective preferences method proposes an algorithm that allows the array of physical or mental objects according to the importance of their characteristics [20, 32, 33]; the method is based on the following stages [32]:

1) Establish two finite sets of objects and characteristics that are determinant in the decision.
2) For each characteristic, a reciprocal matrix is developed, which is representative of the relative estimation connection of an object concerning the rest.
3) The proper dominant value and the corresponding proper vector are obtained for each characteristic, which is normalized with sum 1.
4) The proper vectors are regrouped, creating a matrix, ordinarily rectangular, with values in [0, 1].

5) A reciprocal square matrix is built based on the relative comparison of the characteristics, indicating that the line's characteristic is better than the one in the column.
6) Considering this matrix, the proper dominant value and the corresponding proper vector are generated, which will help in the weighting once normalized with sum 1.
7) The result of the matrix is obtained with the vector, generating a new and definite vector whose values result in the array.

6 Implementation

6.1 Establish the Finite Sets

The set of objects is composed of the 32 states in the Mexican Republic, which are in alphabetical order (Table 3). The set of characteristics is integrated by the 44 items described in Table 2.

The method of subjective preferences operates on the premise that more is better. This means that if an indicator I_1 has associated a numerical value Y_1 for a region R_1 and the indicator I_1 has a numerical value associated Y_2 for the region R_2 with $Y_1 > Y_2$, for the region R_1 it has a better weighting of indicator I_1 concerning region R_2. In order for all indicators to comply with the latest, in some indicators, the scale was inverted through the statistical variable of stochastic oscillator [32] so that all of the indicators would comply with the premise that a higher value has a better weighting. The indicators in which the scale was modified are identified in Table 4, column 3.

Table 3. Set of objects

Object Id.	Name
E1	Aguascalientes
E2	Baja California
E3	Baja California Sur
E4	Campeche
E5	México City
E6	Coahuila de Zaragoza
E7	Colima
E8	Chiapas
E9	Chihuahua
E10	Durango
E11	Guanajuato
E12	Guerrero

(continued)

Table 3. (*continued*)

Object Id.	Name
E13	Hidalgo
E14	Jalisco
E15	México
E16	Michoacán de Ocampo
E17	Morelos
E18	Nayarit
E19	Nuevo León
E20	Oaxaca
E21	Puebla
E22	Querétaro
E23	Quintana Roo
E24	San Luis Potosí
E25	Sinaloa
E26	Sonora
E27	Tabasco
E28	Tamaulipas
E29	Tlaxcala
E30	Veracruz
E31	Yucatán
E32	Zacatecas

Table 4. Set of characteristics

Characteristics Id.	Indicator	Inverted scale
C1	1.2.1	✓
C2	1.2.2	✓
C3	2.1.2	✓
C4	3.1.1	✓
C5	3.1.2	
C6	3.2.1	✓
C7	3.2.3	✓
C8	3.2.4	

(*continued*)

Table 4. (*continued*)

Characteristics Id.	Indicator	Inverted scale
C9	3.2.5	✓
C10	3.2.6	✓
C11	3.3.3	✓
C12	3.3.7	
C13	3.3.9	✓
C14	3.3.12	
C15	3.7.1	
C16	3.7.2.a	✓
C17	3.7.2.b	✓
C18	4.1.2	
C19	4.1.3	
C20	4.1.4	
C21	4.1.5	
C22	4.1.6	
C23	4.2.2	
C24	4.2.3	
C25	4.3.1	
C26	4.4.1	
C27	4.6.1	
C28	4.c.1.a	
C29	4.c.1.b	
C30	4.c.1.c	
C31	5.2.1.a	
C32	5.2.2.a	
C33	5.6.3	
C34	5.6.4	✓
C35	8.5.2	✓
C36	8.5.5	
C37	8.5.6	✓
C38	8.5.7	
C39	8.7.1	✓
C40	9.1.3	
C41	9.c.2	

(*continued*)

Table 4. (*continued*)

Characteristics Id.	Indicator	Inverted scale
C42	9.c.3	
C43	11.1.3	✓
C44	17.8.1	

6.2 Reciprocal Matrix

According to the number of elements in the two finite groups, a 44 × 32 matrix is generated. Values for the matrix were obtained from the Sustainable Development Goals (INEGI, 2019).

A reciprocal matrix is generated for each characteristic. Therefore 544 matrixes are obtained, which are 32 × 32.

6.3 Dominant Proper Vector and Corresponding Proper Vector

Dominant proper vectors and corresponding proper vectors are generated for each of the 44 matrixes with coherence indexes inferior to 0.10 and normalized with sum 1.

6.4 Proper Vector-Matrix

A 32 × 44 matrix is constructed. Its columns align to the corresponding proper vector of each characteristic, and the lines represent the 32 Mexican states. The dominant value is obtained in this matrix, as well as the proper corresponding value, generating the array of the data with values in [0, 1]. The results of this matrix are presented in Table 5.

6.5 Matrix that Represents the Relevance of the Characteristics Observed

In this study, all indicators have the same relevance, therefore the characteristics of the matrix correspond to a vector size 44 × 1, and the value of each element is 1, meaning that all characteristics have the same weight. Consequently, the resulting array from Table 5 does not suffer any changes, finalizing the process.

7 Conclusions

Table 5 identifies the order in which the 32 states of the Mexican Republic are positioned in terms of sustainable development and the direction in which more efforts should be made to reach the goals of the 2030 agenda. It can be observed that Mexico City is the region with the highest sustainable growth, followed by Nuevo León, Aguascalientes, and Querétaro, located in the north and center of the country. Additionally, the regions with the lowest sustainable development are located in the southern region, in México, Chiapas, Guerrero, Oaxaca, Campeche, and Puebla. It can also be observed that the

Table 5. Resulting array

Object Id.	Name	Value
E5	México City	1
E19	Nuevo León	0.9807
E1	Aguascalientes	0.9332
E22	Querétaro	0.9230
E14	Jalisco	0.9113
E3	Baja California Sur	0.8934
E2	Baja California	0.8926
E25	Sinaloa	0.8887
E26	Sonora	0.8879
E7	Colima	0.8831
E31	Yucatán	0.8499
E6	Coahuila de Zaragoza	0.8488
E24	San Luis Potosí	0.8443
E28	Tamaulipas	0.8430
E32	Zacatecas	0.8412
E17	Morelos	0.8404
E11	Guanajuato	0.8393
E23	Quintana Roo	0.8377
E9	Chihuahua	0.8280
E15	México	0.8278
E10	Durango	0.8177
E18	Nayarit	0.8087
E29	Tlaxcala	0.8071
E13	Hidalgo	0.8055
E16	Michoacán de Ocampo	0.7668
E27	Tabasco	0.7422
E30	Veracruz	0.7351
E21	Puebla	0.7319
E4	Campeche	0.7296
E20	Oaxaca	0.6493
E12	Guerrero	0.6478
E8	Chiapas	0.5485

Table 6. Intervals

Group	Interval
1	=1
2	≥0.95
3	≥0.90
4	≥0.85
5	≥0.80
6	≥0.75
7	≥0.70
8	≥0.60
9	≥0.50

sustainable development that Mexico City presents when compared to Chiapas is almost double. Achieving a balance for all Mexican states is a unique challenge for governors and politicians in developing public policies.

Figure 2 presents the results of grouping the different regions to identify states with a similar level of sustainable development. The array is done considering the results from Table 5, and the intervals that were used are presented in Table 6.

As can be observed in Fig. 2, nine groups are generated. The geographical distribution of each state in the Mexican Republic is analyzed and also the group in which each region is located according to the level of sustainable development it presents. It is clearly visualized that the regions with the highest degree of sustainable development, except for Mexico City, located center of the country, are distributed in regions that belong

Fig. 2. Pooling of regions in Mexico

to the northern and north-central regions of the Mexican Republic. The states with the lowest sustainable development are found in the southwest region of Mexico, integrated by the states of Chiapas, Guerrero, and Oaxaca. In these states, along with Campeche and Puebla, public policies must be developed to improve their inhabitants' quality of life.

8 Future Research Directions

In order to identify areas of opportunity for every region in the Mexican Republic, particularly for the states of Chiapas, Guerrero, Oaxaca, Campeche, and Puebla, further analysis is required on each of the vectors that correspond to each state. This will help identify the direction and the variables that need to be analyzed to consolidate efforts and public policies to improve the quality of life in the short run. This will result in better sustainable development and, therefore, more equality in the country.

References

1. PNUD Homepage. http://www.mx.undp.org/content/mexico/es/home/post-2015/sdg-overview.html. Accessed 3 Oct 2019
2. ONU Mexico Homepage. http://www.onu.org.mx/agenda-2030/objetivos-del-desarrollo-sostenible/. Accessed 3 Oct 2019
3. SIODS Homepage. http://agenda2030.mx/ODSopc.html?ti=T&goal=0&lang=es#/ind. Accessed 3 Oct 2019
4. Aguado, I.: La agenda 21 local como instrumento de Desarrollo Sostenible. Tesis de grado. Universidad del País Vasco, Facultad de Ciencias Económicas y Empresariales, Bilbao (2005)
5. Meadows, D.H., Meadows, D.L., Randers, J., Behrens, W.: Los límites del crecimiento: informe del Club de Roma sobre el predicamento de la humanidad. Fondo de Cultura Económica, México (1972)
6. Pinto-López, I., Montaudon-Tomas, C., Malcón-Cervera, C.: Aportaciones académicas en la investigación del desarrollo sostenible: un análisis bibliométrico. Ciencias Empresariales **31**(1) (2018)
7. Web of Science WoS. https://www.webofknowledge.com. Accessed 30 Sept 2019
8. CEPAL: Agenda 2030 y los Objetivos de Desarrollo Sostenible. Una oportunidad para América Latina y el Caribe. Naciones Unidas (2016)
9. ONU Homepage. https://www.un.org/sustainabledevelopment/es/. Accessed 25 Sept 2019
10. Bjegovick-Mikanivic, V., Broniatowski, R., Byepu, S., et al.: A gap analysis of mother, newborn, and child health in west Africa with reference to the sustainable development goals 2030. Afr. J. Reprod. Health **22**(4), 123–134 (2018)
11. Van Beynen, P., Akiwumi, F.A., Van Beynen, K.: A sustainability index for small island developing states. Int. J. Sustain. Dev. World Ecol. **25**(2), 99–116 (2018)
12. Ngunyulu, R.N., Peu, M.D., Mulaudzi, F.M., Mataboge, M.L.S., Phiri, S.S.: Collaborative HIV care in primary health care: nurses' views. Int. Nurs. Rev. **64**(4), 561–567 (2017)
13. Alola, A.A.: The trilemma of trade, monetary, and immigration policies in the United States: accounting for environmental sustainability. Sci. Total Environ. **659** (2019)
14. Ugwu, C.L.J., Zewotir, T.T.: Using mixed-effects logistic regression models for complex survey data on malaria rapid diagnostic test results. Malaria J. **17**(453), 1–10 (2018)

15. Klingberg-Allvin, M., et al.: Co-creation to scale up provision of simplified high-quality, comprehensive abortion care in East-Central and Southern Africa. Global Health Action **11**(1), 1490106 (2018)
16. Lei, Y., et al.: An international program on skill road disaster risk reduction-a belt and road initiative (2016–2020). Journal of Mountain Science **15**(7), 1383–1396 (2019)
17. Akinyemi, B.E., Mushunje, A., Fashogbon, A.E.: Factors explaining household payment for potable water in South Africa. Cogent Soc. Sci. **4**(1), 1464379 (2018)
18. Majinge, R.M., Mutula, S.M.: Access to electronic and print information resources by people with visual impairments in university libraries. A review of related literature. Libr. Manag. **39**(6–7) (2018)
19. Jasmine, B.: Mainstreaming corporate engagement for progress towards the future –Sustainable Development Goals 2030. Curr. Sci. **113**(6), 1027–1028 (2017)
20. Gil-Lafuente, A.M., Torres-Martinez, A., Amiguet-Molina, L., Boria-Reverter, S.: Gender equality index of the autonomous communities of Spain: a multidimensional analysis. Technol. Econ. Dev. Econ. **25**(5), 915–933 (2019). https://doi.org/10.3846/tede.2019.10288
21. ONU1 Homepage. https://ods.org.mx/docs/doctos/McoIndMun_es.pdf. Accessed 26 Sept 2019
22. INEGI (2019). http://agenda2030.mx/#/home. Accessed 26 Sept 2019
23. Pinto-López, I.N., Botello, J., Montaudon-Tomas, C.: Gross domestic product deficiencies as an indicator of sustainable development. The case of Mexico. Bus. Manag. Rev. **9**(4), 362–372 (2018)
24. Grzegorzewski, P.: Distances between intuitionistic fuzzy sets and/or interval-valued fuzzy sets based on the Hausdorff metric. Fuzzy Sets Syst. **148**(2), 319–328 (2004)
25. Dubois, D., Prade, H.: Fuzzy-sets in approximate reasoning. 1. Inference with possibility distributions. Fuzzy Sets Syst. **40**(1), 143–202 (1991)
26. Sacconi, L.: A social contract account for CSR as an extended model of corporate governance (II): compliance, reputation, and reciprocity. J. Bus. Ethics **75**(1), 77–96 (2007)
27. Luo, X.D., Zhang, C.Q., Jennings, N.R.: A hybrid model for sharing information between fuzzy, uncertain, and default reasoning models in multi-agent systems. Int. J. Uncertainty Fuzziness Knowl.-Based Syst. **10**(4), 401–450 (2002)
28. Dodis, Y., Reyzin, L., Smith, A.: Fuzzy extractors: how to generate strong keys from biometrics and other noisy data. In: Cachin, C., Camenisch, J.L. (eds.) EUROCRYPT 2004. LNCS, vol. 3027, pp. 523–540. Springer, Heidelberg (2004). https://doi.org/10.1007/978-3-540-24676-3_31
29. Merigó, J., Gil-Lafuente, A.: New decision-making techniques and their application in the selection of financial products. Inf. Sci. **180**(11), 2085–2894 (2010)
30. Xu, Z.S., Chen, J.: An overview of distance and similarity measures of intuitionistic fuzzy sets. Int. J. Uncertainty Fuzziness Knowl.-Based Syst. **16**(4), 529–555 (2008)
31. Kaufmann, A.: Les Expertons. Hermes, Paris (1987)
32. Kaufmann, A., Gil Aluja, J.: Nuevas técnicas para la dirección estratégica. Universitat de Barcelona, Falta infrmación (1993)
33. Gil-Aluja, J.: Elementos para una teoría de la decisión en la incertidumbre. Milladoiro (1999)
34. Saaty, T.L.: A scaling method for priorities in hierarchical structures. J. Math. Psychol. **15**(3), 234–281 (1977)
35. Yager, R.R.: Families of OWA operators. Fuzzy Sets Syst. **59**(2), 125–148 (1993)
36. Yager, R.R.: Quantifier guided aggregation using OWA operators. Int. J. Intell. Syst. **11**(1), 49–73 (1996)

37. Merigó, J.M., Gil-Lafuente, A.M., Yager, R.R.: An overview of fuzzy research with bibliometric indicators. Appl. Soft Comput. **27**, 420–433 (2015)
38. Kacprzyk, J., Yager, R.R., Merigó, J.M.: Towards human centric aggregation via the ordered weighted aggregation operators and linguistic data summaries: a new perspective of Zadeh's inspirations. IEEE Comput. Intell. Mag. **14**(1), 16–30 (2019)

Check for updates

Leadership and Employees' Satisfaction in Contemporary Organizations

Montserrat Crespi-Vallbona[1(✉)], Xavier Llopart Perez[1], Sefa Boria-Reverter[1], and Oscar Mascarilla-Miró[2]

[1] Department of Business, Faculty of Economics and Business, Universitat de Barcelona, Diagonal, 690, 08034 Barcelona, Spain
{mcrespi,xllopart,jboriar}@ub.edu
[2] Department of Economics, Faculty of Economics and Business, Universitat de Barcelona, Barcelona, Spain
omascarilla@ub.edu

Abstract. The aim of this article is to identify the key attributes in recent leadership nature and to review existing research with regard to this topic, as leadership directly relates to job satisfaction and staff's well-being. Based on a survey, there were collected 568 employees' answers of different restaurant business in Barcelona (Spain) treated with structural equation modelling to confirm if the job satisfaction is positively related to the following variables: empowerment, team building, development and recognition, leader's personal values, and vision engagement. Current tendencies and results show especial emphasis into team building constructs.

Keywords: Job satisfaction · Leadership · Team working · Engagement · Restaurant business

1 Introduction

Leadership and employee satisfaction are two of the main topics in the field of human resource management. Leadership refers to the ability to influence, motivate, and enable individuals towards the achievement of a vision or set of common goals (Madlock 2008; House et al. 2004; Robbins and Judge 2016), while satisfaction refers to the attitude and pleasure derived while doing a job (Locke 1969; Weiss 2002; Aziri 2011; Asghar and Oino 2017). The scientific literature suggests that leadership characteristics have a positive relationship with employee satisfaction (Barling et al. 2002; Saleem 2015). In that sense, Asghar and Oino (2017) argue that an effective leadership style is necessary for improving job satisfaction.

Contemporary organization concerns are related to talent retention and engagement, focusing on motivation, wellbeing and satisfaction of employees. Submissive and subservient employees of old times have disappeared. Today, workers have more training and leisure options that mean the need of new management or leadership toward

A. M. Gil-Lafuente et al. (Eds.): REDCID 2019, LNNS 388, pp. 46–68, 2022.
https://doi.org/10.1007/978-3-030-93787-4_3

follower-centered approaches (Bligh 2011), hybrid configurations (Gronn 2011), complexity theory (Uhl-Bien and Marion 2011), and a variety of other areas (McCleskey 2014). Although sometimes leadership or management are used as synonyms, the reality is that these are concepts with different meanings and results. Management consists of the allocation of scarce resources against an organization's objective, the setting of priorities, the design of work, the achievement of results and the control of all these processes (Zaleznik 1989). Leadership focuses on the creation of a common vision. It means motivating people to contribute to the vision and encouraging them to align their interests with those of the organization. It means persuading, not commanding (Armstrong 2006; Storey 2007; Beardwell and Claydon 2007). It means a people-oriented manner to guide the workforce of an organization. Conversely, management involves an organization-oriented manner to lead the staff. Therefore, and theoretically, leaders care about employee satisfaction. It is significant that employees' dissatisfaction has a lot of impact on organizations. Just to mention some of them that affect directly to individuals: turnover, absenteeism, job stress, sick leave, humiliation, burnout, unproductiveness, and lack of commitment (Ghiselli et al. 2001; Dolan et al. 2007). Furthermore, the dissatisfaction has other organizational consequence, that is, a direct effect on the profitability and quality of the product or service provided (Clark and Oswald 1996; Benjamin et al. 2014).

By doing so, this study aims to identify the leadership attributes that influence job satisfaction in three ways. The first part of the paper is theoretical with regard to leadership as a significant part of job satisfaction. Therefore, this research tries to analyze job satisfaction, focusing on the following leadership variables: empowerment, team building, development and recognition, leader's personal values, and vision engagement. They take theoretically part in the employees' well-being at work, and our objective is state the significance of each leadership component in the satisfaction of employees in restaurant sector in Spain. Then, this paper focuses on the empirical analysis. A wide survey (568) of waiters, sumillers, and chefs in restaurants in Barcelona (Spain) is the illustrative case to analyze the leadership factors that influence their professionals' perception of job satisfaction. Finally, the conclusions provide a range of information to organizations related to their behavior and organizational leadership that will allow them to seek out and boost their employees' satisfaction.

In short, the purpose of this study is to consider the importance of leadership in the employee's wellbeing and confirm whether restaurants in Spain take into account and follow the current approaches to lead a group of workers. Following this objective, this paper analyses the significance of those variables that generate satisfaction to employees in the XXI century. Therefore, the first research objective is to know how employees value empowerment. The second research objective focuses on the importance of team building given by employers. In third place, the objective is to known how employees' value career plans and recognition in relation to their performance efforts. Subsequently, is important to know how workers value the personal values of the leader. Finally, the last objective is analyses how employees feel the importance of the alignment of business mission.

2 Job Satisfaction

Many scholars have theorized about job satisfaction or happiness at work, and in general, they define it as the positive affect towards employment, as a collection of feelings or positive responses associated with the job situation (Gruneberg 1979; Spector 1997). These academics have also analyzed the consequences. Job satisfaction leads individuals to be more productive and innovative, to perform their job adequately and effectively, to be more committed, engaged, and enthusiastic, and to be more involved with the organization (Saari and Judge 2004; Mohammad Mosadegh Rad and Hossein Yarmohammadian 2006).

Alexander et al. (1998) state that the absence of job satisfaction entails employees' relinquishment of organizational goals. Therefore, job satisfaction should be the strategic concern and main objective in any enterprise. If organizations focus on their human capital as their added and competitive value, they necessarily have to protect the employees' well-being. Job satisfaction of employees extensively contributes to the achievement of the organizational objectives, and therefore, stimulates positive attitudes such as work engagement, enthusiasm or organizational commitment (Schaufeli and Bakker 2004). Different theorists argue thrill and enthusiasm measured by engagement or organizational commitment (Salas-Vallina et al. 2017; Hakanen et al. 2008).

Engagement involves enthusiasm, energy, activation, dedication, efficacy, pride, alertness, commitment (Macey and Schneider 2008). According Meyer and Allen (1997), this commitment has three components –affective, continuance and normative- and refers to personal involvement, identification with the organization and value congruence. Engagement also refers to the degree to which the job becomes an essential part in an individual's life, that is, a result of meaningfulness at work (Salas-Vallina et al. 2017). These last theorists conclude that happiness at work or job satisfaction has a unique effect on individual's feelings, attitudes and responses about working conditions, opportunities and outcomes. Avoiding brain drain, talent retention focused on motivation and promoting satisfaction of employees is necessary for companies. In that sense, leadership nature and attributes arise as a crucial key issue.

3 The Changing Approach of Leadership

Theories and practices of leadership are decisive to better understand how to lead individuals in an efficient way in new social, cultural and organizational contexts. As McCleskey (2014) states there is a vital challenge to the academic leadership field to develop leaders and leadership. Leader's roles are changing to meet the demands of greatly accelerated technological growth, increasing workforce diversity and multiculturalism, and diverse leisure options and social interests among employees. Businesses have substantially changed nowadays (Sarkar 2016).

The term VUCA (which stands for volatility, uncertainty, complexity and ambiguity) is more and more present in the corporate world due to the rapid changes that take place in the organizational world and reflect an unstable, turbulent and changing business environment. It also involves leadership styles. The notion of VUCA was introduced in leadership by Horney et al. (2010) as the need to make continuous shifts in people, process, technology, and structure, that is, the need to have flexibility and quickness in decision making.

Therefore, throughout XX century, leadership theories have evolved from an approach based on the personal traits to a leadership based on behaviors, environment and contingency characteristics. At the beginning of XXI century, resonance-building styles emphasize on that leadership that supports commitment, involvement, empowerment, active pursuit of the vision, healthy, collaborative, and productive work relationships (Goleman et al. 2002). Therefore, nowadays resonant leadership or emotional leadership (Goleman 1998; Boyatzis and McKee 2005; Sy et al. 2006) invade research debates and insist on how these characteristics differ from traditional or previously existing leadership theories.

A brief review of previous models of leadership, highlights as scholars focused on studying the leader's behavior and his/her leadership styles. Situational leadership theory proposes that effective leadership requires a rational understanding of the situation and an appropriate response, rather than a charismatic leader with a large group of dedicated followers (Graeff 1997; Grint 2011). Contingency or situational theories of leadership state that there is no "one best way" to lead an organization, to organize a cooperation or to make a decision. Contingency theory states that these actions are dependant (contingent) to the internal and external factors. Situational leadership evolved from a task-oriented versus people-oriented leadership continuum (Bass 2008; Graeff 1997; Lorsch 2010). The continuum represented the extent that leader focuses on the required tasks or focuses on their relations with their followers.

Therefore, the effectiveness of leaders' behaviors depends on contextual variables, such as societal values, the organizational culture, the nature of the tasks and the characteristics and maturity of the followers (Hershey and Blanchard 1979; Fiedler 1967; House 1971; Blake and Mouton 1982; Tannenbaum and Schmidt 1973; Ayman 2004). Depending on the situation, leaders vary their behaviours within the boundaries of their style. From a functional perspective, team leaders or supervisors have to manage personnel resources and their progression toward task accomplishment. Therefore, in the long tradition of studying leadership style, most research distinguished between *task-oriented style* or *initiation of structure* and *interpersonally oriented style* or *consideration.* These two constructs originated in the Ohio State and Michigan State approaches to leadership (Stogdill 1974). Initiating structure refers to task oriented and directive behavior. Consideration refers to friendly and interpersonally supportive supervisory behavior (House 1971; Judge and Piccolo 2004; Yammarino et al. 2005; Wendt et al. 2009). Consideration and initiating structure are relatively independent leadership behaviors. As Eagly and Carly (2003) asserted, this distinction means one group of leaders that behave democratically and allow subordinates to participate in decision-making and another group that behave autocratically and discourage subordinates from such participation.

In Leader-Member Exchange theory (LMX), high-quality LMX relationship builds on trust, respect and mutual obligation. Thus, feedback, clarification, recognition and praise for the employees make them feel a sense of obligation to their leader (Graen and Uhl-Bien 1995; Janssen and Yperen 2004; Ferris et al. 2013). Some personal values relate to LMX. For instance, agreeableness positively relates to cooperation, helping behavior, and adaptive social behaviors. In addition, agreeableness is positively associated to reciprocity behavior, which is central to social exchange relationships (Gouldner 1960) and LMX. Therefore, LMX theory posits that over time a leader will exchange a relationship with each subordinate (Mahsud et al. 2010). In this line, *transactional leadership* focuses on the exchanges that occur between leaders and followers (Bass 1985, 1990, 2000, 2008; Burns 1978).

Transactional leaders appeal to subordinates' self-interest by establishing exchange relationships with them. Transactional leadership theory described by Burns (1978) posited the relationship between leaders and followers as a series of exchanges of gratification designed to maximize organizational and individual gains. Transactional leadership involves managing in the conventional sense of clarifying subordinates' responsibilities, rewarding them for meeting objectives, and correcting them for failing to meet objectives (Burns 1978; Kuhnert and Lewis 1987). It is worth to mention that rewarding employees for good performance especially predicts their satisfaction (Hater and Bass 1988).

However, *transformational leadership* is one of the most inspiring leadership styles (Burns 1978; Bass 1985; Avolio 1999) due to leaders and followers achieve higher levels of morality, trust, confidence, and motivation. It generates positive attitudes at work (George and Jones 1997) and higher levels of employee engagement with organizational objectives, even beyond their own self-interests. Burns (1978: 141) defined a transformational leader as "one who raises the followers' level of consciousness about the importance and value of desired outcomes and the methods of reaching those outcomes". Transformational leadership is future oriented rather than present oriented and that strengthens organizations by inspiring followers' commitment and creativity (Burns 1978). Accordingly, transformational leaders promote participative management practices by allowing their subordinates to be involved in the decision-making processes, empowering followers to develop their potential and thus to contribute more effectively to their organization.

As Silins (1994) indicated, transformational leadership bonds leader and followers within a collaborative change process and thus contributes to the performance of the whole organization. They encourage them to take on personally challenging, demanding tasks, to solve problems and to produce innovative ideas and solutions (Dulewicz and Higgs 2005). The transformational leader defines as that leader able to encourage and motivate employees in addition, to earn their trust and respect (Tung 2016). He or she regularly reminds subordinates of the purpose of their work and stir the emotions of the employees. Consequently, some researchers introduced *charismatic leadership* (Conger and Kanungo 1998) due to the relevance of personal traits of leaders to achieve this commitment, enthusiasm, empowerment and engagement of employees.

At present, to get committed people, leadership needs a new focus, based on humbleness, engagement, adaptability, and vision (Bass 2008; DeRue and Wellman 2009) to coach and motivate employees, to bring their talent to work, to achieve a successful organizational transformation. With regards on resonant leadership, a leader defines as a person who goes with someone in order to achieve his/her objective. A person who builds motivation to carry out actions, designing paths/itineraries, projects, pushing a group towards an objective.

These recent approaches recommend a change in hierarchical structures and an increase of leaders in the role of coach or teacher (Eagly and Carli 2003), as their patterns of behaviors influence the followers' actions and team functioning (Demmy et al. 2002; Fisher 2010; Breevaart *et al.* 2014). Therefore, transformational, resonant, emotional and ethical leadership that is collaborative, cooperating and mentoring leadership is more effective than a highly authoritarian person who merely tells others what to do (Mahsud et al. 2010; Mohanty and Mohanty 2016). Ethical leadership is positively associated with idealized influence, individualized consideration, leader effectiveness, employees' willingness to put in extra effort, and satisfaction with the leader (Brown and Treviño 2006; Sharma 2018). According to Mintzberg (2010) true leaders engage others with their consideration and modesty because they involve themselves in what they are actually doing not for individual gains.

To sum up, in recent decades, leadership researchers have attempted to identify the theoretical perspectives of leadership that are most appropriate under the conditions that are common in contemporary organizations. Most valuable and valid theories are transformational leadership, emotional and resonant leadership. The skills and abilities leaders once needed to help their organizations thrive are no longer sufficient.

Today, more strategic, complex critical-thinking skills are required of business leaders. HR and talent management professionals can help their organizations succeed in today's VUCA environament. According to these approaches, in a design of high commitment human resource practices, there are five variables related to these leadership characteristics that can explain the employees' satisfaction: empowerment, team building, development and recognition, leader's personal values, and vision engagement.

Empowerment

There are different approaches for the definition of empowerment as a sharing knowledge, and as a source of motivation that depends on the leadership style. Also, different factors such as psychological and behavioral empowerment affect job satisfaction. According to Kanter (1979), power within an organization can be either productive or oppressive and can determine the effectiveness of a system. She argues that power reflects the ability to mobilize resources as opposed to dominance and influence.

As general literature confirms, leadership that encourages their team members to have self-initiatives, ideas and problem solving, they trust more their own criteria providing more satisfaction. This is a perceptual outcome of LMX, and it is composed of individuals' cognitions that have about their work roles, including meaning of work, competence, self-determination, and impact (Spreitzer 1995). Long et al. (2014) suggest that transformational leadership and employee perceived empowerment increase the employee job satisfaction. According to Ejere and Abasilim (2013), transformational leadership essentially emphasizes the importance of leaders to motivate their subordinates to carry out their responsibilities. In that sense, Liden, et al. (2000) examined the importance of empowerment on the relations between the job, interpersonal relationships, and work outcomes. Therefore, when leaders empower their employees, this empowerment, this confidence generates satisfaction.

Team Building

The quality of coworkers' relationships is a crucial aspect of employee satisfaction. According to social exchange theory (Blau 1964) and expectancy theory (Vroom 1964) the responsibility of a leader in an organization is to create fair and healthy atmosphere that fulfills the needs and expectations of the employees. There are different aspects that contribute to generating favorable teamwork: cohesion, attractiveness, coordination and interaction, frequent and open communication, and psychological safety. Cohesion is the tendency of a set of individuals to stay harmoniously joined to achieve the group goals. According to Birx et al. (2011) group cohesion is an outcome of team-building activities and is associated with positive feelings toward coworkers, which are an aspect of job satisfaction.

Attractiveness has to do with the satisfaction due to the sense of belonging, the support of the rest of the employees, the feeling of being part of something more important, greater. The level of work coordination and interaction also has effects on teamwork behavior. As Yang (2016: 4) insists: "teams can maximize members' cooperation with colleagues, volunteering for tasks that go beyond their formal work requirements and exhibiting helping behaviors toward others". Porath and Pearson (2010) describe the importance of psychological safety that creates a team, which guarantees engagement in interactions with other individuals, help and feedback. Thus, a positive teamwork atmosphere provides meaningful insights into organizational sustainability and well-being.

Sohmen (2013) underlines that leaders must create a work environment that is conducive to team members working together in cooperative and goal-oriented efforts. Thus, effective leadership is clearly imperative in order to induce team building. Even if the project team is high performing with the right capabilities, it will not be successful in the absence of effective leadership. According to Jibao (2018), transformational leaders stimulate their subalterns to prefer the interests of the team and organization to personal interests. Therefore, when leaders focus their attention on team building creation, looking for the cooperation and coordination of their members, they achieve employees' satisfaction.

Development and Recognition

Recognition refers to the training and promotion opportunities that organizations offer its staff. These opportunities are linked to the performance assessment of each employee. This is the contingent reward dimension of transactional leadership, which suggest that the leaders give special recognition at the good performance of employees (Asghar and Oino 2017).

The organizations use training and development practices to improve the ability of the workforce to implement the firm's business strategy. Improving the competence of the employees is one way to create competitive advantage, organizational cohesiveness and employee commitment. By offering training and development opportunities, employers help employees develop their own personal competitive advantage and ensure their long-term employability (Tannenbaum 2002; Salas and Kozlowski 2009). A common source of employee dissatisfaction is the lack of career advancement opportunities.

When employees do not receive such opportunities, they are likely to not feel committed, loyal, enthusiastic to the organization, and they constantly think to leave. Training and development opportunities are a form of attractive recognition of talented workers. Competitive models often guide the design of career development activities. Organizations that invest in career management are more likely to increase employees' job satisfaction, and organizational commitment (Lee 2000). Adeloka (2011) proposes that career development has a direct influence on the achievement of job satisfaction and career commitment.

Therefore, to reach employees' satisfaction, leaders have to plan professional careers for their subordinates and recognize their performance efforts and commitment.

Personal Values

Another leadership characteristic that generates satisfaction to employees are the personal and moral values (Robbins and Judge 2016). The moral concepts and ethical behavior of leaders and managers have a wide effect on the moral concepts and ethical behavior of their subordinates (staff), and consequently on the organizational processes and results (Boria Reverter et al. 2013; Crespi-Vallbona and Mascarilla-Miró 2018). Therefore, leaders are key organizational figures to embody a caring attitude towards the rest of employees. The personal values embody assumptions of goodness, integrity, honesty, humbleness, civility and virtue. On the other hand, moral wrongdoing may undermine the morality of an organization and legitimize all kinds of (negative) behavior.

Personal characteristics have to do with humbleness, when leaders accept different opinions and recognize that others can know more than they can. In addition, this is achieved with empathy, when leaders show the will to active listening, interaction and communication with the internal and external interested parts. Furthermore, they accept that changes are a constant issue in organizations and working conditions, and therefore they constantly adapt to new changes. That is, they have integrated adaptability, flexibility and openness to work.

As Schminke et al. (2005) say, the actual congruence between the leader's moral development and the employee's moral development affects job satisfaction, commitment, and turnover intentions. Some evidences also show that the employees who perceived their supervisor as more approachable and responsive were more likely to voice their concerns (Aydogdu and Asikgil 2011). Leadership style is understood as a series of managerial attitudes, behaviors, characteristics and skills based on individual and organizational values, leadership interests and reliability of employees in different situations.

This factor captures the extent to which management respects workers, operates with honesty and integrity, promotes efficiency, and has open lines of communication with employees (Aronson et al. 2006). Therefore, the leader's ethical and personal values (such as humbleness, sense of honor, transparency, adaptability, flexibility) influence the employees' satisfaction.

Vision Engagement

According to Bass (1985), charisma is the ability of individuals to arouse people and bring them to follow the leader's mission and vision. Transformational managers make subordinates aware of how important their jobs are for the organization and how necessary it is for them to perform those jobs as best they can, so the organization can attain its goals. When employees are satisfied with their jobs and view their work as important, they are much more likely to provide high-quality service, to improve their working conditions (Tims et al. 2011). Therefore, "leadership is more than simply having a vision and mission and communicating them to the team" (Levin 2005: 666), which means that the leader needs to encourage employees to become good team players. There are practices that encourage employees to get involved in firm activities and create conditions in which employees feel trusted to carry freely out tasks that are in line with the organizational goal (Nam and Lee 2018). Employees who trust their leader are more likely to accept and contribute to the work process and display full commitment to accomplish organizational objectives (Boria Reverter et al. 2013; Long et al. 2014; Crespi-Vallbona and Mascarilla-Miró 2018). Therefore, the alignment of business mission that leader transmits to his/her subordinates influence their satisfaction and wellbeing (Simone et al. 2016).

4 Conceptual Framework and Research Hypothesis

Considering the literature revised, the leadership nature that generate satisfaction to employees in the XXI century are: empowerment, team building, development and recognition, personal values and vision commitment. All of them are determinants of employees' satisfaction (Fig. 1):

H1. When leaders empower their employees, this empowerment, this confidence generates satisfaction.
H2. When leaders focus their attention on team building creation, looking for the cooperation and coordination of their members, it achieves employees' satisfaction.
H3. When leaders plan professional careers for their subordinates and recognize their performance efforts, employee' satisfaction is achieved.
H4. The leader's ethical and personal values (such as humbleness, sense of honor, transparency, adaptability, flexibility) influence the employees' satisfaction.
H5. The alignment of business mission that leader transmits to his/her subordinates influence their satisfaction and wellbeing.

5 Leadership and Job Satisfaction in the Restaurant Sector: The Case Study

The restaurant sector in Spain is one driving force of the economy, since 2015 (Table 1), thanks to tourism, consumers' new tastes and habit evolution. In 2018, restaurant sector increased its sales 2′1% last year (NDP Group 2019), leaving crisis behind and reinforcing three consecutive years of growth. According to Spanish National Statistics (INE 2019), in 2017, 8'7% of employed population works at restaurant sector and this represents 6′8% GPD. At present, there are nearly 280.000 establishments of catering (restaurants and bars), most of them micro and small businesses that have from one to 5 employees. Therefore, the main aim of the research is to determine the leadership factors that influence the perception of job satisfaction among the staff in restaurants, one of the most active sector in Spanish economy.

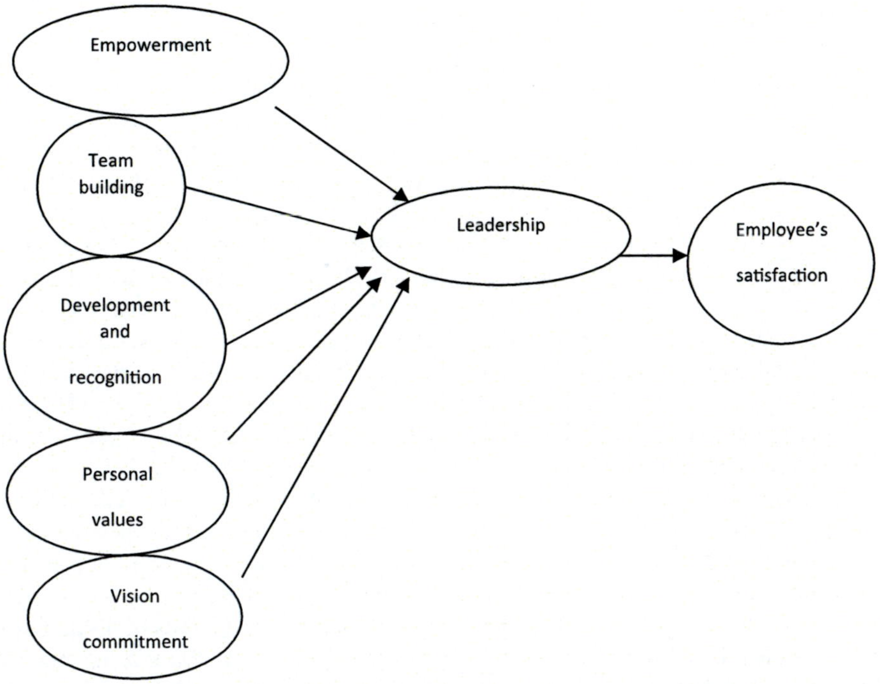

Fig. 1. Current leadership nature that generate satisfaction to employees. Source: Own elaboration

Table 1. Sale evolution in % of the catering sector in Spain (2009–2018)

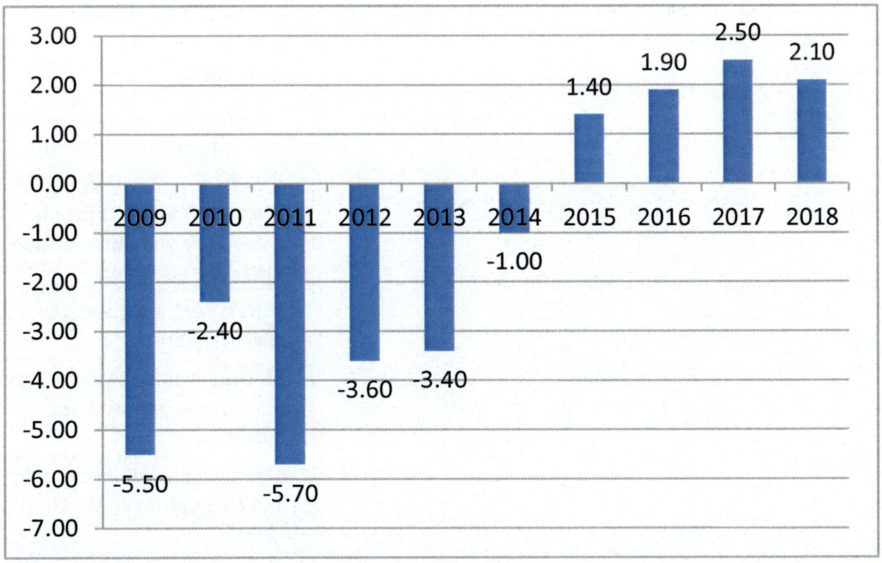

Source: Own elaboration from NDP Group Data (2019)

6 Methodology

In order to be able to carry out an in-depth analysis of the employees' satisfaction related to leadership nature; a survey was prepared addressed to employees in restaurant sector in Barcelona. Data were collected in the Barcelona area between November 2017 and March 2018, visiting 150 business that include gastronomy and food services. 90 of them agreed to participate in this research. These chosen organizations locate in the central and tourist area of the city. They have hierarchical organizational structures; even they are small size organizations. For data collection, these companies were contacted and informed about the purpose of the research, confidentiality issues, and the reporting of the results. Then, after the approval of the business manager, a questionnaire was administered to all staff members at the beginning of each employee's labour day. More than 600 employees (waiters, sumillers, and chefs) answered questions related to leadership variables that lead to job satisfaction, but just 568 questionnaires were valid for the study as they were completely answered.

The questionnaire bases on the aforementioned dimensions of current leadership: empowerment, team building, development and recognition, leader's personal values and vision engagement. The questionnaire structure and questions were developed by the own researchers based on the previous review of the leadership theories and those academics who previously studied each variable (Table 2). It consists of 43 items rated on a 5-point Likert scale (1 "almost always untrue ", 2 "often untrue", 3 "sometimes true/sometimes untrue", 4 "often true", 5 "almost always true"). Job satisfaction was measured through "job satisfaction measuring scale (JSS)" developed by Spector (1997) consisting of 10 questions.

Table 2. Prepared questions to leadership dimensions related to job satisfaction

Dimension	References	Questions
Empowerment (8)	Liden et al. (2000) Kanter (1979)	Management does a good job of assigning and coordinating people
		Management trusts people to do a good job without watching over their shoulders
		People here are given a lot of responsibility
		Management recognizes honest mistakes as part of doing business
		Management genuinely seeks and responds to suggestions and ideas

(*continued*)

Table 2. (*continued*)

Dimension	References	Questions
		Management involves people in decisions that affect their jobs or work environment
		My work has special meaning: this is not "just a job."
		I can be myself around here
Team building (11)	Braun et al. (2013) Hunjra et al. (2010) Acuña et al. (2009)	I am able to take time off from work when I think it's necessary
		People are encouraged to balance their work life and their personal life
		Management shows a sincere interest in me as a person, not just an employee
		When I look at what we accomplish, I feel a sense of pride
		People here are willing to give extra to get the job done
		People celebrate special events around here
		People care about each other here
		This is a fun place to work
		There is a "family" or "team" feeling here
		We are all in this together
		You can count on people to cooperate
Development and Recognition (8)	Danish and Usman (2010) Boon Ooi et al. (2007)	I am offered training or development to further myself professionally
		I am given the resources and equipment to do my job
		Management shows appreciation for good work and extra effort
		Everyone has an opportunity to get special recognition

(*continued*)

Table 2. (*continued*)

Dimension	References	Questions
		I am treated as a full member here regardless of my position
		Promotions go to those who best deserve them
		I feel I make a difference here
		I can be myself around here
Personal Values (11)	Mancheno-Smoak et al. (2009) Krishnan (2001)	Management keeps me informed about important issues and changes
		I can ask management any reasonable question and get a straight answer
		Management is approachable, easy to talk with
		Management delivers on its promises
		Management's actions match its words
		I believe management would lay people off only as a last resort
		Management is honest and ethical in its business practices
		Managers avoid playing favorites
		People avoid politicking and backstabbing as ways to get things done
		If I am unfairly treated, I believe I'll be given a fair shake if I appeal
		When you join the company, you are made to feel welcome

(*continued*)

Table 2. *(continued)*

Dimension	References	Questions
		When people change jobs or work units, they are made to feel right at home
Vision Commitment (5)	Slack et al. (2010) Boria Reverter et al. (2013)	Management makes its expectations clear
		Management is competent at running the business
		Management has a clear view of where the organization is going and how to get there
		Management hires people who fit in well here
		I feel good about the ways we contribute to the community

Source: Own elaboration

Prior to distribution, the questionnaire was preliminarily tested in order to formulate questions that would be clear and understandable to the respondents. Therefore, a preliminary survey was performed on a small group of 30 randomly selected respondents. By analyzing the suggestions obtained from the pilot sample, minor modifications were made to the initially formulated statements, shaping the final version of the questionnaire.

Data analysis was conducted using the Statistical Package for Social Sciences (SPSS) (version 21.0) and the Analysis of a Moment Structures (AMOS-in) (version 18.0) in order to determine the statistically significant drivers of employees' satisfaction and to study the relationships between the aforementioned different variables and the weight that each of them places on the others. Moderated regression analysis was used to test the main and interaction effects. In addition, the independent variables were first centered and then multiplied by each other, in order to avoid a multicollinearity problem that may occur when calculating interaction effects.

7 Results and Discussion

The structure of the sample consists of 613 respondents. However, only 568 questionnaires were completely answered by respondents. 50,3% of respondents were male and 49,7% were female.

To determine the influence of the five leadership dimensions (empowerment, team building, development and recognition, leader's personal values and vision engagement) on employees' satisfaction, a multiple linear regression model was performed. Furthermore, it was conducted a reliability analysis and confirmatory factor analysis. Values of Cronbach's alpha for all five variables are shown in Table 3. All variables have an adequate level of reliability as the coefficient of Cronbach's alpha is over the required threshold of 0.7 (Nunnally 1978).

As results suggest, the highest level of reliability is manifested in the "Development and recognition" variable (Cronbach's alpha = 0.928), whereas in the case of the personal values variable a slightly lower level of internal consistency of the statements is present (Cronbach's alpha = 0.857). Furthermore, the proposed leadership dimensions also have an acceptable level of convergent validity, as the standard deviation (SD) is close to the mean. Those results confirm that in a disruptive labour world, staff assesses positively the personal recognition and development.

Table 3. Means, standard deviations and correlations among leadership dimensions

Variable	Mean (*DT*)	*Alfa*	1	2	3	4	5
1. Satisfaction	4,19 (0,78)	0,889	1				
2. Empowerment	4,09 (0,72)	0,92	,822**	1			
3. Team Building	4,1 (0,75)	0,89	,876**	,868**	1		
4. Development and Recognition	4,14 (0,74)	0,928	,843**	,893**	,879**	1	
5. Personal vàlues	4,57 (0,76)	0,857	,822**	,904**	,879**	,909**	1
6. Vision engagement	4,2 (0,74)	0,909	,829**	,854**	,860**	,860**	,902**

****p < 0,001

According to the results of the regression analysis, there is a statistically significant relationship between all independent variables of leadership and employees' satisfaction (F (7,430 = 256,83, p < 0,001) explaining the 80,6% of the satisfaction variability of all variables of the proposed model. They also show the rank of importance of all of them (see Table 4). Thus, regarding the predictive effect of each of the independent variables, the satisfaction score observed was significantly associated with high levels of empowerment, team building, development and recognition and vision commitment.

With respect to the predictive weight of the significant variables (partial r in absolute value), team building is the best predictor of satisfaction (partial r = 0.373) followed by development and recognition (partial r = 0.192), vision commitment (partial r = 0.179) and empowerment (partial r = 0.101), at a great distance. Personal values show the less statistically significant effect on employees' satisfaction (partial r = 0.065). Therefore, employees' answers reflect their strong interest in working with consolidated, collaborative, cohesive teams.

According to the results of the study, there is a statistically significant relationship between all independent variables and employee satisfaction related to proposed leadership dimensions. A detailed analysis of each hypothesis is presented, according to the significance of each leadership dimension.

Table 4. Effect of leadership variables to employees' satisfaction

Variable	B (ET)	Beta	$IC_{95\%}$ B	*t*	*p*-valor	*r* parcial
Empowerment	0,14 (0,07)	0,13	−0,04; 0,08	2,1	**0,037**	0,101
Team Building	0,45 (0,05)	0,43	0,01; 0,27	8,28	**<0,001**	0,373
Development and Recognition	0,26 (0,06)	0,24	0,34; 0,56	4,01	**<0,001**	0,192
Personal Values	−0,09 (0,07)	−0,09	0,13; 0,38	−1,34	0,181	−0,065
Vision Commitment	0,21 (0,06)	0,21	−0,22; 0,04	3,75	**<0,001**	0,179
Model summary						
R^2 (%)	80,6					
Model	$F(7{,}430 = 256{,}83, p < 0{,}001$					
Assumptions						
Normality†	p = 0,341					
Independence‡	1,99					
Homocedasticity+	p = 0,767					

B: non-standardized coefficients. ET: typical error. Beta: standardized coefficients. †Kolmogorov-Smirnov normality test of residues. ‡Durbin-Whatson test+ +Levene test between residual and predicted values. *p < 0.05 **p < 0.01 ***p < 0.001.

H1. **Empowerment** generates satisfaction. This hypothesis in not supported. Restaurant employees consider empowerment as the less relevant satisfaction variable. Once the menus are designed, plates are thought joined with its accompanying beverage (often wine in Spanish culture), their elaboration follows a very repetitive patterns, which do not leave margin to innovation, creativity or imagination. Thus, this empowerment, this delegation in the restaurant sector is not too valued as satisfactory element due to the short margin of freedom in the standardized processes.
H2. **Team building** provides employees' satisfaction. This hypothesis is supported, it means that team building is one of the best employee satisfaction predictors. That is, in restaurant sector, employees consider that a cohesive, joined-up, coordinated team must be the most important goal for a leader. In the restaurant sector, the quality and good service to diners are the main objectives. They refer to and require an adequate attention of waiters, a short waiting time between the consumer's order and servings and beverages served on table, a quality of food and a nice presentation of cooked plates. It necessarily involves that all the staff must focus on their tasks, working comfortably with all the mates to prepare the food and serve it on the right time, in a calm and ease work atmosphere. All the personnel should collaborate and coordinate, being cohesive and complemented one each other. Leader is responsible to build this team atmosphere as respondents assess it. That is, aspects such as cohesion, attractiveness, coordination, interaction with colleagues, and open communication are also decisive to guaranteeing employees' satisfaction. The effort of leaders must therefore focus on designing very

cohesive work teams, so that the atmosphere of cooperation, friendship, collaboration, helping behavior, mutual understanding, and positive and fluent communication are the daily way to act.

H3. **Development and recognition** generate employees' satisfaction. This hypothesis is also supported, as when leaders plan professional careers for their subordinates and recognize their performance efforts, employees' satisfaction is achieved. Training and career plans offer opportunities to the employees' development and those are relevant variables to restaurant employees. Employees respond with support, engagement and commitment to the organizational tasks when the organization in turn shows them support, respect and recognition. This recognition depends on their performance assessment, and the professional opportunities to promotion. Unfortunately, the organizational structure and size of most restaurant businesses cannot offer real professional promotions and career plans. Then, leaders must focus on recognition, that is, give personal value to each employee, assessing each personal effort and performance by days off, licenses, permissions, training courses of specialized food, wine tasting, etc. Prior studies have found that individual consideration (a trait of transformational and resonant leadership styles) tends to be more effective in the short term as compared to the long term. Leaders who are more considerate tend to enhance employees' performance in the short term (Epitropaki and Martin 2005).

H4. The leader's **personal values** influence the employees' satisfaction. This hypothesis is not supported as respondents consider leader's personal values dimension has no effect on job satisfaction. However, they insist on leadership's main goal is looking after and even pampering their subordinates or collaborators. Any individual has peculiar interests, own inquisitiveness and professional motivations. The employees' perception facing the presence of unfair treatment among workers, the lack of recognition, the disinterest in employees' health benefits, etc. yield labor environments without personal and organizational well-being. The general management of the organization, and in particular middle line managers have to focus on these aspects and pay attention to employee satisfaction.

H5. Leader's **vision commitment** influences employees' satisfaction and wellbeing. This hypothesis is supported as vision engagement is an important leadership dimension for employee's satisfaction in restaurant sector. Shared values, committing with vision of the organization is key to staff satisfaction and to achieve organizational objectives. When there is no alignment by both parts, employee and employer, the discomfort, unrest appear. People works in an annoying way. Service sector has an especial consideration as workers have to show a sense of vocation to serve, help, like the customers, feel the prepared meal.

8 Conclusions

The aim of this article is to identify the current personal and professional leader characteristics in restaurant sector in SME Spanish organizations, as leadership directly relates to job satisfaction and staff's well-being. Following this objective, this paper analyses five variables as determinants of satisfaction to employees in the XXI century: empowerment, team building, development and recognition, leader's personal values and vision engagement. Not all of them generate employees' satisfaction. Employees consider team building, as the worthiest variable leaders have to take into account. Cooperation, collaboration and coordination between kitchen and dinner service are essential to offer a good and quality service to dinners. That is, to work in a good conditions and atmosphere, do tasks correctly and to provide the best catering service.

Furthermore, employees deeply appreciate their professional development (whenever the organization structure affords it) and the employer's recognition of their performance efforts. In addition, it is worth to mention that employees respond with support, engagement and commitment to the organizational tasks, vision and mission when the organization in turn shows them support, respect, fair treatment and recognition.

Finally, employees in restaurant sector state empowerment and personal values dimensions of leadership are less and not relevant, respectively in their job satisfaction.

9 Practical Implications

As results show, catering industry should pay special attention on team group cohesiveness and cooperation, as it is the must variable related to job performance and satisfaction. Organizations do not have to neglect it, if they are interested in their staff wellbeing. Team building states as an important tool to improve the organizational behavior in all terms. Considered as a way that tasks and responsibilities are been assigned through internal processes, team building pursues dealing in the best way with the internal conflicts within the teams; solve ambiguity in tasks as the commitment to achieve the organization expectations.

This is, to generate an effective environment from the beginning. Being part of the formation or decision making about the team membership, employees feel better climate in the working environment, they develop better interpersonal and participation skills and they are more orientated to objectives, as part or job satisfaction. To sum up, team group cohesiveness is the must related to job satisfaction.

Teams characterize by a common purpose, clear goals, competent members, a unified commitment, complementary skills, a collaborative climate, mutual accountability, standards of excellence, and principled leadership.

Limitations and Future Research

The study bases on leadership dimensions as central components of employee satisfaction. However there are external factors that can also affect personnel wellbeing. The empirical study is focused on Barcelona, a touristified area where consumers come from around the world, with peculiar expectations. Future research should compare these results with other areas less touristic and with more number of local consumers, just to analyze if this variable can mediate in the final job satisfaction.

References

Acuña, S.T., Gómez, M., Juristo, N.: How do personality, team processes and task characteristics relate to job satisfaction and software quality? Inf. Softw. Technol. **51**(3), 627–639 (2009)

Adekola, B.: Career planning and career management as correlates for career development and job satisfaction. A case study of Nigerian Bank Employees. Aust. J. Bus. Manag. Res. **2**, 100–112 (2011)
Alexander, J.A., Lichtenstein, R.L., Oh, H.J., Ullman, E.: A causal model of voluntary turnover among nursing personnel in long-term psychiatric settings. Res. Nurs. Health **21**(5), 415–427 (1998)
Armstrong, M.: A Handbook of Human Resource Management Practice. Kogan Page Publishers (2006)
Aronson, Z.H., Reilly, R.R., Lynn, G.S.: The impact of leader personality on new product development teamwork and performance: the moderating role of uncertainty. J. Eng. Technol. Manag. **23**(3), 221–247 (2006)
Asghar, S., Oino, D.: Leadership styles and job satisfaction. Market Forces **XIII**(1), 1–13 (2017)
Avolio, B.J.: Full Leadership Development: Building the Vital Forces in Organizations. Sage (1999)
Aydogdu, S., Asikgil, B.: An empirical study of the relationship among job satisfaction, organizational commitment and turnover intention. Int. Rev. Manag. Mark. **1**(3), 43–53 (2011)
Ayman, R.: Situational and contingency approaches to leadership (2004)
Aziri, B.: Job satisfaction: a literature review. Manag. Res. Pract. **3**(4), 77–86 (2011)
Bass, B.M.: Leadership and Performance beyond Expectations. Free Press, New York (1985)
Bass, B.M.: From transactional to transformational leadership: learning to share the vision. Organ. Dyn. **18**(3), 19–31 (1990)
Bass, B.M.: The future of leadership in learning organizations. J. Leadersh. Organ. Stud. **7**(3), 18–40 (2000)
Bass, B.M.: The Bass Handbook of Leadership: Theory, Research, & Managerial Applications, 4th edn. Free Press, New York (2008)
Barling, J., Loughlin, C., Kelloway, E.K.: Development and test of a model linking safety-specific transformational leadership and occupational safety. J. Appl. Psychol. **87**(3), 488–496 (2002)
Beardwell, J., Claydon, T. (eds.): Human Resource Management: A Contemporary Approach. Pearson Education (2007)
Benjamin, D.J., Heffetz, O., Kimball, M.S., Szembrot, N.: Beyond happiness and satisfaction: toward well-being indices based on stated preference. Am. Econ. Rev. **104**(9), 2698–2735 (2014)
Birx, E., LaSala, K.B., Wagstaff, M.: Evaluation of a team-building retreat to promote nursing faculty cohesion and job satisfaction. J. Prof. Nurs. **27**(3), 174–178 (2011)
Blake, R., Mouton, J.S.: Theory and research for developing a science of leadership. J. Appl. Behav. Sci. **18**(3), 275–291 (1982)
Blau, P.M.: Power and Exchange in Social Life. Wiley, New York (1964)
Bligh, M.C.: Followership and follower-centered approaches, pp. 425–436 (2011)
Boon Ooi, K., Abu Bakar, N., Arumugam, V., Vellapan, L., Kim Yin Loke, A.: Does TQM influence employees' job satisfaction? An empirical case analysis. Int. J. Qual. Reliab. Manag. **24**(1), 62–77 (2007)
Boria Reverter, S., Crespi Vallbona, M., García González, A., Vizuete Luciano, E.: Los Valores Compartidos en la Empresa Española. Universia Bus. Rev. **37**(1), 68–85 (2013)
Boyatzis, R.E., Boyatzis, R., McKee, A.: Resonant Leadership: Renewing Yourself and Connecting with Others Through Mindfulness, Hope, and Compassion. Harvard Business Press (2005)
Braun, S., Peus, C., Weisweiler, S., Frey, D.: Transformational leadership, job satisfaction, and team performance: a multilevel mediation model of trust. Leadersh. Q. **24**(1), 270–283 (2013)
Breevaart, K., Bakker, A., Hetland, J., Demerouti, E., Olsen, O.K., Espevik, R.: Daily transactional and transformational leadership and daily employee engagement. J. Occup. Organ. Psychol. **87**(1), 138–157 (2014)

Brown, M.E., Treviño, L.K.: Ethical leadership: a review and future directions. Leadersh. Q. **17**(6), 595–616 (2006)
Bryman, A., Collinson, D., Grint, K., Jackson, B., Uhl-Bien, M. (eds.): The SAGE Handbook of Leadership. Sage, Thousand Oaks (2011)
Burns, J.M.: Leadership. Harper and Row, New York (1978)
Clark, A.E., Oswald, A.J.: Satisfaction and comparison income? J. Public Econ. **61**(3), 359–381 (1996)
Conger, J.A., Kanungo, R.N.: Charismatic Leadership in Organizations. Sage Publications (1998)
Crespi-Vallbona, M., Mascarilla Miró, O.: Job satisfaction. The case of information technology (IT) professionals in Spain. Universia Bus. Rev. **58**, 36–51 (2018)
Danish, R.Q., Usman, A.: Impact of reward and recognition on job satisfaction and motivation: an empirical study from Pakistan. Int. J. Bus. Manag. **5**(2), 159–167 (2010)
Demmy, T.L., Kivlahan, C., Stone, T.T.: Physicians' perceptions of institutional and leadership factors influencing their job satisfaction at one academic medical center. Acad. Med. **77**(12), 1235–1240 (2002)
DeRue, D.S., Wellman, N.: Developing leaders via experience: the role of developmental challenge, learning orientation, and feedback availability. J. Appl. Psychol. **94**(4), 859–875 (2009)
Dolan, S., Shuler, R., Valle, R.: La gestión de los recursos humanos. McGraw-Hill, Madrid (2007)
Dulewicz, V., Higgs, M.: Assessing leadership styles and organisational context. J. Manag. Psychol. **20**(2), 105–123 (2005)
Eagly, A.H., Carli, L.L.: The female leadership advantage: an evaluation of the evidence. Leadersh. Q. **14**(6), 807–834 (2003)
Ejere, E.I., Abasilim, U.D.: Impact of transactional and transformational leadership styles on organisational performance: empirical evidence from Nigeria. J. Commer. **5**(1), 30–41 (2013)
Epitropaki, O., Martin, R.: The moderating role of individual differences in the relation between transformational/transactional leadership perceptions and organizational identification. Leadersh. Q. **16**(4), 569–658 (2005)
Ferris, D.L., Johnson, R.E., Rosen, C.C., Djurdjevic, E., Chang, C.H.D., Tan, J.A.: When is success not satisfying? Integrating regulatory focus and approach/avoidance motivation theories to explain the relation between core self-evaluation and job satisfaction. J. Appl. Psychol. **98**(2), 342 (2013)
Fiedler, F.E.: A Theory of Leadership Effectiveness. McGraw-Hill Series in Management (1967)
Fisher, C.: Happiness at work. Int. J. Manag. Rev. **12**(4), 384–412 (2010)
George, J., Jones, G.: Experiencing work: values, attitudes, and moods. Hum. Relat. **50**(4), 393–416 (1997)
Ghiselli, R.F., La Lopa, J.M., Bai, B.: Job satisfaction, life satisfaction, and turnover intent: among food-service managers. Cornell Hotel Restaurant Adm. Q. **42**(2), 28–37 (2001)
Goleman, D., Boyatzis, R., McKee, A.: The emotional reality of teams. J. Organ. Excell. **21**(2), 55–65 (2002)
Goleman, D.: La inteligencia emocional en la práctica. Editorial Kairós, Barcelona (1998)
Gouldner, A.W.: The norm of reciprocity: a preliminary statement. Am. Sociol. Rev. 161–178 (1960)
Graeff, C.L.: Evolution of situational leadership theory: a critical review. Leadersh. Q. **8**(2), 153–170 (1997)
Graen, G.B., Uhl-Bien, M.: Relationship-based approach to leadership: development of leader-member exchange (LMX) theory of leadership over 25 years: applying a multi-level multi-domain perspective. Leadersh. Q. **6**(2), 219–247 (1995)
Grint, K.: A history of leadership. In: Bryman, A., Collinson, D., Grint, K., Jackson, B., Uhl-Bien, M. (eds.) The SAGE Handbook of Leadership, pp. 3–14. Sage, Thousand Oaks (2011)

Gronn, P.: Hybrid configurations of leadership. In: Bryman, A., Collinson, D., Grint, K., Jackson, B., Uhl-Bien, M. (eds.) The SAGE Handbook of Leadership, pp. 437–454. Sage, Thousand Oaks (2011)

Gruneberg, M.M.: Understanding Job Satisfaction. Macmillan, London (1979)

Hakanen, J.J., Schaufeli, W.B., Ahola, K.: The job demands-resources model: a three-year cross-lagged study of burnout, depression, commitment, and work engagement. Work Stress **22**(3), 224–241 (2008)

Hater, J.J., Bass, B.M.: Superiors' evaluations and subordinates' perceptions of transformational and transactional leadership. J. Appl. Psychol. **73**(4), 695–702 (1988)

Hersey, P., Blanchard, K.H.: Life cycle theory of leadership. Train. Dev. J. **33**(6), 94 (1979)

Horney, N., Pasmore, B., O'Shea, T.: Leadership agility: a business imperative for a VUCA world. People Strategy **33**, 4 (2010)

House, R.J., Hanges, P.J., Javidan, M., Dorfman, P.W., Gupta, V. (eds.): Culture, Leadership, and Organizations: The GLOBE Study of 62 Societies. Sage Publications (2004)

House, R.J.: A path goal theory of leader effectiveness. Adm. Sci. Q. **16**(3), 321–339 (1971). https://doi.org/10.2307/2391905

Hunjra, A.I., Chani, D., Irfan, M., Aslam, S., Azam, M., Rehman, K.U.: Factors effecting job satisfaction of employees in Pakistani banking sector. Afr. J. Bus. Manag. **4**(10), 2157–2163 (2010)

Janssen, O., van Yperen, N.W.: Employees' goal orientations, the quality of leader-member exchange, and the outcomes of job performance and job satisfaction. Acad. Manag. J. **47**(3), 368–384 (2004)

Jibao, G., Wang, G., Liu, H., Song, D., Changqing, H.: Linking authoritarian leadership to employee creativity: The influences of leader-member exchange, team identification and power distance. Chin. Manag. Stud. **12**(2), 384–406 (2018)

Judge, T.A., Piccolo, R.F.: Transformational and transactional leadership: a metaanalytic test of their relative validity. J. Appl. Psychol. **89**(5), 755–768 (2004)

Kanter, R.M.: Power failure in management circuits. Harv. Bus. Rev. **57**(4), 65–75 (1979)

Krishnan, V.R.: Value systems of transformational leaders. Leadersh. Org. Dev. J. **22**(3), 126–132 (2001)

Kuhnert, K.W., Lewis, P.: Transactional and transformational leadership: a constructive/developmental analysis. Acad. Manag. Rev. **12**(4), 648–657 (1987)

Lee, S.H.: A managerial perspective of the objectives of HRM practices in Nigeria: an exploratory study. Nigeria Manag. Rev. **22**(1), 65–82 (2000)

Levin, R.: Leadership and team building. J. Am. Dent. Assoc. (1939), **136**(5), 666–667 (2005)

Liden, R.C., Wayne, S.J., Sparrowe, R.T.: An examination of the mediating role of psychological empowerment on the relations between the job, interpersonal relationships, and work outcomes. J. Appl. Psychol. **85**(3), 407–416 (2000)

Locke, E.A.: What is job satisfaction? Organ. Behav. Hum. Perform. **4**(4), 309–336 (1969)

Long, C.S., Yusof, W.M.M., Kowang, T.O., Heng, L.H.: The impact of transformational leadership style on job satisfaction. World Appl. Sci. J. **29**(1), 117–124 (2014)

Lorsch, J.: A contingency theory of leadership. Handb. Leadersh. Theory Pract. **13**(1), 411–432 (2010)

Macey, W.H., Schneider, B.: The meaning of employee engagement. Ind. Organ. Psychol. **1**(1), 3–30 (2008)

Madlock, P.E.: The link between leadership style, communicator competence, and employee satisfaction. J. Bus. Commun. 1973 **45**(1), 61–78 (2008). https://doi.org/10.1177/0021943607309351

Mahsud, R., Yukl, G., Prussia, G.: Leader empathy, ethical leadership, and relations-oriented behaviors as antecedents of leader-member exchange quality. J. Manag. Psychol. **25**(6), 561–577 (2010)

Mancheno-Smoak, L., Endres, G.M., Polak, R., Athanasaw, Y.: The individual cultural values and job satisfaction of the transformational leader. Organ. Dev. J. **27**(3), 9–21 (2009)
McCleskey, J.A.: Situational, transformational, and transactional leadership and leadership development. J. Bus. Stud. Q. **5**(4), 117 (2014)
Meyer, J.P., Allen, N.J.: Commitment in the Workplace. Sage, Thousand Oaks (1997)
Mintzberg, H.: Developing leaders? Developing countries? Oxford Leadersh. J. **1**(2) (2010)
Mohammad Mosadegh Rad, A., Hossein Yarmohammadian, M.: A study of relationship between managers' leadership style and employees' job satisfaction. Leadersh. Health Serv. **19**(2), 11–28 (2006)
Mohanty, S., Mohanty, K.: Human asset management, a challenge for service sector: an analysis from employee retention perspective. Int. J. Appl. Bus. Econ. Res. **14**(119), 8271–8298 (2016)
Nam, J., Lee, H.: High commitment human resource practices and employee behavior: a multi-level analysis. Int. J. Manpow. **39**(5), 674–686 (2018)
Nunnally, J.: Psychometric Methods, 2nd edn. (1978)
Porath, C.L., Pearson, C.M.: The cost of bad behavior. Organ. Dyn. **39**(1), 64–71 (2010)
Robbins, S., Judge, T.A.: Essentials of Organizational Behavior. Pearson Education Inc. (2016)
Saari, L.M., Judge, T.A.: Employee attitudes and job satisfaction. Hum. Resour. Manag.: Published in Cooperation with the School of Business Administration, The University of Michigan and in Alliance with the Society of Human Resources Management **43**(4), 395–407 (2004)
Sarkar, A.: We live in a VUCA World: the importance of responsible leadership. Dev. Learn. Organ. Int. J. **30**(3), 9–12 (2016)
Salas, E., Kozlowski, S. (eds.): Learning, Training and Development in Organizations. Routledge Academic, London (2009)
Salas-Vallina, A., López-Cabrales, A., Alegre, J., Fernández, R.: On the road to happiness at work (HAW). Transformational leadership and organizational learning capability as drivers of HAW in a healthcare context. Pers. Rev. **6**(2), 314–338 (2017)
Saleem, H.: The impact of leadership styles on job satisfaction and mediating role of perceived organizational politics. Procedia Soc. Behav. Sci. **172**, 563–569 (2015)
Schaufeli, W.B., Bakker, A.B.: Job demands, job resources, and their relationship with burnout and engagement: a multi-sample study. J. Organ. Behav. Int. J. Ind. Occup. Organ. Psychol. Behav. **25**(3), 293–315 (2004)
Schminke, M., Ambrose, M.L., Neubaum, D.O.: The effect of leader moral development on ethical climate and employee attitudes. Organ Behav. Hum. Decis. Process. **97**(2), 135–151 (2005)
Sharma, D.: When fairness is not enough: impact of corporate ethical values on organizational citizenship behaviors and worker alienation. J. Bus. Ethics **150**(1), 57–68 (2018). https://doi.org/10.1007/s10551-016-3107-9
Slack, F.J., Orife, J.N., Anderson, F.P.: Effects of commitment to corporate vision on employee satisfaction with their organization: an empirical study in the United States. Int. J. Manag. **27**(3), 421–436 (2010)
Silins, H.C.: The relationship between transformational and transactional leadership and school improvement outcomes. Sch. Effectiveness School Improv. **5**(3), 272–298 (1994)
Simone, R., Barakat, G.I., Gama, B., Mazzon, A.: The influence of corporate social responsibility on employee satisfaction. Manag. Decis. **54**(9), 2325–2339 (2016)
Sohmen, V.S.: Leadership and teamwork: two sides of the same coin. J. IT Econ. Dev. **4**(2), 1–18 (2013)
Spector, P.E.: Job Satisfaction: Application, Assessment, Causes, and Consequences. Sage Publications (1997)
Spreitzer, G.M.: Psychological empowerment in the workplace: dimensions, measurement, and validation. Acad. Manag. J. **38**(5), 1442–1465 (1995)
Stogdill, R.M.: Handbook of Leadership: A Survey of Theory and Research. Free Press (1974)

Storey, J.: Human Resource Management: A Critical Text. Cengage Learning EMEA (2007)
Sy, T., Tram, S., O'Hara, L.A.: Relation of employee and manager emotional intelligence to job satisfaction and performance. J. Vocat. Behav. **68**(3), 461–473 (2006)
Tannenbaum, S.: A strategic view of training and learning. In: Kraiger, K. (ed.) Creating, Implementing and Managing Effective Training and Development: State-of-the-Art Lessons for Practice, pp. 10–52. Jossey-Bass, San Francisco (2002)
Tannenbaum, R., Schmidt, W.H.: How to choose a leadership pattern. Harvard Bus. Rev. 3–12 (1973)
Tims, M., Bakker, A.B., Xanthopoulou, D.: "Do transformational leaders enhance their followers' daily work engagement? Leadersh. Q. **22**(1), 121–131 (2011)
Tung, F.C.: Does transformational, ambidextrous, transactional leadership promote employee creativity? Mediating effects of empowerment and promotion focus. Int. J. Manpow. **37**(8), 1250–1263 (2016)
Uhl-Bien, M., Marion, R.: Complexity leadership theory. In: Bryman, A., Collinson, D., Grint, K., Jackson, B., Uhl-Bien, M. (eds.) The SAGE Handbook of Leadership, pp. 468–482. Sage, Thousand Oaks (2011)
Vroom, V.H: Work and Motivation (1964)
Weiss, H.M.: Deconstructing job satisfaction: separating evaluations, beliefs and affective experiences. Hum. Resour. Manag. Rev. **12**(2), 173–194 (2002)
Wendt, H., Euwema, M.C., Van Emmerik, I.H.: Leadership and team cohesiveness across cultures. Leadersh. Q. **20**(3), 358–370 (2009)
Yammarino, F.J., Dionne, S.D., Chun, J.U., Dansereau, F.: Leadership and levels of analysis: a state-of-the-science review. Leadersh. Q. **16**(6), 879–919 (2005)
Yang, J.: Thriving organizational sustainability through innovation: incivility climate and teamwork. Sustainability **8**(860), 1–12 (2016)
Zaleznik, A.: The mythological structure of organizations and its impact. Hum. Resour. Manag. **28**(2), 267–277 (1989)

Design of an Employer Branding Model

Mª Luisa Rayón(✉), Marina Romeo, Montserrat Yepes-Baldó, and Sefa Boria-Reverter

University of Barcelona, 08007 Barcelona, Spain
mrayon@grupoechevarne.com

Abstract. The objective of this research is the design of an *Employer Branding* (EB) model as a tool that facilitates the attraction and retention of talent.

To this end, the model configuration elements are identified from literature, based on the 4 Ps described by McCarthy [45], thus developing a new model, the EB Mix. From this, we establish what the Product offered by the company as an employer is, the Price to be paid by the employee to obtain it, how it is made accessible or distributed (Place), and how this product is promoted among its target audience. It is, therefore, to define the elements that the EB unites, in order to propose a model for the implementation of the EB strategy in organizations. The theoretical model developed has been validated from the inter-rater adjustment. The questionnaire was administered to 30 experts from the field of organizational management who assessed the relevance (from a 5-point Likert scale) and the order of importance of each of the 4 Ps components. An Aiken V > 0.6 and a 90% confidence interval were used as criteria. The order of importance of each dimension was established using the range analysis method. The results show a high level of agreement of the experts in the nominal definition of all the components, the "Product" is considered the most important to establish an EB policy in an organization.

Keywords: Employer branding · Inter-rater adjustment · Marketing · Talent

1 Introduction

The rigorous and systematic study of Marketing began in the field of Social Psychology in the 60s. Since then, in the scientific environment, numerous theories have been developed with a two-sided orientation. On the one hand, those that explain consumer behavior [4, 34] and, on the other, those that focus on the design and establishment of communication strategies [33, 44, 49, 67].

However, there are still many areas to explore. The Employer Branding (EB) study, for example, is one of the areas that, if studied rigorously, can be fruitful, as Collings and Mellahi [20] have pointed out, observing that companies need to maximize the differential that its human capital brings. Having groups comprising of the best people, gives companies a competitive advantage [47], so it is convenient to address this issue.

Kotler y Keller [40, p. 287] point out that "branding is about providing products and services with the power of a brand, and it is essentially about creating differences". These differences are provided by a "mixture of attributes, tangible and intangible, symbolized

A. M. Gil-Lafuente et al. (Eds.): REDCID 2019, LNNS 388, pp. 69–85, 2022.
https://doi.org/10.1007/978-3-030-93787-4_4

in a trademark, which, properly managed, create value and influence" [71, p. 14]. The potential of EB as co-builder of the corporate brand lies in its ability to clarify its essential aspects, that is, its organizational identity, culture, and values [7, 12, 74], enhancing its image as an employer. Furthermore, it clarifies the company's unique value proposition, differentiating it from the competition. If it is managed correctly, it will create value and influence and will attract the best candidates.

The objective of this research is to design an Employer Branding (EB) model. The development of this model is especially relevant in the current socio-labor context, because, after the recent financial crisis and the uncertain economic scenario, attracting and retaining the best employees is a fundamental task to maintain the competitiveness of any company [15].

To achieve the described objective, the configuration elements of the model are identified, having a conceptual basis the 4Ps proposed by McCarthy [45], also known as Marketing Mix. To do this, the Product offered by the company as an employer, the Price to be paid by the employee to obtain it, how it is made accessible or distributed (Place), and how this product is Promoted among its target audience, must all be established.

2 Theoretical Framework

2.1 Brand Concept

The brand is one of the most valuable assets of a company and its management has become a key activity in many organizations. We can define the brand as the set of attributes, tangible and intangible, that identify their products and services and differentiate them from the competition. These differential attributes generate a mental image in your target audience, which, if managed correctly, adds value to the product or service it represents.

Identity is a key element in the strength of a brand that enhances customer appeal [26, 14, 37, 38]. The impact of the corporate brand on the attitudes and behaviors of employees is what has led to the development of the "internal brand" and EB [31, p. 401]. As noted above, if an organization is able to correctly manage its brand image, thereby enhancing its employer image, it will create value and influence and be able to attract the best candidates [51, 61, 72].

2.2 Employer Branding

Ambler and Barrow [3] coined the term Employer Branding in 1996, which is identified as the development and communication of an organization's culture. These same authors specify the employer's branding as the package of functional, economic and psychological benefits provided by the employer, identified with the employer. In other words, the company's value proposition is transferred to its current or potential employees through EB. If a company manages to identify its unique and specific value proposition for the satisfaction of its employees, this assumes an element of differentiation from the competition, which in turn will imply a greater degree of retention, science, and productivity.

Therefore, the EB is presented as a process of identification and creation of a company brand message [7], applying traditional marketing principles, to position itself as an employer of choice, attracting and retaining the most qualified employees [36, 52, 53, 70]. As Barrow and Mosley [9] affirm, EB serves to adapt, with a marketing approach, the tools, and techniques used to hire and link employees. Similarly, Edwards refers to it as "an activity where the principles of marketing, in particular brand science, are applications applied to human resources activities affected to current or potential employees" [26, p. 6]. To develop an EB model, it is useful and enriching to seek help in areas and tools that were previously considered domains of the marketing department, which offers, from a methodological point of view, valid techniques and models. Levionnois [42] defines internal marketing as a set of methods and techniques, put into action in a certain order, allowing the company to increase its level of success while affecting the interest of its customers and its collaborators [21, 65]. Internal marketing, therefore, offers organizations a valid management method or technique to serve both internal and external customers.

Over recent decades, the way of accessing the information of the products has been amended and, with this, the power of the client in the decision-making process has increased. In this environment, brands need to be transparent in communicating the values that accompany their product, share them with their customers to create an emotional bond with them, and collaborate with an increasingly informed customer who claims to be a more active part of the process of Exchange with brands. This reality has made it necessary to make updates, both to marketing and to companies, and also to the 4Ps model. But in neither case, does it make the model obsolete Moreover, any evolving model is built and evolves on a base, which is the 4Ps model [40]. This is the reason why, in the present investigation, it has been chosen as a reference for the development of an EB model according to the principles of marketing [25].

The identification of the four elements of the 4Ps model from which we start is essential for the development of an EB plan, as it will condition the company's strategy to position the product it offers as an employer in front of its target audience, so that it is perceived as a desirable place to work [8, 28].

2.3 The 4 Ps Model and the EB Mix

The 4Ps model considers the relationship between the company and its customers as an exchange relationship [11, 16, 35, 73]. The adequacy of the 4Ps model to an EB model is represented in Fig. 1.

Therefore, a company offers its employees a product as an employer, consisting of a set of tangible and intangible elements (offering) that materialize in a unique and specific experience of working in that company, in exchange for which employees contribute their work. Through the communication channels, the company communicates the particularities of the product and makes it accessible to all through the organizational systems and the leadership style.

Thus, based on the 4Ps of Marketing Mix model, the constituent elements of the new model, which we can call EB Mix, are the following:

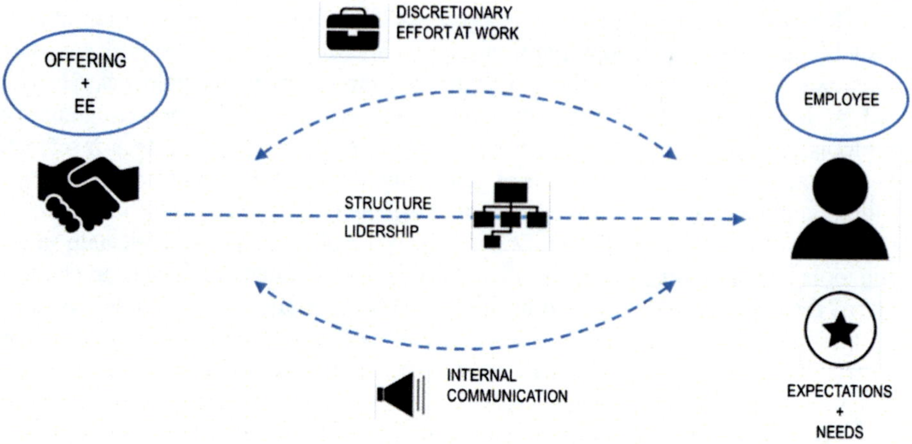

Fig. 1. EB exchange ratio. Own elaboration.

- The Product to promote and distribute through the EB is the work experience (*Employment Experience*); an inseparable consequence of what the company offers its employees (*offering*). This is *The Unique Employment Experience Elements* [26, 43, 48, 58].
- The Price, that is, what the employee contributes to enjoy said product (*Employment experience + offering*), is the discretion in the effort or in the work performed by the employee [26], whose level of specialization, intensity and / or difficulty, will be proportional to the benefits associated with the acquisition of said product [1, 54, 57].
- The Distribution or dissemination of the product is facilitated by the "flexibility of the organizational structure, the overlapping of levels and working groups, and the generalization of participatory leadership" [63, p. 200] [13, 62, 64]. This is signified in having a leadership style that facilitates it, that is, a transformational leadership, whose characteristics are charisma and influence, motivational inspiration, attention to individual differences and intellectual stimulation [10, 29, 32].
- Finally, Promotion implies communicating the particularities of the product, which means taking into account the communication systems and practices used by the organization internally. That is the management of internal communication [5, 22, 55].

The model we are going to develop seeks to identify the value proposition for the employee, which Kotler and Keller [40, p. 288] define as "the added value that is provided to products and services". This intangible asset contributed by the company as an employer, adds or subtracts value from its image in the market, and will determine the balance of the proposed exchange relationship [1, 58].

If the *branding* consists of providing products and services with the power of a brand and is essentially about creating differences, the value proposition for the employee must seek to impact how they think, feel and act with respect to the brand, differentiating from "*The unique employment experience elements*" competition, or what is the same, identifying the employer brand as a unique and differential product.

The model proposed here is intended to clarify the EB Mix as a tool for implementing the specific EB plan of each company. The correct choice and combination of these elements, its coherence with the identity and values of the company [24], the sensitivity to the environment in which it develops and the harmony with the expectations and needs of its customers (employees) is the basis for creating a chain of brand value [39], or what in marketing is called brand capital [40]. Especially important is the consistency of the model against the messages transmitted and the experiences lived by the different interest groups that have "lived the brand" in some way because, otherwise, the candidates will have the perception that the employer's promise is not truthful and consequently the company will seem less attractive [30].

These elements are not unalterable, nor is their combination static, which forces us to reformulate the EB plan when there are variations in the elements that configure it, as represented in Fig. 2.

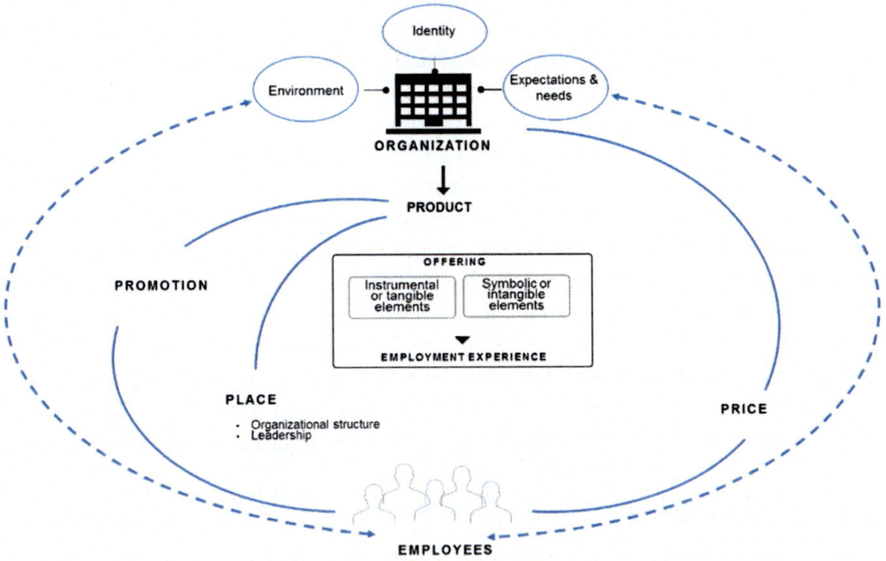

Fig. 2. Graphical representation of the EB model. Own elaboration.

The interrelationships between the different elements of the model recommend addressing each of them from a global perspective, since the modification of one of them will have an impact on the others and will force the redesign of the proposal or, at least, some of its elements [66, 69].

Therefore, an EB plan requires "a meticulous, lengthy and mature process, and includes the combination of time, changes in the definition of policies, and well examined marketing and human resources strategies" [68] (Serna, 2009, p. 38). Hence the importance of measuring, maintaining and optimizing the management of the employer brand, to verify its effectiveness and introduce progressive improvements. One way to monitor its implementation and evolution are climate surveys, the establishment, and review of turnover and retention ratios, hiring costs, market positioning rankings, number of job applications received and accepted, and ability to fill positions [19, 27, 41, 56, 60].

3 Method

First, the elements of the model were identified through the literature review of the existing literature on EB. Although there are no theoretical models that develop the variables proposed here, these variables were identified in different models and approaches of the authors reviewed.

From the different contributions of the authors, the theoretical and operational conceptualization of the elements that were presented in a questionnaire were specified, for validation by 30 experts.

For the analysis of the relevance of the items, the method of inter-rater adjustment from the Aiken validity ratio V was used based on the confidence intervals according to the scoring method [23, 59, 75] and using the program of Merino and Livia [50]. The use of the quotient V of Aiken [2] was chosen because it allows the evaluation of the relevance of each item concerning its construct, taking into account not only the number of categories presented to the judges but also the number of participating experts. In this way, the level of agreement is established based on the normal distribution and obtaining, from it, a probability associated with each item.

The interpretation of the values obtained can be carried out following a liberal criteria [18] of $V = 0.5$, or a more demanding one with $V = 0.7$ or higher [17]. The choice of one or the other criteria depends on the investigative process. In initial moments and, if the number of judges is low, the liberal criteria can be chosen [62]. In our investigation, we have chosen an intermediate criterion, with a $V = 0.6$. The items with values $V < 0.6$, for a 90% confidence interval, were reviewed based on the opinions expressed by the experts.

For the consideration of the absolute importance of the dimensions, the choice was made to analyze the average and the standard deviations of the scores granted by the experts. Dimensions with a standard deviation greater than 1 are analyzed to assess their adequacy in terms of their importance. Finally, the relative importance of the dimensions was analyzed using the range analysis method.

4 Results

4.1 Inter-rater Settings: Instrument Validation

Next, the results obtained for each of the dimensions of the model, of the validation of the instrument using the method of inter-rater adjustment are set out.

Product - The Unique Employment Experience Elements

The responses to the 5 items of this dimension exceeded the desirable agreement level value ($V = 0.60$) for a 90% confidence interval. The results of this dimension are shown in Table 1.

The item of this dimension that reached the highest level of agreement among experts, ($V = 0.92$ with a level of agreement between 0.84 and 0.96 with a confidence interval of 90%) was 1.2, worded as follows: "The work environment, the level of demand and internal relations are elements that identify a company and are valued as much as the economic offer".

Table 1. Results operational definition PRODUCT dimension

Item	$\bar{x}$	*Lower limit*			*Upper limit*		
		IC 90%	*IC 95%*	*IC 99%*	*IC 90%*	*IC 95%*	*IC 99%*
1.1 The identity and values of the company are identifiable and are present in the goods and services that it sells and in what it offers to its employees	1.5	.64	.63	.84	.88	.84	.86
1.2 The work environment, the level of demand and internal relations are elements that identify a company and are valued as much as the economic offer	1.83	.84	.82	.78	.96	.96	.97
1.3 The package of economic and social benefits offered is identifiable	1.47	.63	.61	.57	.81	.83	.85
1.4 A company's offer is all the more attractive the more sensitive it is to the needs and expectations of the employees	1.7	.76	.74	.70	.91	.92	.93
1.5 The configuration of what the company offers workers must be flexible, adaptable to changes in the environment and the needs and expectations of employees	1.7	.76	.74	.70	.91	.92	.93

At the other end, the lowest level of agreement was obtained by item 1.3, phrased as "The package of economic and social benefits offered is identifiable." Some comments that support this result are collected. Its importance is recognized, but it is relativized, considering the fulfillment of other factors to be more important.

The consideration by some judges to incorporate other aspects that were not mentioned in the value proposition was also provided as an open comment, such as: "The existence of a career plan is not explicitly mentioned in the questionnaire and the criteria with which the company executes it. This, in my opinion, is quite important, both when you are looking for work (especially the first jobs) and when you are already in the company "(P27).

In the comments to this item, the importance given to flexibility was repeated as an indicator of personalization of the value proposition, that expresses an interest in the individual, which facilitates the development of a stronger connection with the company.

Price - Quantity/Quality of Work

The price dimension breaks down into five items the fundamental aspects of the quantity and quality of work required by a company of its employees as a key element of the model.

The item of this dimension that reached the highest level of agreement among experts (V = 0.92 with a level of agreement between 0.84 and 0.96 with a 90% confidence interval) was 2.5, whose statement is "The quantity and quality of work depend on both tangible (e.g., economic) and intangible rewards (e.g., feeling proud of being a member of a prestigious company)."

The experts considered the importance of knowing the terms of the agreement, among which the importance of recognition and its influence on the pride of belonging is mentioned, as well as the impact of both types of rewards over time. In summary, they state, in the words of one of the experts, "(…) *agree that a fair reward facilitates the good atmosphere and better performance*" (P28).

Items 2.1 and 2.2 did not reach the desired level of agreement for the lower limit of the 90% confidence interval. The specific indicator of agreement in both cases was V = 0.69, although the 90% confidence interval placed the cut-off points between 0.58 and 0.77. The formulation of these items is: "The employee can modulate the quantity and quality of his work based on the attractiveness of the reward associated with it" (item 2.1) and "The employee can adjust the quantity and quality of his or her work according to the possibilities that he or she considers they have to obtain the reward associated with it "(item 2.2).

The experts considered the items as relevant or pertinent, although not essential when describing the price. Some focused on the temporary effectiveness of the rewards and their long-term effect as a function of their consolidation or not in the remuneration. The results obtained in this dimension are shown in Table 2.

Table 2. Results operational definition PRICE dimension

Item	$\overline{x}$	*Lower limit*			*Upper limit*		
		IC 90%	*IC 95%*	*IC 99%*	*IC 90%*	*IC 95%*	*IC 99%*
2.1 The employee can adjust the quantity and quality of his work based on the attractiveness of the reward associated with it	1.37	.58	.56	.52	.77	.79	.81
2.2 The employee can adjust the quantity and quality of his work according to the possibilities he considers he has to obtain the reward that he associates with it	1.38	.58	.56	.52	.77	.79	.81

(*continued*)

Table 2. (*continued*)

Item	$\bar{x}$	*Lower limit*			*Upper limit*		
		IC 90%	*IC 95%*	*IC 99%*	*IC 90%*	*IC 95%*	*IC 99%*
2.3 If the relationships between employees and superiors are governed by the principle of equity, the level of intensity and quality of work improves	1.6	.7	.68	.64	.87	.88	.9
2.4 The adaptation of the economic and social benefits package to the employee's expectations improves the quantity and quality of their work	1.4	.6	.57	.54	.79	.80	.83
2.5 The quantity and quality of work depend on both tangible (economic) and intangible (reputation) rewards	1.83	.84	.82	.78	.96	.96	.97

According to the experts, although an adequate design of the policies and content of the rewards is fundamental, they do not fully explain the employee's response regarding the quantity and quality of the work offered. The employee can respond autonomously, regardless of the amount and success of the reward offered.

Distribution- Leadership Structure/Style

The distribution dimension breaks down into five items the impact of the structure and leadership style as a key element of the model. As can be seen in the results presented in Table 3, all the items exceeded the minimum agreement level with a 90% confidence interval.

The highest level of agreement is observed in item 3.2 expressed as "The clarity in the structure and the description of positions and functions facilitates the transmission of operational and corporate information, as well as the values and objectives of the company". In this sense, the experts agreed to consider its relevance underlining the idea that "*the values and objectives of the company must be present in all areas*" (P28).

Regarding decision-making (Item 3.1), it is significant to point out that experts with commercial profiles influence the importance of speed, while technical profiles focused the weight on the success of decisions.

Concerning the characteristics of leadership, experts consider that "*what is essential is the ability to generate confidence and synergies to overcome obstacles*" (P10) since "*the employee feels safe if their managers respond adequately to hazards*" (P28). That confidence is generated through its management: "Employees appreciate management leadership *as the ability to face, advise and decide on a problem, since "it projects solvency and therefore generates stability in the project. Build trust"* (P2).

Table 3. Results operational definition DISTRIBUTION dimension

Item	$\bar{x}$	*Lower limit*			*Upper limit*		
		IC 90%	*IC 95%*	*IC 99%*	*IC 90%*	*IC 95%*	*IC 99%*
3.1 A good system for making quick and effective decisions is essential for employees to have a positive experience of working for a company	1.6	.7	.68	.64	.87	.88	.9
3.2 The clarity in the structure and the description of positions and functions facilitates the transmission of operational and corporate information, as well as the values and objectives of the company	1.93	.9	.88	.85	.99	.99	.99
3.3 Coordination between services and departments is key to the work of employees	1.77	.8	.78	.74	.94	.94	.95
3.4 The ability to overcome obstacles of the managers of a company impacts the commitment and identification of employees	1.47	.63	.61	.57	.81	.83	.85
3.5 Transformational leadership is essential when transmitting the company's values	1.53	.67	.64	.6	.84	.85	.87

But not only does it matter what a leader does and how they do it but what they are like and their characteristics, hence it is pointed out that "*managers have to be inspiring leaders*" (P4) and they will be as long as they are consistent: "*when transmitting values, the key* **is** *the facts, not the speeches* " (P5) so that " those values are *represented and demonstrated on a day-to-day basis* "(P9) and they are "*close by and know how to listen and set an example* "(P26).

Promotion – Internal Communication

The promotion dimension presents aspects related to internal communication through 5 items for the assessment of experts as a key element of the model. The highest level of agreement was reached in item 4.3 and expressed as "Communication should encourage feedback between management and employees". Experts consider that "*if the employee is not aware that through their hierarchical structure they communicate and study their valuations, I think that the whole model would not make sense. The fundamental basis of communication is reciprocity*" (P1). This is considered "*determinant to identify needs*" (P2) in addition to considering that "*it is essential that communication goes both ways*"

(P28). However, other experts focus primarily on descending communication and conceive participatory processes as a threat. In this context, it is noted "*I think that the communication must be, basically, from the company to the employee. Feedback in the other direction can be very conflictive, and cause the company's value proposition to be undermined or damaged*" (P5).

Item 4.4 did not reach the desired level of agreement for the lower limit of the 90% confidence interval. The punctual agreement indicator was V = 0.69 although the limits of the confidence interval were between 0.58 and 0.77. Its formulation is "The company should provide information by means of internal communication regarding aspects of the corporate brand and the employer's brand". The results are shown in Table 4.

Table 4. Results operational definition PROMOTION dimension

Item	$\overline{x}$	*Lower limit*			*Upper limit*		
		IC 90%	*IC 95%*	*IC 99%*	*IC 90%*	*IC 95%*	*IC 99%*
4.1 Internal communication must convey information about the identity and values of the company	1.7	.76	.74	.7	.91	.92	.93
4.2 Internal communication should clarify the elements that constitute the company's offer as an employer	1.53	.67	.64	.6	.84	.85	.87
4.3 Communication must provide feedback between management and employees	1.77	.8	.78	.74	.94	.94	.95
4.4 The company must provide information through internal communication regarding aspects of the corporate brand and the employer's brand	1.37	.58	.56	.52	.77	.79	.81
4.5 The company must channel the information in different ways depending on the audience it wants to reach or the message it must convey	1,57	.69	.67	.63	.86	.87	.89

After analyzing the experts' responses and comments, it is confirmed that communication about the corporate brand is considered more effective when it comes through the reality of the company rather than in a formal way. Feedback given by the experts suggests that brand values are perceived better and more strongly when they are part of life or the experience of working in a company than when they arrive through a declaration of intent. To this effect, formal communication will serve to clarify these values, but not as the only or primary way to deliver them. This feedback is in line with the approach of the model in which the values are part of the value proposition of a company, which is made accessible to employees through the structure and style of leadership.

4.2 Absolute and Relative Importance of Dimensions

Firstly, the experts were asked to assess the importance of each of the dimensions when establishing an EB policy in organizations. Even if all of them score higher than 4.27 on a 5-point scale, which qualifies them as important or very important, it should be noted that the highest level is obtained by the product dimension, that is, the differential offer that a company makes to its current or potential employees. The means and standard deviations for each dimension are indicated in Table 5.

Table 5. Means and standard deviations

Dimension	Product	Price	Place	Promotion
$\bar{x}$	4,77	4,27	4,47	4,53
DE	.4	.7	.9	.7

Subsequently, the experts were asked to order the dimensions according to their relative importance. It should be noted that the responses of 28 out of the 30 experts have been taken into account, as some of them did not match what was requested.

The result obtained is as follows:

1st - Value proposition of the company
2nd - Design of the structure and leadership style
3rd - Quantity and quality of work required- desired
4th - Internal corporate communication

It is observed that it is coincidental that the dimension in which experts showed a higher level of agreement in their nominal definition (product) is also the one they consider most important when establishing an EB policy in an organization. The attractiveness of a company as an employer is directly related to what it offers. However, the order is not the same starting from the second element in importance.

Secondly, in importance, is the structure and style of leadership, which, in the proposed model, is how the company makes this value proposition accessible.

Thirdly, in order of importance, they indicate the quantity and quality of the work required to enjoy the value proposition. We believe that this dimension has been better valued by the experts for omitting, in the questionnaire presented to the experts, the equivalence of the "price to pay" dimension of the 4Ps model, as well as for the effort to clarify the nominal definition. The importance attached to this dimension was accompanied by comments that pointed out the difficulty of achieving that balance or proportionality, and its relationship with the value proposition presented by the company. In this regard, it is worth noting the observation of one of the experts, who pointed out that the importance of each of the elements varies over time.

Although at first the order of importance establishes it as:

1. Value proposition
2. Quantity/quality of work
3. Internal structure and leadership
4. Internal corporate communication

In the course of time and the actual experience of working in a company, this order is modified by the following:

1. Internal structure and leadership
2. Internal corporate communication
3. Quantity/quality of work
4. Value proposition

He justifies it in the following way: "*Classifying it in two different ways (...) is in line with the fact that the employer's brand must change over time from the employee's perspective. Without losing sight of the "initial contract", (the employer needs what the employee can give him, and he gives it depending on the agreement they have reached) the approach can and should change. The internal structure and leadership must be strong, increasing in importance over time, to strengthen the employee's confidence towards the employer's value proposition and this is not possible without an agile, simple and direct internal communication*" (P10).

It should be noted that the expert who makes this assessment is a middle manager who has been working in the same organization for more than 20 years. His reflection raises a deeper analysis of how the company's value proposition becomes confused with the channels through which it materializes and becomes real. The proposal of theoretical value is, according to the criteria of this expert, relegated to the least important place, because the experience of what has been lived in front of theoretical approaches is imposed.

Least important was the role of internal communication. The explanation of this position in the order of importance is reflected in feedback given on its instrumental role and also in some items of the operational definition of this dimension. Some experts qualified the role of communication by subtracting value in the configuration of EB policies and assigning a purely informative role.

5 Conclusions

The results show a high level of agreement between the experts in the nominal definition of all the components, resulting in the "product" being the most important aspect in establishing an EB policy in an organization. The importance given to this element of the model is shared equally between tangible and intangible elements. Without denying the importance of the economic offer, aspects such as reputation, flexibility or career plan stand out as essential to determining the attractiveness of the employer brand.

This overall valuation of the *offering* also affects the consideration of the "price" element, since it is considered that the reward affects the quality and quantity of work, provided that it is well designed, is known by the employee and takes into account both tangible and intangible aspects. This, as some observations indicate, cannot be established solely in economic terms, since aspects such as "*The relationship, closeness and ability to listen have a strong effect on the loyalty of the employee*" (P1). In summary, an adequate design of the reward systems is considered key, despite considering that its impact on the quantity and quality of work was not decisive, resulting in the ability of the employee to respond autonomously, regardless of the amount and success of reward offered.

Regarding the most desirable characteristics of the organizational structure and leadership style as "distribution channels" (Place) of the company's value proposition, clarity (positions and functions) and coherence are emphasized. Coherence refers both to internal coherence with the values and objectives of the company, as well as coherence between discourses and facts. This differentiates leaders from simply managers, since the former can build confidence to overcome obstacles and are a source of inspiration for members of an organization.

The aspect in which the highest level of agreement was reached when assessing the "promotion" element was the existence of feedback. Reciprocity in communication, and participation, is valued as an essential feature in developing an EB plan. However, the "instrumental" role of communication was highlighted, whose fundamental mission is to clarify aspects of the value proposition that must come through the reality lived in the company, especially with regard to intangible aspects of the organisation. That is, values are perceived more strongly through structure and leadership than by messages or slogans launched by formal channels of communication. This aspect links to the coherence demanded of leadership as a way to disseminate the company's value proposition. Furthermore, in internal communication, "the reality" prevails over communication considered to be propaganda.

From this last reflection, the results obtained when assessing the order of importance of the dimensions are consistent, since, after the product or value proposition, the second element is considered to be distribution, thirdly, the price, and finally communication by assigning it an informative and clarifying role.

As stated above, the EB Mix model provides tools to Human Resources professionals to establish key considerations when designing an organisational EB plan.

References

1. Adams, J.S.: Inequity in social exchange. Adv. Exp. Soc. Psychol. **2**, 267–299 (1965)
2. Aiken, L.R.: Three coefficients for analyzing the reliability and Validity of ratings. Educ. Psychol. Measur. **45**, 131–142 (1985)
3. Ambler, T., Barrow, S.: The employer brand. J. Brand Manag. **4**(3), 185–206 (1996)
4. Argo, J.J., White, K., Dahl, D.W.: Social comparison theory and deception in the interpersonal exchange of consumption information. J. Consum. Res. **33**(1), 99–108 (2006)
5. Asif, S., Sargeant, A.: Modelling internal communications in the financial services sector. Eur. J. Mark. **34**(3/4), 299–318 (2000)
6. Avolio, B.J., Walumbwa, F.O., Weber, T.J.: Leadership: current theories, research, and future directions. Ann. Rev. Psychol. **60**, 421–449 (2009)
7. Backhaus, K., Tikoo, S.: Conceptualizing and researching employer branding. Career Dev. Int. **9**(5), 501–517 (2004). https://doi.org/10.1108/13620430410550754

8. Bagozzi, R.P.: Marketing as exchange. J. Mark. **39**, 32–39 (1975)
9. Barrow, S., Mosley, R.: Bringing the Best of Brand Management to People at Work. Wiley, NewYork (2005)
10. Bass, B.M.: Does the transactional–transformational leadership paradigm transcend organizational and national boundaries? Am. Psychol. **52**(2), 130 (1997)
11. Blau, P.: Exchange and Power. Wiley, New York (1964/1986)
12. Borden, N.H.: The concept of the marketing mix. J. Advert. Res. **4**, 2–7 (1964)
13. Bower, G.H.: Imagery as a relational organizer in associative learning. J. Verbal Learn. Verbal Behav. **9**(5), 529–533 (1970)
14. Burmann, C., Zeplin, S.: Building brand commitment: a behavioural approach to internal brand management. J. Brand Manag. **12**(4), 279–300, 374–393 (2005)
15. Carrière, J., Bourque, C.: The effects of organizational communication on job satisfaction and organizational commitment in a land ambulance service and the mediating role of communication satisfaction. Career Dev. Int. **14**(1), 29–49 (2009). https://doi.org/10.1108/13620430910933565
16. Cavusgil, S.T.: International marketing strategy. In: Thorelli, H.B. (ed.) Pergamon Press (1984)
17. Charter, R.A.: A breakdown of reliability coefficients by test type and reliability method, and the clinical implications of low reliability. J. Gener. Psychol. **130**(3), 290–304 (2003)
18. Cicchetti, D.V.: Guidelines, criteria, and rules of thumb for evaluating normed and standardized assessment instruments in psychology. Psychol. Assesments **6**, 284–290 (1994)
19. Cohen, A.: Organizational commitment and turnover: a meta-analysis. Acad. Manag. J. **36**, 1140–1157 (1993)
20. Collings, D.G., Mellahi, K.: Strategic talent management: a review and research agenda. Hum. Resour. Manag. Rev. **19**(4), 304–313 (2009)
21. Cooper-Hakim, A., Viswesvaran, C.: The construct of work commitment: testing an integrative framework. Psychol. Bull. **131**, 241–259 (2005)
22. Crable, R.E., Vibbert, S.L.: Public Relations as Communication Management. Bellwether Press (1986)
23. Cronbach, L.J.: Coefficient alpha and the internal structure of test. Psychometrika **16**, 297–334 (1951)
24. Dutton, J.E., Dukerich, J.M., Harquail, C.V.: Organizational images and member identification. Adm. Sci. Q. **39**(2), 239–263 (1994)
25. Edwards, M.R.: An integrative review of employer branding and OB theory. Pers. Rev. **39**(1), 5–23 (2010). https://doi.org/10.1108/00483481011012809
26. Edwards, M.R., Edwards, T.: Employee Responses to changing aspects of the employer brand following a multinational acquisition: a longitudinal study. Hum. Resour. Manag. **52**(1), 27–54 (2013)
27. Elangovan, A.R.: Causal ordering of stress, satisfaction and commitment, and intension to quit: a structural equations analysis. Leadersh. Org. Dev. J. **22**(4), 159–165 (2001)
28. Emerson, R.: Exchange theory: a psychological basis for social exchange. Sociol. Theor. Progress **2**, 38–57 (1972)
29. Emrich, C.G., Brower, H.H., Feldman, J.M., Garland, H.: Images in words: presidential rhetoric, charisma, and greatness. Adm. Sci. Q. **46**(3), 527–557 (2001)
30. Fernández-Barcala, M., González-Díaz, M.: Brand equity in the European fruit and vegetable sector: a transaction cost approach. Int. J. Res. Mark. **23**(1), 31–44 (2006)
31. Foster, C., Punjaisri, K., Cheng, R.: Exploring the relationship between corporate, internal and employer branding. J. Prod. Brand Manag. **19**(6), 401–409 (2010)
32. García-Guiu, C., Molero, F., Moriano, J.A.: Authentic leadership and its influence on group cohesion and organizational identification: the role of organizational justice as a mediating

variable/El liderazgo auténtico y su influencia sobre la cohesión grupal y la identificación organizacional: el papel de la justicia organizacional como variable mediadora. Revista de Psicología Social **30**(1), 60–88 (2015)
33. Hastings, G., Stead, M., Webb, J.: Fear appeals in social marketing: strategic and ethical reasons for concern. Psychol. Mark. **21**, 961–986 (2004)
34. Hess, R.L.J., Ganesan, S., Klein, N.M.: Interactional service failures in a pseudorelationship: the role of organizational attributions. J. Retail. **83**(1), 79–95 (2007)
35. Homans, G.C.: The Nature of Social Science. Harcourt, New York (1961)
36. Kim, S.H., Laffranchini, G., Wagstaff, M.F., Jeung, W.: Psychological contract congruence, distributive justice, and commitment. J. Manag. Psychol. **32**(1), 45–60 (2017). https://doi.org/10.1108/JMP-05-2015-0182
37. King, C.: 'One size doesn't fit all': tourism and hospitality employees' response to internal brand management. Int. J. Contemp. Hosp. Manag. **22**(4), 517–534 (2010)
38. King, C., Grace, D.: Internal branding: exploring the employee's perspective. J. Brand Manag. **15**(5), 358–372 (2008)
39. Kotler, P., Armstrong, G.: Principles of Marketing. Pearson Education, Madrid (2009)
40. Kotler, P., Keller, K.L.: Dirección de Marketing. Pearson Prentice Hall, Madrid (2009)
41. Larson, E., Fukami, C.: Relationships between worker behavior and commitment to the organization and union. In: Proceedings of the Academy of Management, pp. 222–226 (1984)
42. Levionnois, M.: Marketing interno y gestión de recursos humanos. Díaz de Santos, Madrid (1992)
43. Lievens, F., Highhouse, S.: The relation of instrumental and symbolic attributes to a company's attractiveness as an employer. Pers. Psychol. **56**, 75–102 (2003)
44. Littlejohn, E.W., Foss, K.A.: Theories in Human Communication. Waveland Press, Inc., Illinois (2010)
45. McCarthy, J.E.: Basic Marketing. A Managerial Approach. Richard D. Irwin, Homewood (1960)
46. Manríquez, M.R., Rendón, L.P., Sánchez-Fernández, M.D., Guerra, J.F.: Estrategias de compensación y su relación con el compromiso organizacional en los docentes de las Instituciones de Educación Superior en México. Revista Ibero Americana de Estratégia **16**(2), 90–103 (2017)
47. Martin Molina, A.: El Marketing interno de los recursos humanos. Cómo vendernos. Capital Humano. Año **XXI**(221), 58–63 (2008)
48. Mathieu, J.E., Zajac, D.: A review and meta-analysis of the antecedents, correlates, and consequences of organizational commitment. Psychol. Bull. **108**, 171–194 (1990)
49. Augusto, C., de Matos, C., Rossi, A.V.: Word-of-mouth communications in marketing: a meta-analytic review of the antecedents and moderators. J. Acad. Mark. Sci. **36**(4), 578–596 (2008). https://doi.org/10.1007/s11747-008-0121-1
50. Merino Soto, C., Livia Segovia, J.: Intervalos de confianza asimétricos para el índice la validez de contenido: Un programa Visual Basic para la V de Aiken. Anales de Psicología **25**(1), 169–171 (2009)
51. Merz, M., Herbst, U., Yavuz, O.: Measuring employer brand value. In: Proceedings of the EMAC Conference (2009)
52. Meyer, J.P., Allen, N.J., Topolnytsky, L.: Commitment in a changing world of work. Can. Psychol. Assoc. **39**(1–2), 89–93 (1998)
53. Meyer, J.P., Maltin, E.R.: Employee commitment and well-being: a critical review, theoretical framework, and research agenda. J. Vocat. Behav. **77**, 323–337 (2010)
54. Mladinic, A., Isla, P.: Justicia organizacional: Entendiendo la equidad en las organizaciones. Psykhe **11**(2) (2011)
55. Morgan, R.M., Hunt, S.D.: The commitment-trust theory of relationship marketing. J. Mark. **58**(3), 20–38 (1994)

56. Moroko, L., Uncles, M.D.: Characteristics of successful employer brands. J. Brand Manag. **16**(3), 160–175 (2008)
57. Navarro, J., De Quijano, S.D.: Dinámica no lineal en la motivación en el trabajo: propuesta de un modelo y resultados preliminares. Psicothema **15**(4), 643–649 (2003)
58. Osorio, M.: El impacto de la remuneración en la Gestión de Recursos Humanos/The impact of compensation on Human Resources Management. Boletín de Estudios Económicos **71**(217), 85 (2016)
59. Penfield, R.D., Giacobbi, P.R., Jr.: Applying a score confidence interval to Aiken's item content-relevance index. Meas. Phys. Educ. Exerc. Sci. **8**(4), 213–225 (2004)
60. People Matters employer branding en España 2011: estado de situación. Estudio de People Matters (2011)
61. Piehler, R., King, B., Xiong, L.: The importance of employee brand understanding, brand identification, and brand commitment in realizing brand citizenship behaviour. Eur. J. Mark. **50**(9/10), 1575–1601 (2016)
62. Quijano, S., Navarro, J., Yepes, M., Berger, R., Romeo, M.L.: auditoría del sistema humano (ASH) para el análisis del comportamiento humano en las organizaciones. Papeles del psicólogo: revista del Colegio Oficial de Psicólogos **29**(1), 92–106 (2008)
63. Quintanilla, I.: Recursos humanos y Marketing interno. Pirámide, Madrid (1991)
64. Quintanilla, I.: Psicología social del trabajo. Ediciones Pirámide, Madrid (2014)
65. Riketta, M.: Attitudinal organizational commitment and job performance: a meta-analysis. J. Organ. Behav. **23**, 257–266 (2002)
66. Ríos, M.F., Sánchez, J.C., Muñoz, R.R.: Procesos estratégicos y estructura organizacional: implicaciones para el rendimiento. Psicothema **13**(1), 29–39 (2001)
67. Schivinski, B., Dabrowski, D.: The effect of social media communication on consumer perceptions of brands. J. Mark. Commun. **22**(2), 189–214 (2016)
68. Serna, C.C.: Nuevas tendencias en la retención y mejora del talento profesional y directivo. EOI Escuela de Organización Industrial (2009)
69. Susaeta, L., Pin, J.R., Belizón, M.J.: Factores determinantes de la reputación del CEO: Un análisis sectorial entre las principales empresas españolas. IESE Business School (2008)
70. Sutherland, M.M., Torricelli, D.G., Karg, R.F.: Employer of choice branding for knowledge workers. S. Afr. J. Bus. Manag. **33**(4), 13–20 (2002)
71. Swystun, J. (ed.): The Brand Glossary. Palgrave Macmillan, New York (2007)
72. Tett, R.P., Meyer, J.P.: Job satisfaction, organizational commitment, turnover intention, and turnover: path analyses based on meta-analytic findings. Pers. Psychol. **46**, 259–293 (1993)
73. Thibaut, J.W., Kelley, H.H.: The Social Psychology of Groups. Wiley, New York (1959)
74. Van Riel, C.B.M., Balmer, J.M.T.: Corporate identity: the concept, its measurement, and management. Eur. J. Mark. **31**(5/6), 340–355 (1997)
75. Wilson, E.B.: Probable inference, the law of succession, and statistical inference. J. Am. Stat. Assoc. **22**(158), 209–212 (1927)

Check for updates

Investment Evaluation Proposal with Fuzzy Tools: Case Molecular Immunology Center

Onailis Oramas Santos(✉), Lourdes Souto Anido, and Maritza Ortiz Torres

Faculty of Economics, Havana University, 10400 Havana, Cuba
{onailisos,lourdes,maritza}@fec.uh.cu

Abstract. Companies are currently developing in a changing environment, which has led to more complex and uncertain environments and greater difficulty in their ability to make decisions.

Among the most important decisions taken in the business management process is the approval, evaluation and execution of investments, since these generally involve high capital disbursements and even the participation of debt.

The Molecular Immunology Center (CIM using its acronym in Spanish), a Cuban company belonging to the BIOCUBAFARMA group, is not exempt from this situation. INIM is the area in charge of Research and Development within the CIM and currently has a slow development of products from the patent until its commercial phase.

This research aims to address the evaluation of an investment in the INIM from the optimum of fuzzy mathematics, due to the characteristics of the assets (intangible) and / or the sources of income in general that this entity presents, pretending that the technique addressed is applied to other investments and other areas of business decision making.

Project supported by "Red Iberoamericana para la Competitividad, Innovación y Desarrollo" (REDCID), project number 616RT0515, in "Programa Iberoamericano de Ciencia y Tecnología para el Desarrollo" (CYTED).

Keywords: Decision making · Fuzzy mathematics · Intangibles · Investment valuation

1 Introduction

Business systems are currently facing a type of competition in global markets in which there is an accelerated development of new technologies, new process, information and management techniques arise that, in some way, imply a challenge for any entrepreneur. Today, moreover, consumers are more demanding, the scenarios much more dynamic than in previous decades and all this indicates that the environment is not static.

This changing environment has led organizations to more complex and uncertain environments and greater difficulty in their ability to make decisions. Therefore, the current business challenge is to develop skills that facilitate adaptation and adaptation to new demands and market conditions, in order to compete successfully.

A. M. Gil-Lafuente et al. (Eds.): REDCID 2019, LNNS 388, pp. 86–99, 2022.
https://doi.org/10.1007/978-3-030-93787-4_5

This research was carried out at the Center for Molecular Immunology (CIM using its acronym in Spanish), a Cuban biotechnology institution founded in 1994. The area responsible for Research and Development within the CIM is INIM, in addition to being the place where most of it is created and patented of product development. Therefore, it is considered that the income generated by the provision of intangible services (patent license, charge for milestone in technology transfer, sale of assets of mixed companies) is the main source of resources generated by this area.

Currently, the generation of new products and income of the INIM is limited by the slow development of the products from the patent until its commercial phase. This disagreement is caused by the internal company due to the lack of pilot production facilities and insufficient medical diagnostic technology for the stratification of patients. The above difficulties also affect the medium and term increase in CIM exports.

In order to solve the problem detected, this research evaluates the feasibility of investment in equipment for the Product Development Department of INIM, taking into account the tools of fuzzy mathematics as an alternative to uncertainty.

The application of this technique provides the entity with a useful tool in decision making that will improve business management.

2 Theoretical and Methodological Framework

In terms of business management, the acquisition of the necessary resources to carry out the business mission is essential; but not only those resources that guarantee the operation of the business, but also those that endow it with new capabilities that allow it to meet the unsatisfied demands of the market.

Faced with a context of constant change, the information that provides an estimate in terms of certainty will be more inaccurate when assuming that all the elements involved in the projects are known and true. However, using the fundamentals of fuzzy logic, "originated almost 50 years ago with the seminal paper of Lofti A. Zadeh" (Merigó et al. 2015), a solution is obtained that takes into account different levels of presumption, and shows ranges that, if not so accurate, do incorporate and allow all uncertainty in each of the variables of a draft. Therefore, in relation to the modeling and resolution of problems in uncertain environments, fuzzy mathematics "will allow us, in the absence of being more exact, to be more honest" (Kaufmann et al. 1987) improving the information necessary to decision making (Alvarado and Menjivar 2017).

This research applies part of the tooling of fuzzy logic to the evaluation of investment projects. To do this, it proposes the following order of activities:

1. Execution of a market study that validates the need for an investment, identifying the demanded products, their characteristics and / or properties, the quantities required and the prices of similar products. It is important at this stage to be able to identify the possible states of nature that can be presented for each of the variables, their probabilities of occurrence not being necessary.
2. Identify alternative decisions. At this time the board of directors of the organization, with the participation of the rest of the workers and external experts, must determine if it is in a position, with its available productive capacity, to meet that demand and,

if not, they must explore the options available. Such options could be more or less expensive depending on the size of the investment. It is recommended to evaluate more than one option, to achieve effective decision making.

3. Evaluate alternative decisions. Once the different alternative decisions have been identified, these should be evaluated at the opportunity study level, considering not only the investment costs in these alternatives, but also the operating benefits or losses that would be obtained for each of the states of nature. For the valuation of the options, the investment evaluation indicators will be used Net Present Value (NPV), Internal Rate of Return (IRR) and Discounted Recovery Period (DRP); in addition to social benefits and costs in case they can be identified. The mathematical operations to be carried out would be carried out as the fuzzy logic puts it, given the different scenarios of the states of nature:

Assuming any two confidence intervals: $A^\alpha = \left[a_1^\alpha, a_2^\alpha\right]$ y $B^\alpha = [b_1^\alpha, b_2^\alpha]$ with $a_1^\alpha, a_2^\alpha, b_1^\alpha, b_2^\alpha \in \mathbb{R}$, the operations addition, subtraction, multiplication, division and filing will be executed as follows (Oramas and Ortiz 2015):

- Sum:

$$[a_1^\alpha, a_2^\alpha]\,(+)\,[b_1^\alpha, b_2^\alpha] = [a_1^\alpha + b_1^\alpha,\ a_2^\alpha + b_2^\alpha]$$

- Subtraction:

$$[a_1^\alpha, a_2^\alpha]\,(-)\,[b_1^\alpha, b_2^\alpha] = [a_1^\alpha - b_2^\alpha,\ a_2^\alpha - b_1^\alpha]$$

- Multiplication:

$$[a_1^\alpha, a_2^\alpha]\,(*)\,[b_1^\alpha,\ b_2^\alpha] = [a_1^\alpha * b_1^\alpha,\ a_2^\alpha * b_2^\alpha],$$

if the fuzzy set is defined in R +, if it would not be as follows (Morillas Raya 2006):

$$\left[a_1^\alpha, a_2^\alpha\right](*)\left[b_1^\alpha, b_2^\alpha\right] = \left[a_1^\alpha * b_1^\alpha \wedge a_1^\alpha * b_2^\alpha \wedge a_2^\alpha * b_1^\alpha \wedge a_2^\alpha * b_2^\alpha, a_1^\alpha * b_1^\alpha \vee a_1^\alpha * b_2^\alpha \vee a_2^\alpha * b_1^\alpha \vee a_2^\alpha * b_2^\alpha\right]$$

Symbols ∧y∨ mean minimum and maximum respectively. There is a neutral for multiplication: 1 = [1; 1].

- Division:

$$[a_1^\alpha, a_2^\alpha]\,(/)\,[b_1^\alpha, b_2^\alpha] = [a_1^\alpha / b_2^\alpha,\ a_2^\alpha / b_1^\alpha]$$

if the fuzzy set is defined in R +, if it would not be as follows:

$$\left[a_1^\alpha, a_2^\alpha\right](/)\left[b_1^\alpha, b_2^\alpha\right] = \left[a_1^\alpha / b_1^\alpha \wedge a_1^\alpha / b_2^\alpha \wedge a_2^\alpha / b_1^\alpha \wedge a_2^\alpha / b_2^\alpha, a_1^\alpha / b_1^\alpha \vee a_1^\alpha / b_2^\alpha \vee a_2^\alpha / b_1^\alpha \vee a_2^\alpha / b_2^\alpha\right]$$

This is not true if $a_1 \leq 0 \leq a_2$ (Morillas Raya 2006).

- Radiation:

$$\surd A^\alpha = [\surd a_1^\alpha,\ \surd a_2^\alpha], \text{ if the fuzzy set is defined in R } +.$$

- Reverse:

$$\left[a_1^\alpha, a_2^\alpha\right]^{-1} = \left[1/a_2^\alpha, 1/a_1^\alpha\right], \text{ if } 0 < \mathrm{a}_1 \leq \mathrm{a}_2$$

There is another set of mathematical operations and restrictions that were not exposed in this section for not making relevant contributions to this research.

The following mathematical expression will be used to determine the NPV:

$$NPV = -C_0 + \sum\nolimits_{t=1}^{n} \frac{C_t}{(1+k)^t} \quad (1)$$

Being:

C_0 = Initial investment cost
Ct = Cash Flow in period t
t = Lifetime
k = Capital discount rate.

The selection criteria will be:

- NPV > 0, the investment exceeds its cost and the project will be attractive to investors. While its value indicates how much you earn with the business and how much capital is valued in t.
- NPV = 0, no profits or losses will be obtained and investors will be indifferent to the execution of the project. In this case the company will materialize the project if it has any other objective that is not only to create value.
- NPV < 0, the investment does not exceed its cost so the project will not be attractive to investors, who will reject it.

In the case of social investments, even if the resulting NPV is negative, they are sometimes executed for their relevance to other policies or areas of society.

The IRR is defined as the type of discount that zeroes the net present value, that is, that discount rate that equals the updated value of cash inflows with the updated value of expected payments. (Peña et al. 2016).

Its mathematical expression responds to the following equation:

$$IRR = k_1 + \frac{NPV_{k_1} \times (k_2 - k_1)}{NPV_{k_1} + /NPV_{k2}/} \quad (2)$$

Being:

$NPV_{k_1} = NPV$ Positive at the update rate k_1
$k_1 = \mathit{Positive\ NPV\ discount\ rate}$
$NPV_{k_2} = NPV$ negative at the update rate k_2
$k_2 = \mathit{Negative\ NPV\ discount\ rate}$

Regarding the selection criteria, you have to:

- IRR > k, the expected profitability of the project exceeds that required so that it will be attractive to investors and will accept it.
- IRR = k, the expected profitability of the project is equal to that required and investors will be indifferent to its execution.
- IRR < k, the expected profitability of the project is lower than that required and therefore this will not be attractive to investors who will reject the project.

4. Select the best decision alternative. The selection of the best decision alternative will be made by comparing the results of the NPV, IRR and DRP indicators, the latter being just one more argument, that is, the indicators that ultimately determine the NPV and IRR for the economic significance. That enclose: gain or loss in net terms.
5. Evaluate the feasibility of the selected decision alternative. Once the best alternative has been selected (which will be the one with the most benefits for both the company and the company), it should be evaluated in greater depth, determining as accurately as possible all the elements related to the management of the project itself, say, project manager or director, legislature to be governed by, necessary budgets, labor, permit, internal and external customers, calendar and critical path, etc.
6. Determine the balance of payments effect for the country and the profitability threshold. Once everything related to the project has been known, the impacts on the country's currencies must be determined, that is, if the project will generate more or less foreign exchange than the one that it issues and, in general, what the amounts or prices or costs would be minimums that make the project generate neither losses nor gains; extremely important elements in business management and strategies

Each of the calculations carried out in the activities indicated above takes into account the values of the states of nature previously defined for the variables at the different presumption levels. Then a fuzzy number will be formed that will be represented through the segments formed by "cutting" (assigning a value) the membership function at certain levels α (presumption level), as illustrated in the following Fig. 1:

As the presumption level α decreases, the segments obtained are progressively fitted. Level 0 always corresponds to the referential and the maximum presumption level (1) is possible that the subset is empty.

As is appreciable, the proposed tool presents a reasonable and applicable form to both commercial and service entities and productive companies, allowing to manage investment projects efficiently, and working with variables such as demand, prices, unit costs, etc. in conditions of uncertainty.

In this way, it is possible to simplify the amount of information and time needed when working with probabilistic models, which require time series that allow you to make forecasts with a high degree of certainty, and reliable information related to standard deviations.

Next, the application of these theories to the evaluation of the feasibility of an investment carried out by the Molecular Immunology Center, Cuba, will be presented.

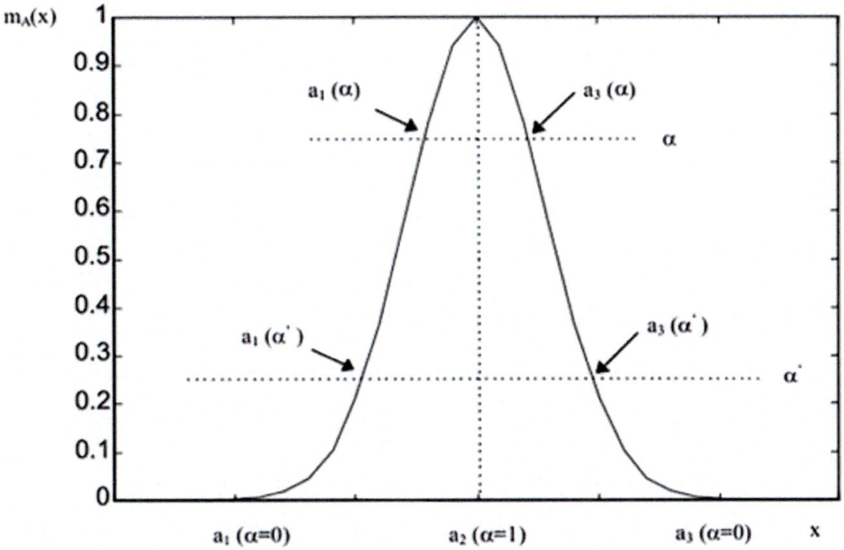

Fig. 1. Normal and convex diffuse number with α-cuts. obtained from (Morillas Raya 2006).

3 Results

3.1 Need for Investment

Following the XX Anniversary of the creation of the CIM, a diagnosis was made for the identification of the main bottlenecks or problems in the center that greatly impede the development and efficiency of various pillars of its business management system. One of the bottlenecks in the center is in product development.

The transfer to facilities with the capacity to generate enough product to address clinical trials has been identified as one of the longest stages in the context of the CIM.

Among the fundamental causes that have influenced this were insufficient capacity in terms of physical space, equipment and work infrastructure, which does not allow to address several products a year and at the same time work on the development of platform technologies such as: high producing and stable strains, culture media, formulations, etc.

With a view to responding to the needs raised and solving the identified difficulties, an investment in new capacities of the INIM building to which the Product Development Management belongs is in progress.

This investment will triple the physical area (from 146 m^2 to 500 m^2), increase the capacity of engineering services and update the regulatory level of the Pilot Plant of products in development, because with it the conditions are created to make a qualitative leap in the INIM product development work and expand its scope and quality.

3.2 Identify Alternative Decisions

In this case, an exhaustive search of the assets that would be necessary to buy, as well as their suppliers, was carried out. Then it was continued to identify, of the possible suppliers, those that belonged to the portfolio of suppliers approved for the center. Once this information was collated, the evaluation of the options began.

3.3 Evaluate Alternative Decisions

For the evaluation of alternative decisions, a project team was formed that assessed each of the suppliers' offers together with the states of nature provided for the demands, prices, unit costs, financing, etc. Then the tender continued.

3.4 Select the Best Decision Alternative

Once the bidding process was completed, the most suitable supplier was selected, resulting in the winner of a direct producer capable of supplying all the necessary assets.

3.5 Evaluate the Feasibility of the Selected Decision Alternative

For the evaluation of the investment project, the following elements were taken into account in the cash flow conformation:

- The total investment amount will include the necessary reinvestments after the total depreciation of certain assets during the evolution period.
- In consultation with senior management of the company and other people considered knowledgeable about the subject, in addition to the project team, it was concluded that the variables that generate more uncertainty and that according to their opinion can be expressed through Triangular blurry numbers are: the expected income and the discount rate.
- They will be included in the cash flow:
  - Sales taxes
  - The incremental operating costs
  - Other taxes, fees and contributions
  - Depreciation of the equipment to be purchased
  - Income taxes
  - The item previous expenses
  - The variation of working capital
- The investment will be financed with a loan and the evaluation time horizon will be ten years.
- The total amount of equipment is 7,497,100 CUC, which includes freight expenses amounting to 22,000 CUC, and for testing and commissioning of 37,000 CUC.

Projected Fuzzy Income. Due to the difficulty in estimating income accurately, since they have their own instability due to their nature and depend largely on what scientists are capable of creating and / or the results that can be obtained from a laboratory test, which presents high uncertainty, aims to work the revenue line as a triangular fuzzy number.

For year 1, 2,000,000 CUC was assumed as a pessimistic value, an amount that coincides with one of the lowest obtained by the center; As a higher presumption value, 3,800,000 CUC was assumed, the average balance of income for intangible services of the last 9 years, because although it is expected that greater product development has been preferred to be austere due to the instability of income of the previous years, as an optimistic value, 4,500,000 CUC was assumed, an amount even less than the difference between the minimum value and the highest presumption value, but which coincides with the maximum results obtained.

For the 3 years after the first equipment was put into operation, an annual income growth rate of 2% was set, since the percentage of revenue growth was analyzed for this type of service that has historically presented the center from one year to another, and the same rate was assumed on average with respect to the maximum presumption level value. From the fifth year, once all the equipment is installed and fully operational, this rate is increased by 3 percentage points taking into account the criteria of the CIM experts and the characteristics specified by the producers of the equipment to be imported.

Fuzzy Discount Rate. The discount rate to be used to calculate the current value of future cash flows will depend, among other factors, on the risks assumed that are inherent to the investment. In this case it is considered that the rate will be given by a triangular fuzzy number due to the uncertainty, instability and risk of investments in technology.

Thus, since the investment is medium-high risk, the maximum value of 15%, the highest presumption value of 13% and a minimum value of 12% will be assumed, leaving the discount rate conformed by the number triangular blur (12, 13, 15), values that were established taking into account the criteria of the CIM experts and the study of international bibliographic sources for this type of investment.

It is assumed that the opportunity cost of capital will fluctuate in these values throughout the evaluation time horizon, that is, that interval will remain constant and a different interval will not be established for each year.

Taxes Company Pays: Sales tax: 1% of the sales revenue generated; Income tax: 35% and Other taxes, fees and contributions: 1% which are, among others, contributions to territorial development.

Operating Cost. The operational costs to consider are the incremental investment, which are composed of raw materials and materials (reagents, which are imported by 80%), salary and energy are estimated at 1,564,684.38 CUC. Taking into account the criteria of the CIM experts, for the 3 years after the first equipment was put into operation, an annual growth rate of operating costs of 2% was set, corresponding to the expected revenue growth in that period, and after the fifth year, once all equipment is installed and fully operational, this rate is increased by 3 percentage points also in correspondence with the expected growth of revenues in that period.

Depreciation. According to (Ministry of Finance and Prices 2012), the depreciation rates of fixed assets that the company will acquire will be 6% in the case of machinery in general, 10% for furniture and shelves, 15% for Office equipment and equipment and 25% for computer equipment.

Previous Expenses. The previous expenses for technical task, market study, feasibility study, permits and licenses granted by the consulting bodies and agencies are estimated at 3,000 CUP, which depreciate at a rate of 25%.

Working Capital. For the present investment, it was considered that the working capital with which it should begin would be equal to the operating costs of the first semester of exploitation, this value amounts to 782,342.19 CUP.

It was determined to calculate the working capital in this way taking into account the type of investment being evaluated, whose results are influenced by long research cycles, hence the calculation of the working capital is not adjusted by the period method of financial maturation. In addition, it was determined that it was the value of the operating costs of the first semester in order to minimize risks given the long collection cycles and short payment cycles of the company.

Financing. Although the investment is planned to be executed in 3 stages with different amounts in each one, the entire loan is requested in year zero, because in the first place, it is very cumbersome and complicated to request 3 loans from the bank in 3 years separately, since it requires a lot of information that justifies the use of money, delays in giving answers and would oppose granting 3 credits for 3 consecutive years to the company. Secondly, sometimes the suppliers of the equipment to buy offer discounts if purchased in quantities or different types of equipment simultaneously, and it would be convenient to have money in hand to take advantage of the discounts in a timely manner if they were granted, despite that customs regulations do not allow imports of equipment for such high amounts.

The financing source considered was a bank loan in the amount of 7,497,100 CUC at an interest rate of 7%, as stipulated in Circular No. 5/2012 of the General Treasury Department of the Central Bank of Cuba. Thanks will be considered 3 years, taking into account the trustworthiness of the center for being a leading company in Cuba, in addition that after these years, the investment would be completed and in the fourth year it would begin with the return of the principal that would be amortized by a amount of 1,499,420 CUC during the following 5 years.

With the previous data and using operations with fuzzy numbers, a 10-year cash flow in total currency (CUC + CUP) was prepared, assuming an exchange rate of 1CUC = 1CUP (see Table 1).

The indicators were then calculated to determine whether or not the investment is profitable, starting with the Fuzzy Net Present Value and the Fuzzy Internal Rate of Return.

- Fuzzy Net Present Value (NPV)

Table 1. Summary of fuzzy cash flow.

Year	Cash flow		
	Min value	Highest presumption value	Max value
0	−785.342,19	−785.342,19	−785.342,19
1	−23.727,05	1.122.872,95	1.568.772,95
2	39.267,05	1.208.799,05	1.663.617,05
3	85.862,89	1.278.785,53	1.742.699,89
4	−1.434.210,39	−217.429,30	255.763,35
5	−1.326.615,13	−48.994,99	447.857,29
6	−1.244.075,67	97.425,48	619.120,37
7	−1.160.820,42	247.755,79	795.535,43
8	−1.102.813,59	376.191,44	951.360,06
9	507.402,41	2.060.357,68	2.664.284,73
10	4.302.465,61	5.933.068,64	6.567.192,05

With the results of the fuzzy cash flow, the net present value formula (1) is applied and under these conditions the Fuzzy Net Present Value is:

$$NPV_{Fuzzy} = (-785.342,19; -785.342,19; -785.342,19) + \frac{(-23.727,05; 1.122.872,95; 1.568.772,95)}{(1,12; 1,13; 1,15)^1}$$
$$\ldots$$
$$+\frac{(4.302.465,61; 5.933.068,64; 6.567.192,05)}{(1,12; 1,13; 1,15)^{10}}$$

Performing the respective operations and respecting the properties of the fuzzy numbers, it is obtained that:

$$NPV_{Fuzzy} = (-2.326.419,49; 4.608.623,52; 7.731.652,82)$$

The result indicates that the project may present negative values, which means that in unfavorable circumstances there will be losses amounting to 2,326,419.49 MT, being the possibilities of obtaining wider favorable results, until obtaining a net capital gain of 7,731,652.82 MT in an optimistic scenario of income and discount rate.

To determine with greater precision the possibilities of favorable results, the NPV is presented graphically with a triangular fuzzy number (see Fig. 2).

First the area of loss is determined with the formula of the area of the

$$A_{triangle} = \frac{(b \times h)}{2}$$

This area is subtracted from the total area obtaining the utility area, and is compared with the total area, which represents all possible project results. The previous calculations allow to determine the percentage of favorable results of the project like in Table 2.

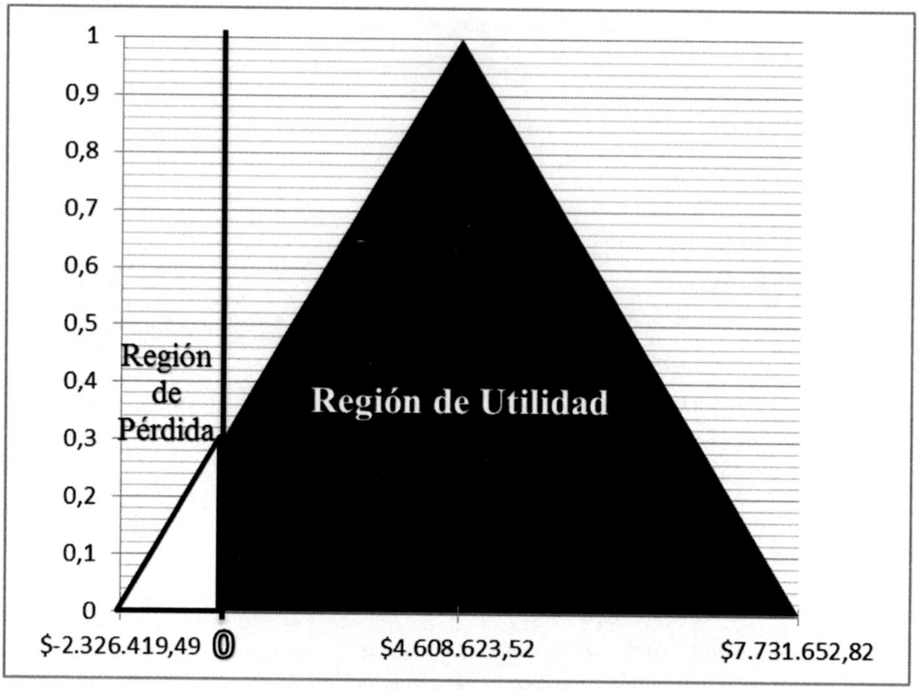

Fig. 2. Fuzzy NPV.

Table 2. Calculation of the regions of loss and profit of the investment.

Triangle area	**5.029.036,16**
Loss area	390.208,66
Profit area	**4.638.827,50**

$$\textit{Percentage of favorable results} = \frac{4.638.827,50}{5.029.036,16} \times 100 = 92,24\%$$

The investment project can be considered as viable, since even under conditions of a complex and uncertain environment the new investment generates sufficient income so that with a 92.24% probability the amount of the investment is recovered, the cost of financing and most importantly, generate a utility for the CIM and for the patients who will benefit from the new products. The level of favorable results is 92.24%, which is considered very acceptable.

- Fuzzy Internal Rate of Return (IRR)

  One of the controversial points in the estimation of project evaluation criteria, under the methodology of fuzzy mathematics, is the calculation of the IRR. Some authors, such as (Buckley 1987) and more recently, (Kahraman et al. 2002) and (Milanesi 2016), argue that a fuzzy IRR cannot be defined, since following its equation, the calculation of the IRR requires equalizing the NPV to zero. According to these authors, zero is a clear number, while the NPV in this context is a fuzzy concept, so equality would be impossible. An alternative interpretation of the IRR was used in this work, using the representation through alpha-cuts proposed by (Buckley 1992) and (Terceño et al. 2003). Therefore, it is assumed that future cash flows have been estimated using fuzzy numbers, and if the IRR of the project is continuously increasing with respect to cash flows, then both rates can be obtained through the corresponding alpha-cuts.

Performing the respective operations respecting the properties of fuzzy numbers and substituting in (2):

$$IRR_{Fuzzy} = (-8\%; 136\%; 198\%)$$

The values shown by the IRR for this case are somewhat inconsistent, because there are sign changes in the cash flows used to calculate them. Therefore, for the purposes of this study, the decision should be taken based on the NPV indicator.

- Fuzzy Discounted Recovery Period (DRP)

  The fuzzy result indicates that the project may have, in favorable conditions, a recovery in 6 months and 21 days, but in unfavorable conditions it does not recover in the evaluation time horizon.

3.6 Determine the Balance of Payments Effect for the Country and the Profitability Threshold

In order to determine the impact of the investment project on the country's foreign exchange balance, the performance of the currency for each of the proposed scenarios was calculated.

The income projected by the exports of intangible services and as exits, the disbursements for the import of equipment and machinery, and the importation of 80% of the raw materials and materials used were taken as foreign exchange entries. This investment does not generate import substitution. Once the calculations were made, a positive net currency in currencies could be observed for any of the three projected scenarios, which evidences the desirability of actually carrying the investment not only for the center and for the patient community that is favored, but also for the national balance of payments.

Therefore, the performance of the currency was calculated and it was obtained that for the pessimistic scenario it will yield 17%, for the highest presumption level 70% and for the optimistic scenario 77%. This is nothing more than the weight gain of the invested currency.

In order to know to what extent the income could vary so that the project did not generate neither losses nor gains, we proceeded with Microsoft Excel and it was obtained that in relation to the pessimistic scenario, the income is below the breakeven point, so that they should increase by 32.5% so as not to have losses as a result, in relation to the scenario of a higher level of presumption and optimism, it is reflected that revenues can decrease up to 31.2% and 42.3% respectively, which in that point will not obtain neither losses nor gains, but any percentage of greater decrease will generate losses for the investment.

4 Final Remarks

Once the investigation was completed, the relevance of the set of actions defined to evaluate the investment projects could be felt using some elements of fuzzy logic.

The theory of fuzzy subsets and within it related to triangular fuzzy numbers, constitutes an area in which business decision making can be inserted very well, an example of this is the economic-financial evaluation presented in this research, which corroborates the convenience of investment in equipment with a 92.24% probability that the amount of the investment will be recovered.

References

Alexander Baranov, E.M.: Valuation of compound real options for investments in innovative projects in pharmaceutical industry. Procedia Econ. Finan. **27**, 116–125 (2015)

Alvarado, H.M., Menjivar, E.E.: Incertidumbre en evaluacion de proyectos de inversion del sector de pequeños hoteles en el area metropolitana de san salvador. Tesis en opción al título de Master en Administración Financiera. El Salvador (2017)

Buckley, J.: The fuzzy mathematics of finance (1987)

Buckley, J.: Solving fuzzy equations in economics and finance (1992)

Durán Herrera, J.J.: Economía y Dirección Financiera de la Empresa. Ediciones Pirámide S.A, Madrid (1992)

Estudio de Factibilidad INIM (2015)

Kahraman, C., Ruan, D., Tolga, E.: Capital budgeting techniques using discounted fuzzy versus probabilistic cash flows (2002)

Kaufmann, A., Gil Aluja, J., Barre, R.: Técnicas operativas de gestión para el tratamiento de la incertidumbre. Editorial Hispanoeuropea, Barcelona (1987)

Madrigal Moreno, F.J.: Los Conjuntos Borrosos en la Administración Financiera. Una Alternativa para el Caos, la Complejidad y la Incertidumbre. Tesis de Doctorado que como parte de los requisitos para obtener el grado de Doctor en Administración. Universidad Autónoma de Querétaro, Facultad de Contaduría y Administración, Santiago de Querétaro (2011)

Merigó, J.M., Gil-Lafuente, A.M., Yager, R.R.: An overview of fuzzy research with bibliometric indicators. Appl. Soft Comput. **27**, 420–433 (2015). https://doi.org/10.1016/j.asoc.2014.10.035

Milanesi, G.S.: La tasa interna de retorno promedio borrosa: desarrollos y aplicaciones. J. Econ. Finan. Adm. Sci. **21**(40), 39–47 (2016)

Ministerio de Economía y Planificación. Decreto 327: Reglamento del proceso inversionista en Cuba. La Habana (2014)

Ministerio de Finanzas y Precios. Resolución No. 47. Gaceta Oficial, La Habana (2012)

Morillas Raya, A.: Indroducción al análisis de datos difusos. Curso de Doctorado en Economía Cuantitativa. Departamento de Estadística y Econometría. Universidad de Málaga (2006)

Oramas, O., Ortiz, M.L.: lógica difusa en la gestión de la cadena de suministros. Gestión y Desarrollo **12**, 53–68 (2015)

Peña, M., et al.: Criterios económicos borrosos para el análisis de factibilidad de proyectos de software en ambientes de incertidumbre. Revista Cubana de Ciencias Informáticas, 117–132 (2016)

Peumans, H.: Valoración de Proyectos de Inversión. Ediciones Deusto, Bilbao (1967)
Rodríguez Mesa, G.M.: Formulación y Evaluación Financiera y Social de Proyectos de Inversión. Universidad de La Habana, La Habana (2007)
Santos Santos, T.: Estudio de Factibilidad de un Proyecto de Inversión: Etapas en su Estudio (2008). Eumed.Net. http://www.eumed.net/ce/2008b/tss.htm. Accessed 5 Mar 2016
Sapag Chian, N., Sapag Chian, R.: Preparación y Evaluación de Proyectos (5ta ed.). Editorial McGraw-Hill Interamericana S.A., Bogotá (2008)
Terceño, A., De Andrés, J., Barberá, M., Lorenzana, T.: Using fuzzy set theory to analyse investments and select portfolios of tangible investments in uncertain environments. Int. J. Uncertainty Fuzziness Knowl.-Based Syst. (2003)
Zadeh, L.: Fuzzy Sets. Information and Control, vol. 8. Department of Electronical Engineering and Electronics Research Laboratory, University of California, Berkeley, California (1965)

Performance Evaluation Procedure Based on Fuzzy Mathematics and OWA Operators

Lourdes Souto Anido(✉)

Faculty of Economics, Havana University, 10400 Havana, Cuba
lourdes@fec.uh.cu

Abstract. The Performance Evaluation (ED) process provides the company with information that is useful for making decisions about training, the granting of remuneration and, in general, the work status of the employees and their motivation. This paper present a new procedure for the Performance Evaluation based on the Theory of Blurry Subsets and OWA operators. Due to the importance of negotiation processes for a country like Cuba, so dependent on the external sector, the proposal will be grounded in the evaluation of the performance of the negotiators. The new procedure points out reducing subjectivity existing in the Performance Evaluation Process and increasing objectivity in evaluations trying to make the decision making process more accurate.

Keywords: Performance evaluation · Fuzzy subsets theory · OWA operators

1 Introduction

In the current overview of the treatment of human resources in which it becomes the key to the success of companies, the evaluation of their performance is a key activity for the improvement of the production process.

Work performance should be measured in organizations to obtain information on the status of staff and achieve effective feedback on work done on human resources. This element allows to know indicators of the different areas that disturb or favor the performance of the workers to have a notion of the area or areas on which effort and money must be invested in order to approximate the efficiency of the personnel to their potential.

Individual criteria and subjectivity are factors that intervene in the ED process with a high frequency. In order to reduce subjectivity, the use of mathematical tools for data processing that supports the decision-making process, specifically the theory of fuzzy subsets, is proposed.

A. M. Gil-Lafuente et al. (Eds.): REDCID 2019, LNNS 388, pp. 100–120, 2022.
https://doi.org/10.1007/978-3-030-93787-4_6

Because of the importance of negotiation processes for a country like Cuba, so dependent on the external sector. This article proposes a procedure to evaluate the performance of Cuban negotiators based on the theory of fuzzy subsets and OWA Operators.

2 Performance Evaluation

Performance should be measured in organizations to obtain information on staff status and achieve effective feedback on the work done. Most of the definitions that can be found in this regard coincide in that it is a process where the work carried out by an employee is checked continuously over time.

Starting with the definition established by (Young and Peñalver 1998): performance evaluation is a "Systematic and periodic process of quantitative and qualitative estimation of the degree of effectiveness of a person in his job". From which it can be extracted that the evaluation must be a systematic and periodic process; it does not matter if the results obtained by these are satisfactory but it is continued evaluating after introducing the changes to determine if they were significant. Therefore, it must be a continuous feedback process where the results are executed and checked to introduce the necessary changes and then start over, with the aim of becoming more effective and efficient.

For its part, (Chiavetano 2000) considers the evaluation of performance as: "The systematic appreciation of the performance of each person in the position or the potential future development. Any evaluation is a process to stimulate or judge the value, excellence, qualities of some person." Remarking that an essential technique is being considered in administrative activity.

(Martínez Gullén 2003) assumes that performance evaluation is the "continuous and systematic analysis procedure that allows judgments about the personnel of a company, in relation to their work, in a given period of time (past and future), whose objective is the integration of the organizational objectives with the individual ones." The peculiarity of this definition is the fact that it considers the evaluation as very important process to align the business objectives of the individuals, in order to be able to link the workers more with the entities and reach the achievement of the objectives with greater measure.

The National Office of Standardization (Cuban Standard: 3000: 2007) defines this activity as the "systematic measurement of the degree of effectiveness and efficiency with which workers carry out their work activities during a certain period of time and their potential development, and constitutes the basis for developing and executing the individual training and development plan".

Due to the relevance of the performance evaluation, several methods have been created to judge how employees carry out their work. According to the authors (Bohlander et al. 2003), most of these techniques constitute an effort to reduce the inconveniences that are evident in other approaches. These methods of performance evaluation are classified according to what they measure: characteristics, behaviors or results.

Characteristic Based Methods
The authors (Bohlander et al. 2003) consider that this method consists in measuring to what extent an employee has certain characteristics such as reliability, creativity, initiative and leadership that would be necessary for the entities at some point in time. They are popular for their simplicity and easy to manage. If the "list" of features is not designed in relation to the position, the result will be far from reality and can give a subjective opinion.

Graphic Rating Scales
This method allows employees to be evaluated using previously defined and analyzed factors. The evaluator indicates on a scale at what level each worker is in relation to a certain characteristic.

Forced Choice
It consists in evaluating the performance of individuals by means of descriptive phrases of alternatives of types of individual performance. In each block there are four sentences in which the evaluator must forcefully choose only one, the one that most applies to the performance of the evaluated employee. Each election is assigned a score previously.

Behavior-Based Methods
Behavior-based methods are based on the description of what actions should or should not be displayed in the position. Its maximum utility is to provide information for the development of human resources.

Critical Incidents
This method arises because in human performance there are behaviors that can lead to the occurrence of positive (successes) or negative results (failures). For its application, the immediate supervisor must observe and record when an employee performs an exceptional event, taking into account those that were positive and negative. In case the behavior is extremely positive it should be enhanced and put into practice and when the negativity is the one that should be corrected and eliminated.

The advantages of this method are that the incidents of human behavior are specific and can facilitate employee feedback and development. However, favorable and unfavorable incidents have to be taken into account, because it is possible that the employees evaluated have negative feelings about this method.

Results-Based Methods
The superior objectivity of this method, with respect to the previous ones presented, represents one of its greatest advantages. In addition to the decrease in influence due to the opinion and subjectivity of the directive when using results observation, such as sales or production figures.

Productivity Measurements
The examples where this method is most used are in which employees can easily align to the objectives of the companies, such as: sellers evaluated according to the volume of sales or production workers based on units produced and at high executives based on profitability. The main problem that can affect it is that it can be distorted by external

factors (shortage of raw material or recession in a given market), with which the process carried out by the workers is demerited and it would be unfair to blame them.

Administration by Objectives (APO)
It arises as an option that overcomes productivity measurements trying to eliminate its main disadvantage through the creation of a format in which both employee and employer contribute to the creation of goals with which the former will be valued.

Other Evaluation Methods
The 360° Performance Evaluation
Yum Brands Inc. employs 850,000 workers in its Pizza Hut, Taco Bell, KFC, A&W and Long John Silver franchises. To reach its goal of fully satisfying the needs of customers, Yum has identified lines that customers prioritize such as speed, cleanliness and hospitality. The company used a 360º web-based system in the appropriate language for the region. The system collects evaluations and information on worker performance. (Gómez-Mejía 2008).

It is an evaluation made in a circular form, by all those who have some kind of interaction with the evaluated one. The superior, other workers, subordinates, internal and external customers, suppliers, generating an evaluation with a scope of 360° around the evaluated are involved. It is a more complete form of evaluation since it is obtained in many ways.

The methods mentioned above are used to assess the performance of a worker regardless of their job, then a procedure is presented for the specific case of the negotiators.

Procedure for Evaluating the Performance of Negotiators Based on Data Analysis
It is a procedure elaborated by (Marrero Ancízar 2014) who focus his research in the performance of the negotiators, differentiating it from the rest outlined above. It is composed of two phases, where at first the efficiency of the negotiators of the same entity is analyzed, and then the negotiating team is compared with that of other entities.

The Data Envelope Analysis is a non-parametric method of estimating boundaries appropriate to assess the technical efficiency of commercial negotiators, as it adapts to the essential characteristics of this research object since it does not require that a functional relationship for development be established of the model; also considering the incorporation of environmental variables and offering as a result complete and individualized information on the efficiency of each negotiator evaluated.

It is a complete method but it can be cumbersome as much information is needed. Its use can sometimes be complex and the company needs to have the software in question. In addition, it lacks a link with the rest of the human resources management processes in order to establish corrective actions.

ED processes are influenced to some extent by subjective factors. In order to mitigate this subjectivity, the use of mathematical tools for data processing that supports the decision-making process, specifically the theory of fuzzy subsets, is proposed.

3 Analysis of the Evaluation Method Currently Used in Cuban Companies to Evaluate Commercial Negotiators

The following describes the performance evaluation procedure currently used by national companies to evaluate commercial negotiators.

The Evaluation of the Results constitutes the basis for the development of the training plans and serves as an instrument for the determination of the permanence, promotion, training, suitability, material stimulation and moral recognition of the workers.

Knowing this, the indicators to be evaluated established by the companies are:

- Work results
- Quality of work done
- Compliance and use of working hours
- Domain and knowledge of the activity that develops
- Prestige and social recognition
- Work behavior and ethical personnel
- Compliance with occupational safety and health standards
- Use and care of material resources
- Compliance with the training and individual development plan

As you can see most of these criteria, without being erroneous, when evaluating them they are qualitative, because they do not have an indicator or a range to assess them, so they have implicit a high degree of subjectivity.

Procedure for Carrying out the Monthly Performance Evaluation
First, the negotiator is responsible for conducting a self-assessment by granting a certain amount of points (up to the maximum established) to each of the indicators listed above. In total, the sum of the values granted may reach 100 points; from which the worker determines whether or not he obtains the total.

Subsequently, his immediate boss issues his criteria on the one evaluated in the same indicators. If there are discrepancies in the results, the value granted by the boss is taken as a reference. However, the assessed negotiator can appeal to these values being subject to review, with the new senior head of the evaluator taking charge of the new evaluation.

Procedure for Conducting the Annual Performance Evaluation
The procedure by which performance is evaluated in the period of one year uses the same variables as the monthly period. Only one essential difference is evidenced: in the monthly evaluation, the results are measured with quantitative values on a scale of 1 to 100; while in the annual evaluation a description of each indicator is appealed, its results being qualitative. The conclusive result of this process is given in the following categories:

- *Superior Labor Performance:* Labor performance and superior performance of the worker, with high economic and social impact, present and future, identified with the labor competencies required in his position.

- *Adequate Labor Performance:* Work performance and worker performance, appropriate to the requirements established for their position.
- *Poor Work Performance:* Work performance and worker performance, below the requirements established for the position.

To be submitted to this process, the worker must have worked at least 70% of the established time. This procedure must be carried out on a mandatory basis every year, taking into account the following logic:

1. Analysis of the results of the monthly evaluations during the year.
2. Assessment of the fulfillment of work functions and their potential development in the field of the company.
3. Analysis of the training actions in which the worker participated during the year.
4. Assessment of the use of the knowledge acquired, in their work functions, which imply an increase in quality, efficiency and effectiveness.
5. Analysis of deficiencies that the worker still presents.
6. Characteristics of their strengths, possibilities and capabilities.
7. Preparation of the evaluation document.
8. Discussion with each worker.
9. Conclusive result.
10. Signature of the evaluation by the evaluated and the evaluator.

As with the monthly evaluation, the worker has the right to discuss the result assigned to him with the boss who evaluated him and express his opinions. In case of disagreement, the evaluated person may claim before the immediate superior to the one who carries out the evaluation within 7 business days after the notification, who decides what proceeds within the next 20 business days. There is no recourse against this decision either in the administrative or judicial proceedings.

Once described the way in which the performance of workers in the company is evaluated, the existence of aspects to improve is evidenced. The first aspect to highlight is the generality of the procedure. This is applied uniformly to all workers in the company, without distinction in the special characteristics of each position. Thus, in the case of commercial negotiators, it does not establish particular indicators that successfully measure their performance.

Secondly, the method poses a high level of subjectivity regarding the criteria of the evaluator, which can vary the result substantially. There is no way that unifies the criteria when choosing an employee to stimulate. Also the fact that the process is not automated and that it needs more time to carry it out prints a certain lack of rigor and gives it a mechanical character. This is evident when the worker is not sincere in his self-assessment and the evaluator is indifferent to the results.

Another aspect to note is the lack of agreement when expressing the results of the evaluations. The monthly one is measured on a 100-point scale and does not have ranges that qualitatively define the results. That is, it is not defined from what amount of points are necessary for a worker to be considered "Good" or "Bad", nor does it end up expressing one of these results, but is limited to using a numerical value. The opposite occurs in the annual evaluation, where the results are expressed only qualitatively and there is no way to validate the criteria of the evaluator.

3.1 Analysis of the Causal Relationships Existing in the Performance of the Negotiators

In order to propose a new procedure to evaluate the negotiators, once the one they are currently applying has been analyzed, the causal relationships that may be affecting the performance of the negotiators are analyzed. With this objective the theory of forgotten effects is used that allows analyzing the causal relationships present as well as the levels of direct and indirect incidents.

The first step was the determination of the possible causes that could be affecting the main performance indicators of the negotiator, for this a brainstorm was carried out, previous bibliographic review on the subject, among the research group being selected those that appear in the following table (Table 1):

Table 1. List of causes and effects to consider.

Variables that influence the negotiator profile	Indicators of negotiator performance
C1. Qualification	E1. Satisfaction of the Negotiator with its Performance
C2. Individual Behavior	E2. Satisfaction of the Organization with the Agreements Achieved
C3. Negotiation Styles	E3. Achievement of the Objectives
C4. Initiative	E4. Impact of the results
C5. Working conditions	
C6. Salary	
C7. Autonomy before negotiation processes	

Once the causes and effects were defined, they were evaluated by a group of six experts composed of specialists from the Negotiating Group of the company. They expressed their valuations through a survey. The results are shown below.

First, the direct incidents between causes and effects were sought (Table 2).

In a second moment, they pass on to express their valuations between causes and causes and of the effects with the effects, looking for the relationships that prevail between them. Once the survey data has been entered in the FuzzyLog software, the results show the existence of ten forgotten effects (Table 3).

Next, the main relationships found are analysed (Fig. 1).

When analyzing the results obtained, it can be seen that there is a direct influence of the Working Conditions on the Satisfaction of the Negotiator with its Performance valued at 0.6. There is also an implicit relationship between the Working Conditions and the Behavior of the Individual. Working conditions promote satisfaction and motivation,

Table 2. Estimated incidence between causes and effects $\underset{\sim}{[M]}$

	E1	E2	E3	E4
C1	0,85	0,75	0,87	0,87
C2	0,87	0,77	0,8	0,8
C3	0,72	0,7	0,8	0,8
C4	0,75	0,73	0,85	0,77
C5	0,6	0,5	0,6	0,58
C6	0,87	0,73	0,87	0,87
C7	0,75	0,7	0,68	0,8

Source: Machine output of the FuzzyLog program.

Table 3. Forgotten effects $\underset{\sim}{[O]} = \underset{\sim}{[M^*]} (-) \underset{\sim}{[M]}$

	E1	E2	E3	E4
C1	0	0	0	0
C2	0	0	0,1	0,1
C3	0,8	0,1	0	0
C4	0	0	0	0
C5	0,3	0,3	0,2	0,8
C6	0	0	0	0
C7	0	0,1	0,8	0

Source: Machine output of the FuzzyLog program.

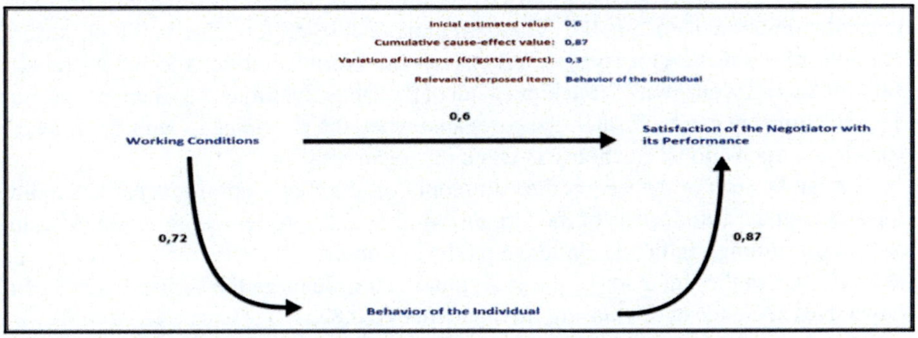

Fig. 1. Variation of incidence (forgotten effect) between the working conditions cause and the satisfaction effect of the negotiator with its performance. Source: Machine output of the FuzzyLog program.

which influences their behavior and generates a relationship that takes influence to 0.87 (Fig. 2).

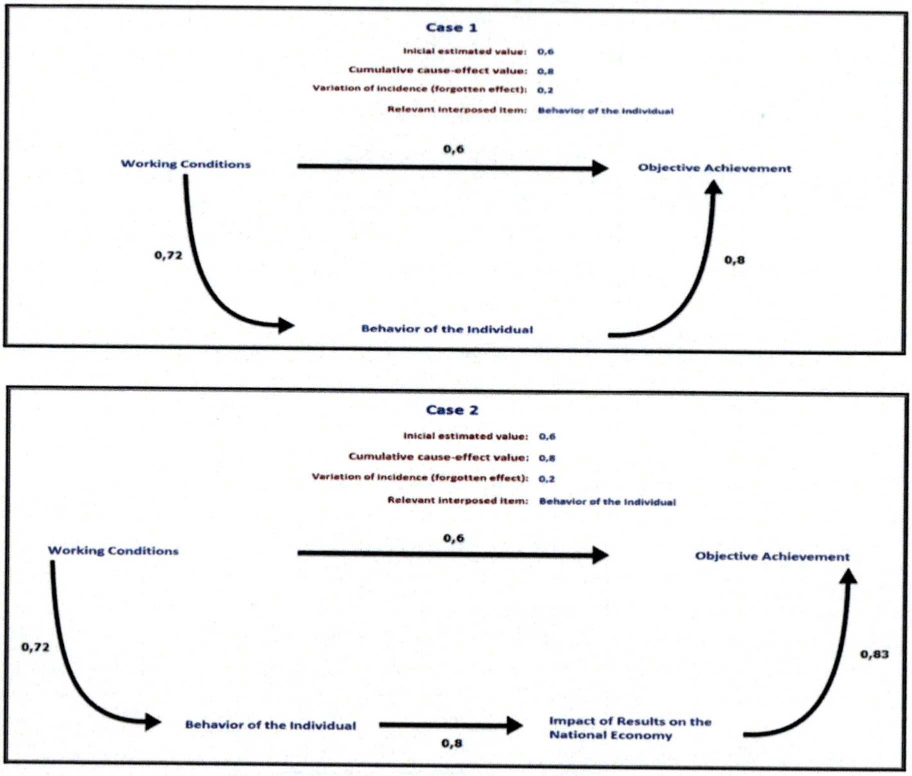

Fig. 2. Variation of incidence (forgotten effect) between the working conditions cause and the objective achievement effect. Source: Machine output of the FuzzyLog program.

In the case of the relationship Working Conditions with Achievement of the Objectives, the direct incidence is 0.6. The cumulative cause-effect value is 0.8 because the relations of indirect incidents present between the Work Conditions-Behavior relationships of the Individual with the achievement of the Objectives must be taken into account. This relationship can be further disaggregated when the relationship with the Impact of Results on the National Economy is taken into account (Fig. 3).

As can be seen in the figure, the Autonomy on the Negotiation process has a direct influence on the Satisfaction of the Organization with the Agreements Achieved valued at 0.7. But through indirect relationships the Autonomy or rather the lack of it generates changes in the Behavior of the Individual that influences the Achievements of the Objectives and ends up increasing the incidence to 0.87.

Another variable that despite having no forgotten effect does have high influences on the effects is the Salary. Logically, this issue is a priority and of vital importance at the national level. Due to this, one of the objectives of this paper is to stimulate it through the improvement of the Performance Evaluation process.

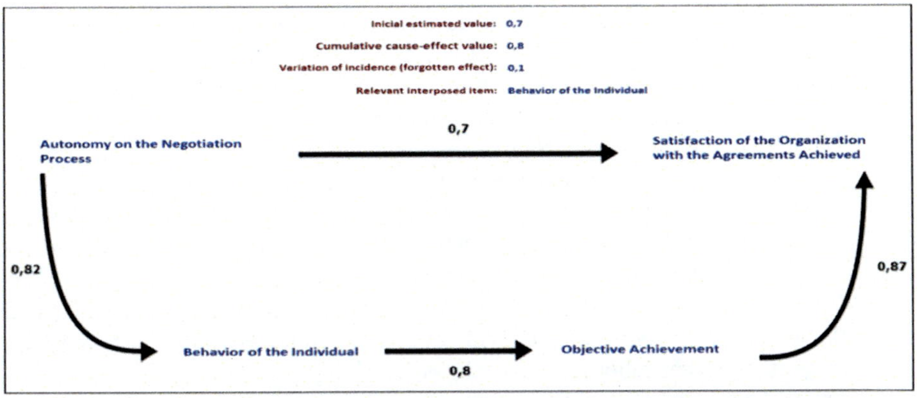

Fig. 3. Variation of incidence (forgotten effect) between the autonomy cause on the negotiation process and the satisfaction effect of the organization with the agreements achieved. Source: Machine output of the FuzzyLog program.

4 Proposal of the Procedure

Once those variables that affect the performance of the negotiator are known, which allows for a more complete view of the situation, the procedure proposal is developed.

With the application of the theory of fuzzy subsets and the mathematical operators OWA and LOWA, the goal is to unify the evaluators' perception of workers with the measurable criteria obtained from the 10 fundamental variables used in the procedure. This information that implies the subjectivity provided by the Criteria of the Evaluators is contained in a range consisting of a minimum and a maximum valuation. In this way, those in charge of the evaluation have a greater margin to express their criteria and those evaluated can obtain higher values than in a static evaluation.

In the selection of the variables to be used for the procedure, some of those currently used by the company were chosen and others from previous research carried out by (Marrero Ancízar 2014) that were analyzed based on the Theory of Forgotten Effects to select the most influencers on the company. With this analysis the decision is made not to choose Autonomy, Salary, Working Conditions, among others because they are not variables in which the negotiator can influence and therefore it affects in a counterproductive way to include it in the study. The list of variables is constituted as follows:

1. Results: Compliance with the objectives set for the negotiation process
2. Individual Behavior: It arises from the merger of the variable Labor and ethical behavior of the company with the criteria of mastery of the emotional competencies of the profile and other elements such as the sensory system, caution, the tendency to ask and the ability of argumentation.
3. Negotiation Style: It is formed by the negotiator's styles or attitudes during a negotiation process, associated with the two basic strategies set forth in Sect. 1.
4. Initiative: Characterizes people who find alternative solutions to conflicts in the negotiation process and are not limited to doing only what is established. This variable has a high correlation with the Flexibility variable that is not included

among those selected for the procedure because it does not depend on the entity's workers.

5. Language: It is a variable that responds to the preparation of negotiators and is essential when conducting international negotiations.
6. Technical knowledge: Mastery of the tasks and functions of the job position it occupies, as well as general aspects of the business sector where it resides.
7. Prestige and recognition: It is related to the image that negotiators leave during the dissimilar negotiation processes in which they have participated and also with the prestige they enjoy inside and outside the organization.
8. Compliance and use of the workday: Responds to the need to perform tasks within the workday for your position. Without wasting time performing acts outside the established by the labor regime of the entity.
9. Safety, health and material resources: Combines elements of the entity's variables Compliance with occupational health and safety standards and use and care of material resources.
10. Compliance with the training and individual development plan: This variable links the cycle of Performance Evaluations. The problems recognized in a previous evaluation must be eradicated through training, otherwise this indicator is affected.

In the proposed procedure, an indefinite scale (valuations with a range between 0 and 10) is used to collect the information. Then the ranges would be shortened to a pentary scale (valuations with a range between 0 and 5), which are the most used in the Cuban business system.

In order to make the procedure more flexible, weights are used in each variable. These are obtained from the criteria of experts who grant different levels of importance within the operator and are used to order the results. When fulfilling a function similar to that of the weights, and being the variables of the procedure as a whole the variables included in the LOWA operator, they are used as weights.

The weights used on the variables to be measured are added by means of the arithmetic mean of the set of valuations granted by 8 experts. The experts who gave their criteria are composed of employees with experience in the subject, who contribute a practical point of view, as well as outstanding academics who contribute with their contributions from the theoretical field. These experts were evaluated for their level of competence, seeking to work with experts with high competition coefficients.

Among the advantages of the mathematical operator OWA, one of the ones that stands out when evaluating performance is the fact of being able to organize the values obtained up or down. In this case, the descending option is chosen, with those with the highest results being reflected first. Its use will be necessary in the 2nd part of the process where a group of workers are evaluated in order to determine those with greater results.

Once the fundamental aspects to perform the procedure have been analyzed, each step is described step by step, which are reflected in the following graph (Fig. 4).

The procedure consists of two fundamental stages, both of equal relevance to the final result. Both are developed with the Microsoft Excel program to facilitate the understanding and use of the new evaluation mode as it is a program commonly used in companies. In the first part the individual evaluations of the employees are carried out concluding with one of the 5 evaluations proposed (Excellent, Good, Regular, Bad, Very bad). The

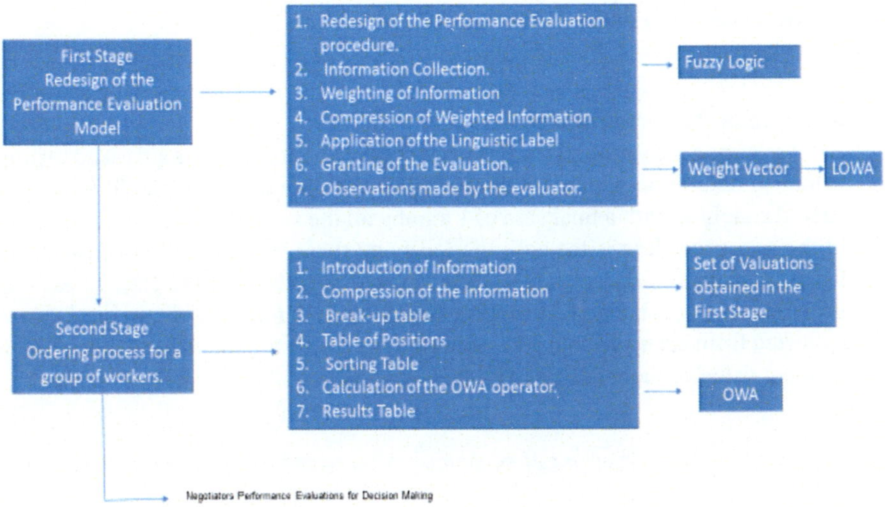

Fig. 4. Procedure for the proposal of the performance evaluation.

second part uses a simple OWA to order the workers already evaluated in descending order, determining which ones obtain the best results than others.

It is important to note that the stages must be separated into different Excel sheets because although the information is similar in both cases the recipient is different. The first provides significant benefits to workers who are evaluated through an effective procedure with few failures, which is impartial to different workers despite containing the subjectivity of the perceptions of the evaluators. The second stage is more useful for the Human Resources department that acquires relevant information such as who needs training and who should be encouraged. Which is necessary for the rest of the department's processes in order to meet their business objectives and increase the level of development of long-term workers.

The software used presents problems when working with undefined variables such as Linguistics, therefore, these are replaced by numerical values in order to facilitate their use as shown below (Table 4).

Table 4. Equivalence of linguistic labels with their numerical values in the procedure.

Linguistic label compressed	Compressed numerical values
Excellent	5
Good	4
Regular	3
Bad	2
VeryBad	1

First Stage: Redesign of the Performance Evaluation Model
The first stage consists of 7 steps in total.

1. Redesign of the Performance Evaluation procedure
After selecting the variables for Performance Evaluation, experts are asked to provide values to determine which are the variables with the most weight. This weighting should conclude with a final sum of 1 among all the variables. One of the advantages of the new procedure is the ease in modifying these weights if it is considered that the significance of the variables varied if the objectives of the Human Resources Department were changed. Once the assignment is completed, the weights that were collected from each variable are added through an arithmetic mean to conclude with a global value of them (Table 5).

Table 5. Weights and their aggregation, granted to the variables by the experts for the initial procedure.

Crit d expert	Expert 1	Expert 2	Expert 3	Expert 4	Expert 5	Expert 6	Expert 7
Variable 1	0.15	0.2	0.15	0.17	0.13	0.12	0.2
Variable 2	0.09	0.14	0.1	0.07	0.11	0.09	0.10
Variable 3	0.12	0.15	0.15	0.13	0.12	0.14	0.1
Variable 4	0.12	0.1	0.12	0.13	0.11	0.14	0.11
Variable 5	0.15	0.1	0.12	0.1	0.12	0.11	0.12
Variable 6	0.15	0.1	0.1	0.1	0.13	0.12	0.12
Variable 7	0.03	0.02	0.03	0.05	0.04	0.04	0.06
Variable 8	0.05	0.06	0.04	0.07	0.09	0.05	0.04
Variable 9	0.02	0.01	0.09	0.08	0.06	0.05	0.02
Variable 10	0.12	0.12	0.1	0.1	0.09	0.14	0.13
Total	1	1	1	1	1	1	1

media Criterio d	
Variable 1	0.16
Variable 2	0.1
Variable 3	0.13
Variable 4	0.118571
Variable 5	0.117143
Variable 6	0.117143
Variable 7	0.038571
Variable 8	0.057143
Variable 9	0.047143
Variable 10	0.114286
Total	1

2. Information Collection
To obtain the necessary information, the self-evaluations of each worker, the fulfillment of the work plan and the perceptions of the immediate bosses must be consulted, which can be determined by direct observations in the workplace or by opinions of the work group.
All the information as a whole is introduced using the theory of fuzzy subsets, so it is given a valuation based on 10 (looking for more options that benefit the evaluated one) that consists of a maximum and a minimum value. Giving rise to a range where the evaluated is.
An example of how these criteria would be collected is presented in the following table (Table 6).

Table 6. Example of valuations introduced for the ten proposed variables.

Indicators	Description	Minimum value	Maximum value
Results	Compliance with the objectives set for the negotiation process	7	8
Individual behavior	It arises from the fusion of the variable "Work behavior and ethical stall of the company" with the criteria of mastery of competences	6	9
Negotiation style	It is formed by the styles or attitudes of the negotiator during the negotiation process, associated with the two basic strategies	5	8

3. Weighting of Information
   To work according to the objectives and needs of the company at the time, the ranges proposed by the evaluator are multiplied with the weights of the experts. The weights used are those derived from the grouping of the experts' criteria. In this way the operator collects both the opinions of the experts and the information gathered by the evaluator (Table 7).

Table 7. Weighted valuations.

Weighted	
11.2	12.8
6	9
6.5	10.4
4.742857	8.3
8.2	11.71429
7.028571	10.54286
2.314286	3.085714
5.142857	5.714286
3.3	4.242857
6.857143	9.142857

4. Compression of Weighted Information
   It aims to bring the results of an indeterminate scale to a pentary to maintain the 5 possible results to the evaluation (Excellent, Good, Fair, Bad, Very bad). This step occurs simultaneously with the following.
5. Application of the Linguistic Label
   Applying this method facilitates the management of variable valuations because its objective is to facilitate work with qualitative information. Its application is possible based on nested conditional functions that verify membership criteria for the minimum and maximum of the Fuzzy numbers adjusted using a pessimistic tie-off criterion in which if the ranges are in different categories, the linguistic label given precedes the maximum.

The function begins by checking for the best linguistic label value that can be given to the corresponding maximums and minimums. If the stated values do not meet the requirements, the maximum value that can be acquired from the linguistic label is assigned through a method of differentiation in the comprehension criteria. Between the maximum and minimum values that are run in the program there must be a numerical difference and the maximum value must be a number greater than the minimum. If it is not in the manner stated above, the result that emerges is ERROR.

6. Granting of the Evaluation.
   The granting of the Evaluation is the last step that uses mathematics in the new procedure. The result is achieved by applying an arithmetic mean of the 10 values that each variable acquires. This average is processed by conditional functions and a text corresponding to one of the 5 values that can be obtained is displayed in the result box. As in the function described in the previous step of not fulfilling the necessary conditions then ERROR is shown as a result.
7. Observations made by the evaluator.
   The evaluator's observations are the judgments issued by him during the process, which influenced the subjectivity of his criteria when assigning a certain value.

   The bases for building this last step are shown below.

- Achievements in the development of work: These are the set of events and goals exceeded by the one evaluated during the period under evaluation.
- Strengths: It is the group of qualities, skills and abilities that the evaluated person possesses that make their performance superior to the standard in their job.
- Weaknesses: These are the qualities, skills and abilities that the evaluated person has not developed and that are necessary for their performance.
- Projections and recommendations: This is the set of goals proposed by the evaluator that must be exceeded to evaluate the performance in relation to the previous period.

If the recommendation in this aspect is exceeded, the following variables arise for the following Performance Evaluation.

Once this step has been completed, the evaluated person can issue his or her judgment if they disagree with any of the above. If the reasons are convincing, they lead to a negotiation between an evaluator and an evaluated person, that if necessary, a second evaluator must be available. Finally, the aspects that varied derived from the negotiation are included in the procedure.

From this first stage of the procedure, the worker obtains the result of his performance evaluation.

Second Stage: Ordering Process for a Group of Workers

The second part of the procedure is in an Excel different from the first, although it works with the evaluations determined in the first. Its quintessential user is the Department of Human Resources because its result helps determine who should be given incentives, who should go through training and the topics of such training. To obtain these results, a simple OWA operator is used to analyze the evaluations as a whole of the workers.

Like the previous stage it is composed of 7 steps:

1. Introduction of Information
   The first step in this stage is very similar to the previous one, it is programmed in Excel so that the information is processed automatically. The evaluator must enter the weights that the experts gave to the variables and valuations of each worker, in correspondence with their name located in the first row (Table 8).

Table 8. Data entry table for the second stage of the procedure.

	Criteria for ordering																			
Workers	Results		Behavior		Negotiation style		Initiative		Language		Technical knowledge		Prestige and recognition		Workday		Safety, health and resources		Training plan	
Leyla	8	10	6	8	8	9	6	7	8	9	7	8	6	8	8	9	7	8	7	8
Roberto	7	9	8	10	9	10	7	9	9	10	8	9	7	8	8	9	6	7	7	8
María	7	9	5	7	7	8	6	8	7	8	7	8	5	7	8	9	6	8	5	7
Lourdes	7	9	8	10	9	10	7	8	6	8	4	6	7	9	8	9	5	7	6	7
Eduardo	6	7	5	6	6	7	7	8	5	6	4	6	5	7	5	7	5	6	7	9
Diego	6	7	7	8	5	6	8	9	6	7	6	7	8	9	6	8	7	9	6	8

2. Compression of the Information.
   Compression is performed in the same way as in the previous stage, using the same function and structure. The difference in this case is that the information is in rows instead of columns and is executed for a group of workers which generates a change in the combination of cells.
3. Tiebreaker table
   The Tiebreaker Table is a strictly mathematical step that has no influence on the evaluation, nor does it measure any parameter of it. It is created by the lack of the program in some way to process two or more equal values, resulting in an alteration in the procedure. The difference established is insignificant being 0.1 * 〖10〗 ^(- 14) and takes as reference the values of the information compression table (Table 9).

Table 9. Example of tiebreaker table.

	Tiebreaker table									
Workers	Results	Behavior	Negotiation style	Initiative	Language	Technical knowledge	Prestige and recognition	Workday	Safety, health and resources	Training plan
Leyla	5	4	5	4	5	4	4	5	4	4
Roberto	5	5	5	5	5	5	4	5	4	4
María	5	4	4	4	4	4	4	5	4	4
Lourdes	5	5	5	4	4	3	5	5	4	4
Eduardo	4	3	4	4	3	3	4	4	3	5
Diego	4	4	3	5	4	4	5	4	5	4

4. Table of Positions
   The Table of Positions gives a hierarchical position to each value of the variables within the results of the same worker, that is, compares the results achieved by the one evaluated in the established variables and orders in descending order (starting with the best result) the same. Thus it can be seen in the main aspects where the employee has deficiencies and train him attending to them. This operation is carried out by means of the function JERARQUIA.EQV (AV47; AV47: BE47) and usually conflicts if it finds two values equal to not being able to determine which one is superior to the other, so the data of the Breakout Table must be used (Table 10).

Table 10. Sample position table.

	Positions table									
Workers	Results	Behavior	Negotiation style	Initiative	Language	Technical knowledge	Prestige and recognition	Workday	Safety, health and resources	Training plan
Leyla	4	10	3	9	2	8	7	1	6	5
Roberto	7	6	5	4	3	2	10	1	9	8
María	2	10	9	8	7	6	5	1	4	3
Lourdes	5	4	3	9	8	10	2	1	7	6
Eduardo	6	10	5	4	9	8	3	2	7	1
Diego	9	8	10	3	7	6	2	5	1	4

5. Sorting Table
   The ordering table allows the hierarchy established in the previous process to order the evaluations from highest to lowest. This allows positions of cells to be matched with positions of other cells that meet a given criteria. This step is essential to achieve the necessary order for the calculation of the OWA.
6. Calculation of the OWA operator.
   To calculate the simple OWA operator, the data collected in the analysis of the experts and the information obtained from the Management Table prepared previously are used. The use of this table facilitates the operator application because the data must be organized in descending order for the calculation to be functional. The calculation of the OWA operator is done by adding the multiplications, in order, of the weights by the evaluations ordered from highest to lowest (Table 11).
7. Results Table
   The Results Table is obtained from applying a function similar to that used in the Ordination Table on the values obtained from the OWA calculation. That is, it is assigned a hierarchical order and the results are displayed in order starting with the one with the highest performance in their performance. In this way, the best performing negotiators can be seen with the naked eye, thus facilitating decision-making when providing a stimulus (Table 12).

The Table shows the value that negotiators achieve in each of the pre-established variables. In addition to the respective values of Arithmetic Average, Weighted Average (WA) and Organized Weighted Average (OWA). The latter is 'by which the negotiators are ordered.

Table 11. OWA operator, weighted average and arithmetic average.

Operators		
Arithmetic Average	WA	OWA
4.4	4.46429	4.50857
4.7	4.8	4.78143
4.2	4.21714	4.26
4.4	4.36857	4.51143
3.7	3.73286	3.90286
4.2	4.07429	4.27571

Table 12. Example of results table.

Organized														
	Criteria for ordering											Operators		
Workers	Results	Behavior	Negotiation style	Initiative	Language	Technical knowledge	Prestige and recognition	Workday	Safety, health and resources	Training plan		Arithmetic Average	WA	OWA
Roberto	5	5	5	5	5	5	4	5	4	4		4.7	4.8	4.781
Lourdes	5	5	5	4	4	3	5	5	4	4		4.4	4.369	4.511
Leyla	5	4	5	4	5	4	4	5	4	4		4.4	4.464	4.509
Diego	4	4	3	5	4	4	5	4	5	4		4.2	4.074	4.276
María	5	4	4	4	4	4	4	5	4	4		4.2	4.217	4.26
Eduardo	4	3	4	4	3	3	4	4	3	5		3.7	3.733	3.903

5 Validation of the Procedure by User Criteria

To be able to apply the procedure continuously in a company, it is necessary to go through a validation process executed by future users. It must be approved by a consensus between workers and the union, in addition to having previous documentation as established in national legislation. 40 members of different commercial entities distributed among executives, specialists and those responsible for performance evaluation were selected.

The technique with which user validation is executed is the IADOV. The result implies the application of a survey consisting of 3 closed and 2 open questions that collect the validation criteria.

The number obtained by each user indicates their level of satisfaction, framed in the following order (Zenea 2011):

- Clear satisfaction
- More satisfied than dissatisfied
- Undefined
- More dissatisfied than satisfied
- Clear dissatisfaction
- Contradictory

The application of the technique showed that 50% considered the procedure very satisfactory, while the remaining 50% liked it more than it dislikes. Based on the results, the Global Satisfaction Index (ISG) is calculated, reaching a value of 0.75, as shown graphically in the following figure (Fig. 5).

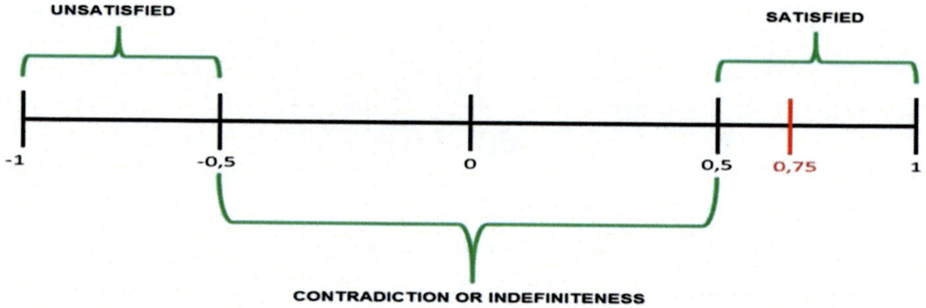

Fig. 5. Graphical result of the Global Satisfaction Index (ISG).

Below are the answers to the open questions that are included in the survey. They are separated into positive and negative aspects.

Positive aspects:

- It is able to define what are the variables in which the negotiators have deficiencies in order to determine their training.
- Use an automated program that generates automatic responses to the evaluation (2)
- You have the possibility to restructure the variables if you want to change the form of evaluation. (one)
- Compare the results obtained by a group of people and define who achieves the best. (2)

Negative aspects:

- It is a cumbersome method and can be complicated when it comes to understanding it. (3)
- It could automatically insert the values of the evaluation of the negotiators of the first stage in the second, in order to increase the speed of response (one)
- The method has not yet been applied so that inconveniences might arise on the part of the evaluators. (2)

The negative aspects provided by the survey are not considered relevant. They arise from the high use of mathematical tools that allow rapid response capacity, but these operators should not be programmed by the company because the final result of the investigation consists of an Excel that includes the processes already introduced. In general, the positive aspects of the negatives are highlighted, confirming once again the high degree of satisfaction that users show about the procedure in the entity and the acceptance of the procedure proposed by the IADOV method.

6 Final Considerations

Performance Evaluation is the way to measure the results and generate decisions about future actions to improve them.

Once the Performance Evaluation process of the negotiators using the Cuban companies was analyzed, deficiencies were found, such as the generality of the procedure, the level of subjectivity regarding the criteria of the evaluator, the lack of automation and the difference when expressing the monthly and annual results.

Based on the Theory of Forgotten Effects, variables such as Autonomy, Working Conditions and Salary were found that, despite being fundamental in the negotiation, should not be used to measure performance since workers do not have the capacity to influence them.

The new procedure created is divided into 2 stages and its advantages are based on the optimization of time, subtracts subjectivity and identifies the points at which each negotiator must be trained.

The proposal was validated through user criteria obtaining a satisfaction index of 0.75.

Acknowledgements. Project supported by "Red Iberoamericano para la Competitividad, Innovación y Desarrollo" (REDCID), project number 616RT0515, in "Programa Iberoamericano de Ciencia y Tecnología para el Desarrollo" (CYTED).

References

Bohlander, G., Sherman, A., Snell, S.: Administración de recursos humanos. International Thomson Editores, Madrid (2003)

Brazzolotto, S.: Aplicación de la evaluación del desempeño por competencias a las organizaciones. Universidad Nacional de Cuyo, Mendoza (2012)

Chiavetano, I.: Administración de Recursos Humanos. Quinta Edición. McGraw-Hill, Bogotá (2000)

Gómez-Mejía. La Evaluacion de Desempeño 360 (2008)

Herrera, F., Martínez, L.: A 2-tuple fuzzy linguistic representation model for computing with words. IEEE Trans. Fuzzy Syst. **8**, 746–752 (2000)

Kacprzyk, J., Yager, R.R., Merigó, J.M.: Towards human centric aggregation via the ordered weighted aggregation operators and linguistic data summaries: a new perspective on Zadeh's inspirations. IEEE Comput. Intell. Mag. **14**(1), 16–30 (2019)

Linares-Mustaros, S., Ferrer-Comalat, J.C., Corominas-Coll, D., Merigo, J.M.: OWA operators in the theory of expertons. Int. J. Intell. Syst. **34**(3), 345–365 (2019)

Liu, P.D., Liu, J.L., Merigó, J.M.: Partitioned Heronianmeans based on linguistic intuitionistic fuzzy numbers for dealing with multi-attribute group decision making. Appl. Soft Comput. **62**, 395–422 (2018)

Marrero Ancízar, Y.: Procedimiento para la determinacion del perfil de los negociadores comerciales cubanos y laevaluación de su eficiencia tecnica. Universidad de La Habana, La Habana (2014)

Martínez Gullén, M.C.: La gestión empresarial: Equilibrando Objetivos y Valores. Ediciones Diaz de Santos, Madrid (2003)

Merigó, J.M., Casanovas, M., Martínez, L.: Linguistic aggregation operators for linguistic decision making based on the Dempster-Shafer theory of evidence. Int. J. Uncertain. Fuzziness Knowl.-Based Syst. **18**(3), 287–304 (2010)

Merigó, J.M., Gil-Lafuente, A.M.: OWA operators in human resource management. Econ. Comput. Econ. Cybern. Stud. Res. **45**(2), 153–168 (2011)

Merigó, J.M., Gil-Lafuente, A.M., Zhou, L.G., Chen, H.Y.: Induced and linguistic generalized aggregation operators and their application in linguistic group decision making. Group Decis. Negot. **21**(4), 531–549 (2012). https://doi.org/10.1007/s10726-010-9225-3

Saavedra Tain, J.: Procedimiento para la Evaluación del Desempeño basado en Matemática borrosa y operadores OWA para la empresa ETECSA. Universidad de la Habana, La Habana (2017)

Yager, R.R.: On ordered weighted averaging aggregation operators in multi-criteria decision making. IEEE Trans. Syst. Man Cybern. B **18**, 183–190 (1988)

Yager, R.R.: Families of OWA operators. Fuzzy Sets Syst. **59**, 125–148 (1993)

Yager, R.R., Kacprzyk, J., Beliakov, G.: Recent Developments on the Ordered Weighted Averaging Operators: Theory and Practice. Springer, Berlin (2011). https://doi.org/10.1007/978-3-642-17910-5

Young, E., Peñalver, O.: Manual del director de Recursos Humanos. Cinco Dias, Madrid (1998)

Zenea, M.M.: Metodología para el análisis del funcionamiento del consejo de dirección en el sistema gerencial del MINAG. Tesis presentada en opción al grado científico de Doctor en Ciencias económicas. Universidad de La Habana, La Habana, Cuba (2011)

Zeng, S.Z., Su, W.: Linguistic induced generalized aggregation distance operators and their application to decision making. Econ. Comput. Econ. Cybern. Stud. Res. **46**, 155–172 (2012)

The Fit Between Corporate Entrepreneurship and Innovation in the Organizational Performance Through SHRM

Montserrat Salvans Blanch[1](✉), Anna Maria Gil-Lafuente[2], M. Luisa Sole Moro[2], and Sefa Boria-Reverter[2]

[1] Barcelona University, Barcelona, Spain
montserrat.salvans@ub.edu
[2] Law Faculty, Avinguda Diagonal 684, C-103, 08034 Barcelona, Spain

Abstract. The objective of this paper is to test the moderating effect of Strategic Human Resource Management (SHRM) in the relationship between organizational innovation performance and corporate entrepreneurship as a measure in technology-based firms. The design and methodology is a postal survey is used in this research. A sample of 250 European technology-based organizations is selected from Amadeus database. The questionnaire will be sent to CEO, finance managers and HR managers. A hierarchical moderated regression will be used as the statistical analyze and we expect all three hypotheses will be supported by results. This paper helps to fill the gap in the relationship between corporate entrepreneurship and Strategic Human Resource Management (SHRM) in order to achieve an innovative organization performance. Besides, it will give company managers some inspirations to develop competitive advantages in technological organizations.

Keywords: Strategic human resource management · Corporate entrepreneurship · Organizational performance · Innovation · Internal innovation · Venturing · Strategic renewal · Strategy · Human capital

1 Introduction

Nowadays, organizations are more and more influenced by the fast technological change and the highlights of importance in innovation to combat the today's increasingly uncertain, changing and complex markets where firms have to possess critical factors that enable them to create value and obtain a competitive edge [1]. In a macroeconomic environment shaped by the global markets and the fast technical development, firms have finally founded the need to innovate their organizations through the SHRM practices to

Project supported by "Red Iberoamericana para la Competitividad, Innovación y Desarrollo" (REDCID), project number 616RT0515, in "Programa Iberoamericano de Ciencia y Tecnología para el Desarrollo" (CYTED).

A. M. Gil-Lafuente et al. (Eds.): REDCID 2019, LNNS 388, pp. 121–143, 2022.
https://doi.org/10.1007/978-3-030-93787-4_7

establish a good correlation between corporate entrepreneurship and the innovative performance organization. Those have become an important source of competitive advantage which could explore new opportunities and increase organizational growth [2, 3]. Therefore, how to achieve organizational goals and develop organizational performance is crucial for firms.

Corporate entrepreneurship refers to an entrepreneurial behavior that through innovation, strategic adjustment enhances organizational performance, dominate new market and find new business opportunities [4–6]. Corporate entrepreneurship is conceptualized as a multidimensional construct with three dimensions: internal innovation, venturing and strategic renewal. Previous literature has showed that corporate entrepreneurship has a positive influence on organizational performance [7].

Strategic Human Resource Management (SHRM) focuses on integrating organizational activities with HRM policies and practices to achieve organizational goals [8]. To success in the entrepreneurial field, firms must use organizational strategies into operations [9]. Therefore, in the managerial field, SHRM also plays a significant role.

Innovation is related to the maintenance of competitive advantage and performance [10]. The assumption is that an organization's capacity to innovate resides in its employees' capabilities and motivation, and that HRM is involved in the whole innovation process, because employees' output is needed for the development and implementation of innovations [11].

In previous literature, papers either talk about the influence of SHRM on organizational performance. Although existing HR research often implies that HR outcomes serve as a key mediator between HR systems and key outcomes, the specific natures of models of this meditation depend on the theoretical perspective researchers have adopted when examining this relationship. On the one hand, several researchers have adopted the behavioral perspective of HRM. According to this perspective, organizations do not perform themselves, but instead use HR practices to encourage productive behaviors from employees and thus to achieve desirable operational and financial objectives [12]. If an organization requires efficient employees, for example, its chosen HR practices and their effectiveness would likely differ from those of an organization that requires employees to be cooperative, to focus on service, or to engage in some other critical role behavior. The effectiveness of HR practices is realized when employees act in ways that are needed for implementing strategies and achieving various business objectives, in this case it is the influence of corporate entrepreneurship on innovation organizational performance. However, few researches that examine both constructs exist [13]. Thus, there is still a gap that needs further research. The role of human capital (HC), it is found to be an important driver of innovative performance [14].

Also, literature is more and more focus in the determinant importance that HRM practices represents of organizational innovation in beneficial to the human capital as a value to sustain competitive advantage and long-term success.

This paper address the following questions: What role does Strategic Human Resource Management (SHRM) play in the relationship between corporate entrepreneurship and innovative organizational performance? Does SHRM have a moderating effect to corporate entrepreneurship and innovative organizational performance?

To deal with these questions, we propose three hypotheses by developing a literature review about the relationship between corporate entrepreneurship and organizational performance through the implementation of SHRM.

We examine separately the moderating effect of SHRM in three dimensions of the corporate entrepreneurship construct (internal innovation, venturing and strategic renewal).

The rest of this paper is constructed as follows: In Sect. 2, the literature review on SHRM, corporate entrepreneurship and innovation organizational performance, along with the development of conceptual model and hypotheses. Section 3 presents the data and methodology we plan to use in our research. Section 4 explains the expected results by using statistical analysis. The last section addresses the discussion, theoretical contributions, managerial implications and lines of future research.

2 Literature Review and Conceptual Model

2.1 Strategic Human Resource Management (SHRM)

Human resource management (HRM) covers all activities about the management of employment relationships in enterprises [15]. Micro HRM (MHRM), international HRM (IHRM) and strategic HRM (SHRM) are three major subfields of HRM.

Human Resource Management (HRM) covers the main functions of HR policies and practices which could be divided into two fields; one is towards employees such as recruitment, selection, induction, training and development, performance management, and remuneration. The other is towards the organization and employee voice systems.

International HRM is about HRM policies and practices in multinational enterprises and Strategic HRM is about using organizational strategies to improve firm performance [16]. SHRM make contributions by focusing on the combination of organizational activities with HRM policies and practices to achieve organizational goals [8].

There are different definitions or approaches on the literature. Some scholars [17, 18], indicate that SHRM combines HRM policies, practices and internal capacity to adapt and adjust to firms' competitiveness. Others conceptualize SHRM as the combination of HRM and corporate strategies [19–21].

Furthermore, [22–24] thought that SHRM is the integration of HRM with business strategy. Wright & McMahan [25] pointed out that SHRM are transformed into activities to enable an organization to achieve goals from the integration with business strategy. In addition, Azmi [26] emphasizes that SHRM has a proactive function.

SHRM theoretical assertion is that HR systems bring competitive advantages that usually result from firm innovation [23].

According to Delery and Doty [27], there are three perspectives for scholars to study the SHRM on organizational performance. Those perspectives are universalistic or best practice approach; contingency or fit approach; and configurational or resource-based view approach [13, 27–30].

Delery & Doty [27] proposed these major SHRM theoretical perspectives and found strong support to all of them. Thought that the universalistic and contingency perspectives are focus on the methods used by HR to keep enterprises' benefits. While configurational perspective supplies the rational framework and SHRM theories [29].

Gomez-Mejia & Balkin [31] propose thought the contingency and configurational perspectives would contribute to firm performance by linking firms' strategies with SHRM practice. Arthur [32] and Huselid [33] suggested that universalistic or best practice approach concentrates on the policies and practices that are more valid and creating more profits to corporations. According to Becker & Gerhart [10], showing abilities to improve corporate performance and generalizability are the key to identify the universalistic or best practice approach.

Many scholars have demonstrated that the universalistic approach has a positive influence on organizational performance. Cook & Ferris [19] pointed out that SHRM has an efficient influence on environmental changes. Martell & Carroll [34] suggested that SHRM could increase operational benefits, reduce costs, and enhance creativity. Gomez-Mejia et al. [35] suggested that SHRM could make firms more proactive, provide clearer objectives and encourage participation to improve firms' performance. Welbourne & Andrews [36] thought that when facing environmental changes, SHRM could produce cohesion to put employees together to promote organizational operation. For that reason, some policies are emphasized by scholars, such as equal opportunities [37], flexible working times and flexible compensation [38, 39]. It is worth mentioning that Pfeffer [40] proposed seven practices to enhance the universalistic or best practice approach: employment security, selective hiring, self-managed teams, high compensation, training, reduction of status differences and sharing information.

The contingency perspective or fit approach puts HR practices into organizational level have found a positive influence on corporate performance [35]. Skaggs & Youndt [41] found strong support for the human capital fit to firm's strategy and the positive influence on organizational performance. Rodwell & Teo [42] found that SHRM has a positive and significant influence on organizational performance.

In that sense, some scholars [19, 43, 44] found a particular set of SHRM practices that enhance firm performance. These set of practices are known as the High Performance Work System (HPWS). HPWS is not an individual practice but bundle of practices [44, 45]. HPWS refers to human resources practices that include comprehensive employee recruitment and selection, intensive compensation and performance management systems, extensive employee involvement and training, teamwork, and employee participation [27, 32, 46–48]. When HPWS fits corporate strategic goals, it may improve organizational performance [27].

However, the approach the configurational or Resource-Based View (RBV) has become more important in recent literatures [47] and it is used as a theoretical foundation to organizations [16, 49]. This approach concentrates on organizational and human capital competencies of HRM [26]. Valuable, rare, inimitable, and non-substitutable resources are the four requirements of RBV approach and when they are used properly with the HR strategies, they could bring in sustainable competitive advantage for the firm [50, 51]. To keep these advantages, HR practices should add value to the firm and make them unique [49, 52, 53].

In summary, systematic and strategic perspectives of SHRM both promote corporate performance. Thus to say, in organizational progress, the HRM function is vital to help to make appropriate strategic decisions [23, 54–56]. Indeed, the evidence supporting a positive and direct relationship between SHRM and firm performance is overwhelming in the literature. Further more, qualitative evidence indicates that HRM can create an enabling environment for technological innovation through improvements and changes in working conditions on implementation of technological in innovation projects facilitating internal innovation.

2.2 Corporate Entrepreneurship and Firm Performance

Corporate Entrepreneurship (CE) is considered as an important factor for the evaluation of firm performance. The entrepreneurial efforts of an established and large company depend on the corporate entrepreneurship, and not the other way around as it provides the output results. The performance of different companies is a major concern nowadays, because these companies are likely to experience challenging situations. There is major decline in the survival rates of the developing companies.

Joseph Schumpeter framed corporate entrepreneurship as the best known economist in the entrepreneurship field in early 20th century in economy in view of changes [57]. He believed that economic development does not rely on capital accumulation but a set of new combinations and innovations which come from market demands. In other words, entrepreneurship is a procedure of creating new innovation activities [58]. There are a set of activities to enhance these combinations or innovations: the creation of new goods or new quality of goods, the creation of new methods of production; the exploitation of a new market, the seizure of a new source of supply of firm, the identification of opportunities [59] development and implementation of a new business concept [60, 61].

As more, Schumpeter's work as the highly influential between on entrepreneur and innovation developed that entrepreneurship have been foundational to an array of issues from this two concepts [62].

There are two groups of entrepreneurships, one has mentioned above as the procedure of bringing new combinations which is corporate entrepreneurship in established firm [63, 64]. Especially at present, we usually consider entrepreneurship as a procedure within firms and focuses on innovation, growth, and uniqueness [65]. The other is about the process of funding new organizations [66] which are called start-ups [64].

Different scholars give their own definition of corporate entrepreneurship. Covin & Miles [65] defined corporate entrepreneurship as the process of innovation with a goal of promoting and redefining enterprises. That corporate entrepreneurship is enabling organizations to innovate, develop new business and renew their activities. It is a constant procedure of improving organizational performance by learning new knowledge [67, 68]. Martín Rojas and others suggest that corporate entrepreneurship is a way to achieve competitive advantage by developing organizational knowledge [69].

To sum up, corporate entrepreneurship is an entrepreneurial behavior that through innovation, strategic adjustment enhances organizational performance, dominate new market and find new business opportunities [4, 5].

Corporate entrepreneurship is conceptualized as a multidimensional construct that has three dimensions, including internal innovation, venturing and strategic renewal activities [7, 58, 71].

- Internal innovation refers to organization's tendency to capture new ideas to develop new product, new market or better organizational performance within the enterprise [72, 73].
- Venturing refers to the willingness to open a new business which is in an unknown [74] or not actively exploited area may not deliver new product or the pressure of exploration of new fields [58, 73].
- Strategic renewal refers to activities that cause changes in strategy or structure and build new relationships inside firms or between firms and its environment [58].

Although these three dimensions work as a combination, they are different from each other [75]. Because of different strategic and structural goal, orientation, newness, ways of creation, opportunity recognition and exploitation, each of them should have their own outcome, promoter and obstacle [76].

Corporate entrepreneurship depends on organizational ability of learning existing knowledge and exploitation of new knowledge, thus, has a significant impact on financial and market performance [71, 77, 78]. Zahra [79] suggest that corporate entrepreneurial activity is a key factor to organizational development and benefit.

Other studies [5, 80] argued that corporate entrepreneurship is positively related to firm performance because it can use organizational resource to increase output, expand market and develop adaptability. In contrast, short of key resources is the main reason leading to the failure of corporate entrepreneurship [81, 82]. What is worse, when firms fail, they may disappear because the failure could bring competitive weakness rather than just not gain competitive advantages [83].

The entrepreneurial attitudes and behaviors are essential for firms of all sizes in the competitive environment. In technology-based organizations, corporate entrepreneurship and risk-taking are really important for success so that firms are focus on the development of product and technology [80, 84].

Corporate entrepreneurialship have a crucial connection with human capital in the enterprise because social capital conditions individual human capital in business systems. Human capital relies heavily on employees' social skills make the efficient functioning of social processes and mechanisms easier [85].

The general notion that CE could has significant and positive performance implication. The reveal differences in the magnitude by which strategic renewal, innovation and corporate venturing impact firms' performance, such that innovation and corporate venturing have stronger performance effects than strategic renewal. The firms have to take into account the advantages and disadvantages of the respective CE activity before engaging in CE [86, 87]. Furthermore, it is important for firms to consider the time frame related to the implementation of the respective CE activities. Innovation and corporate venturing can be put into practice relatively quickly, yielding faster performance effects [87], while strategic renewal activities take more time to take full effect [58].

2.3 The HR Policies Effect on the Innovative Behavior in the Human Capital Enterprise

Innovation has been recognised as a key element of competition and dynamic efficiency in markets. Innovation is a way of facilitating a firm's response to external changes, pressure from competition, changes in customers' demands and the constant requirement for new and better products and services [63]. According this direction the organizational strategy is increasingly focused on increasing its innovative results. It is in this area where the role of human resources and their management have become more important in the last decade [89]. Scholars demonstrate how influence the innovation at the organizational level through human resources practices or human resources systems in the innovative results of the organizations.

According to Chen & Huang [90], innovation is to develop organizational expertise together with human capital which companies have in order to produce new products and service. Companies perform particular SHRM practices by providing organizational development to achieve strategical targets such as improving innovation performance. Organizations explore the expertise and knowledge that they have through these applications. Then they integrate their new capabilities into the production process and in this way SHRM practices contribute innovative activities [91].

To modify employees' behavior in organizations, they must perceive that human resource practices as a signal within an organization [92], as behaviors are valued and they are rewarding, ultimately, the consequences that each behavior has within the organizations. Stimulating in this way the behaviors desired within the organizations through SHRM, in which when employees perceive that the organization as a provider of value, this feeling generates the reciprocity of the workers that will contribute value in order to achieve the established goals [93].

If this perception is focused on SHRM through human resources practices that advocate innovative ideas and in an environment that promotes the generation of new ideas, the consequence will be its participation with innovative behaviors. Proved the effect between the high-commitment work system and the innovative behavior of the workers. Already in the systematic review of Seeck and Diehl [94] the importance of the high commitment of the policies of RRHH _ SHRM for innovation is established. This high commitment of SHRM is advantageous for the results of innovation, [95], since aspects such as occupational safety, psychologically by employees influence their behavior by motivating them to take risks.

Employees who perceive their organization as an information sharing body and which supports them are those that bring innovative behavior towards the organization [96], in those organizations that SHRM have developed in an innovative climate organizational behavior that can be improved by strengthening the contributions of SHRM to organizations in building a climate that supports innovative behaviors. At the same time providing information and being strategic with employees, through transparent and organizational information, motivates employees to innovate behavior, since they perceive the organization as a shared information entity.

Human resources practices and policies based on how they are implemented by their managers are directly influencing the innovative behavior of employees [96, 97].

Multiple strategies are attempted to implement in modern organizations with the purpose of influencing employee behaviors and motivations, on firm's productivity and organizational performance [98], however, is little known about its effect, that would explain the process through which they affect the human capital behaviour. Although there are some studies that account for associations between organizational practices and performance [98–101].

According to Seeck, Marjo, & Diehl [94] human capital activities, in turn, they indicate that practices that foster employee engagement, loyalty, learning and intrinsic motivation are conducive to innovation. At the same time, the evidence warns against the potentially negative effects of some practices applied independently, mainly those related to material incentives and different control and evaluation measures. Also, in the various individual practices, in turn, they indicate that practices that foster employee engagement, loyalty, learning and intrinsic motivation are conducive to innovation.

Lau and Ngo [102] in turn highlight the role of organizational culture, which play a mediation role between SHRM and the results related to the innovation, in the same way especially in the human capital activities' influence.

On consequence, several practical components of high-performance work have different impacts on firm's propensity to be innovative through its impact on the climate of variable mediating innovation, participation and renewal in order to achieve the goals for an innovation organizational performance [103].

2.4 Conceptual Model and Hypotheses

Based on this initial literature review, we can infer that both SHRM and corporate entrepreneurship have positive and direct influence on organizational performance. However, the relationship between these constructs is not researched extensively, both theoretically and empirically. Thus, we are wondering what role SHRM plays in the relationship between corporate entrepreneurship and organizational performance. Does SHRM have a moderating effect to corporate entrepreneurship and organizational performance? With such questions in mind, we propose the following conceptual model (Fig. 1):

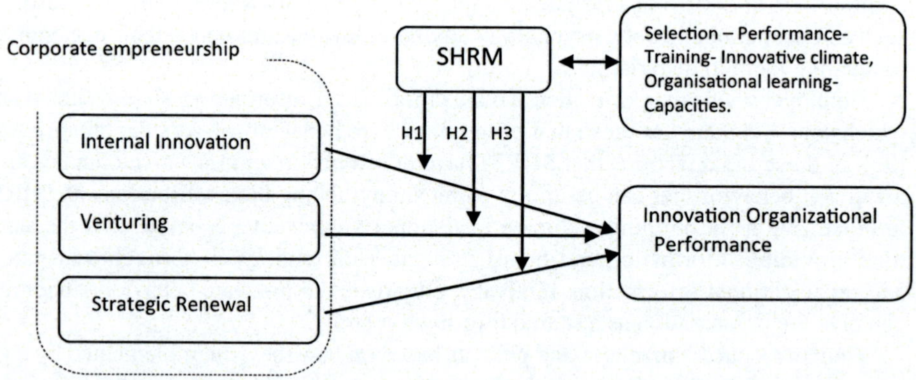

Fig. 1. Conceptual model - the fit between corporate entrepreneurship and organizational performance through SHRM

Internal Innovation. In technology based organizations, competitive advantage is a basic component [97]. In every business, innovation is the precondition of competitive advantage and HR practices could promote opportunities to enhance internal innovation [104]. Matsuno [105] suggest that entrepreneurial organizations would weaken their advantage in existence, economic performance without new product development and innovation. Therefore, entrepreneurial firms should concentrate more on new product development and innovation than others less entrepreneurial.

Meanwhile, Wang & Zang [97] suggest that SHRM could promote both technological innovation and entrepreneurship. Furthermore, a long-term and high-commitment HR practices have a positive influence on innovation. James [106] proposes that SHRM plays an important role on the technology innovation. Marsili [107] thought that with the support of SHRM, there is an effective association between entrepreneurial behavior and innovation. As one component of SHRM, High Performance Work Systems (HPWS) could increase organizational innovation [108]. Accordingly, HPWS could generate opportunities to combine existing knowledge with new knowledge by extensive training which could provide updated knowledge to employees. In technological firms, individual's skills are positively influence the corporate entrepreneurship [109].

Furthermore, innovative environments are created when actions are encouraged. It occurs in an environment, where failure is not shunned and is used as a learning experience. Therefore, it is said that innovation must be rewarded and promoted. There should be an environment that encourages informal communication, where ideas are formed. There must be a suitable atmosphere for the bootlegging of ideas and working on ideas must always be encouraged. Moreover, employees must be organized into small teams for future-oriented projects. Bureaucratic red tapes and rigid procedures are barriers to innovative and breakthrough thinking. There must be substantial investment in entrepreneurial activities that allow new ideas to flourish in an innovative environment. The most positive effects of innovation are seen in high-tech companies rather than in low-tech industries [86].

Besides, HPWS could promote employees' participation to motivate them to search for innovation and opportunities. In technological organizations HR managers should use practices to attract and motivate their employees to promote innovations [110]. Furthermore, building an entrepreneurial atmosphere of trust, collaboration and teamwork could enhance exchange and integration of new ideas and knowledge [111]. In technology based firms, innovation is always intense [112] and as a competitive advantage, it could have a positive influence on productivity [113], sustainable growth [114] and the effect of technology procedure [115]. Thus, the internal innovation could enable economic development [116] and financial benefits [117]. Based on this rational, we hypothesized that:

> ***Hypothesis 1****: In technology-based organizations, SHRM moderates the positive relationship between internal innovation and organizational performance, such as, the higher the SHRM implementation is, the positive relationship will be stronger.*

Corporate Venturing. Corporate venturing is to create new business in existing firms which demands integration of existing knowledge with new knowledge and different working and reward systems [73, 117, 118].

Some scholars have pointed out that SHRM is a key factor in organizational venturing [111]. Kuratko and colleagues, for example, suggest that reward systems which belong to SHRM would encourage venturing and innovation in entrepreneurial behaviors [119].

When organizations fail in their corporate strategy, SHRM could help in decisions making to avoid overregulation in managing employees. With the support of SHRM, corporations would get more useful resources [120]. Wang & Poutziouris [121] suggest that without enough HR practices, it may lead the reduction of organizational business under the pressure of resource shortage and salaries.

When developing new business, firms are facing uncertainty of product, market and technology. In this situation, firms need to increase their risk tolerance with HR strategies [122]. When organizations are in a risk taking situation, they believe that venturing will increase their competitive advantages in the battle with their competitors [123]. Furthermore, with venturing activities, firms would find more opportunities to succeed.

In summary, corporate venturing occurs when processes are institutionalized through development of innovative products and processes with the emphasis on long-term prosperity. Competitive and capable managers have to be employed and developed for corporate entrepreneurship. Apart from this, corporate venture teams also have to be hired and developed, and facing challenges in their operations, firms that willing to take risks will increase benefits and have better performance [124]. Thus, we hypothesized:

> ***Hypothesis 2****: In technology-based organizations, venturing is more positively related to organizational performance with a higher level of SHRM. When firms have higher SHRM implementation, the positive relationship is stronger.*

Strategic Renewal. According to SHRM literature, strategic agility and knowledge management are two mainstays. Strategic agility refers to strategic sensitivity, leadership unity and resource fluidity. Among them, leadership unity includes the ability of HR professionals which may help making decisions faster. With the help of strategic agility, firms could change their strategic renewal to respond to the external environmental changes and give the best usage of human capital [125]. Besides, scholars concluded that strategic agility includes organizational resilience, flexibility, adaptability and dynamic capabilities [126, 127]. They suggested that by all these characteristics mentioned above, organizations could achieve their competitive advantage which is essential to technology based enterprises. Meanwhile, in technology based firms, HR managers are involved in the progress of decision making [128]. Furthermore, improving organizational strategic choices will lead to gain more competitive advantage in the technological field [71, 129, 130] and contribute to firm renewal and new markets in corporate entrepreneurship [131].

Also, strategic renewal efforts such as the redefinition of a company's vision, mission, business concept, reorganization of activities and competitive approach are primary means for firms to adapt to changing environments in order to adapting organizational structures that better align firms' strategies with their organization and their competitive advantages [132].

Strategic renewal enhances firms' performance by increasing their ability to extend firm capabilities and creatively leverage them to add shareholder value [71]. However, strategic renewal also entails potential downsides, such as confusion on the part of the customers and employees. Based on this rational, we hypothesized that:

Hypothesis 3*: In technology-based organizations, strategic renewal is more positively related to organizational performance with a higher level of SHRM. When firms use more SHRM implementation, the positive relationship is stronger.*

3 Methodology

3.1 Samples and Procedure

To test our hypotheses, we will use a sample of 250 European technology-based organizations that occupy various ranges of industries. To select the sample randomly, we will use Amadeus database.

A self-completion questionnaire is proposed to investigate the moderating effect of SHRM in the relationship between corporate entrepreneurship and innovative organizational performance. Two different questionnaires will be prepared to target CEO and/or finance managers and human resource managers.

We use a postal survey procedure. A cover letter be sent with the questionnaire to explain our research objectives and motivate people to complete and return the questionnaire. At the same time, a self-addressed envelope with a stamp on be provided.

3.2 Test for Response Bias

In order to check whether exist response and non-response bias that may occur in our investigation, we will use t-test to exam differences between respondents and non-respondents. In the t-test, we will compare firm size, firm age, firm ownership and industry.

3.3 Measures

3.3.1 Dependent Variable

We will use a 7-point Likert scale to measure organizational performance. Five items developed by Martín-Rojas et al., [69, 133]. Such as, organizational performance measured by return on assets (economic profitability or ROA).

3.3.2 Independent Variables

7-point Likert scale will be used to measure corporate entrepreneurship (1 "totally disagree" to 7 "totally agree"). To survey the multidimensional components of corporate entrepreneurship, we will use three different subscales. To measure internal innovation, four items developed by Li [76]. To measure venturing four items developed by Martín-Rojas. To measure strategic renewal, four items developed by Martín-Rojas et al. [69, 133].

3.3.3 Moderating Variable

To measure SHRM, the nine items adapted from Gurbuz & Ibrahim [27]. Managers were required to answer these items from 1 "totally disagree" to 7 "totally agree".

3.3.4 Control Variable

In this research, 4 different variables will be used as controls: firm size, firm age, type of ownership and industry sector.

The firm size measure was operationalized as the number of employees, because larger firms could use better SHRM [133], perform better in corporate entrepreneurship and have a direct effect to search more market [134].

The firm age was measured by the years the organization has been in business. We will consider the firm age because older firms have more relationships in and outside their industries [68].

The firm ownership has relationship with SHRM and organizational performance and we plan to separate it into public firms, private firms, foreign-invested firms and others. Meanwhile, organizational industries should be considered too, such as manufacturing and service.

4 Analyze and Results

To test our hypotheses, we are going to use hierarchical moderated regression analysis. By this method, we could build interaction between our variables. First, we enter control variables (firm size, age, ownership and industry). Second, we entered our independent variables, namely internal innovation, venturing and strategic renewal. In the third step, SHRM, the moderating variable, is entered. Nest, we test the moderating effect of SHRM in the relationship between internal innovation and organizational performance. Besides, we test the moderating effect of SHRM in the relationship between venturing and organizational performance. Last but not least, we test the moderating effect of SHRM in the relationship between strategic renewal and performance.

In order to test our hypotheses, we can test that support, or reject them, as followed:

- In H1, were that β will be a positive number and p is significant ($p < 0.01$), so that, H1 could be supported which means SHRM moderates the positive relationship between internal innovation and organizational performance in technology-based organizations.

- In H2, were β will be a positive number and p is significant ($p < 0.01$) H2 could be supported which means SHRM moderates the positive relationship between venturing and organizational performance in technology-based organizations.
- In H3, were β will be a positive number and p is significant ($p < 0.01$), so that, H3 could be supported which means SHRM moderates the positive relationship between strategic renewal and organizational performance in technology-based organizations.

5 Discussion

The objective of this study try to test the moderating effect of SHRM in the relationship between corporate entrepreneurship and organizational performance in technology-based firms. To test our hypothesis, we will send questionnaires to CEO, finance managers and HR managers of 250 European technology-based organizations. Hierarchical moderated regression analysis as statistical analysis strategy. We could test that SHRM could interact with corporate entrepreneurship in their three dimensions: internal innovation, venturing and strategic renewal, so that, the organization could generate better performance and increases its competitive advantage.

5.1 Theoretical Contribution

This research try to make several contributions. First of all, we want to address the gap in the relationship between corporate entrepreneurship and organizational performance. We hypothesize that the moderating effect of SHRM will play a major role. Previous studies had been focus on either the direct effect between SHRM and organizational performance [33, 135], or on corporate entrepreneurship and organizational performance [7, 29], or other moderating effect, such as political ties, business ties [75]. However, there are little previous empirical evidences focusing on the SHRM role played on the link between corporate entrepreneurship and firm performance; an interesting avenue for further research development.

Second, in our research, we do not want to compare corporate entrepreneurship as uni-dimensional construct, instead we hypothesized that it has three separate dimensions; internal innovation, venturing and strategic renewal [79]. By this approach, we can determine separately which dimension (internal innovation, venturing and strategic renewal) has a strong effect in explaining the variance.

Furthermore, our study contributes to the literature of SHRM by enhancing its moderating effect on corporate performance. Meanwhile, it highlights the importance of SHRM in the context of corporate entrepreneurship field and innovation organizational performance.

5.2 Managerial Implications

Technology-based organizations should pay attention to corporate entrepreneurship in order to gain more competitive advantages. Without better implementation of SHRM, firms could not incentive so much internal innovation, venturing and strategic renewal.

An excellent technological organization could achieve growth and benefit by using better strategies to motivate their employees, develop organizational structure and enable corporate entrepreneurship. Thus, HR managers should give full play to their role in incentivizing internal innovation, venturing and strategic renewal considering encouraged a suitable atmosphere for the bootlegging of ideas and working on employees' new ideas for any innovitiness organizational performance.

5.3 Limitations and Future Investigations

Our study has several limitations which could lead to future researches. First, we focus on the moderating effect of SHRM, but we tested SHRM as a whole and not only the factors which have a direct influence on CE for innovative organizational performance. Therefore, future study could discuss which SHRM dimension plays a more specific moderating effect on the relationship between corporate entrepreneurship and organizational performance. Researchers in the field of human resources management did not give enough attention to these relationships.

The proposed framework represents a model that tries to contribute to the literature exploring the relationship among influence of SHRM, CE and innovation in the organization. For future researches this model can be completed with a model that tries to contribute to the literature through exploring the connections and relationships among human resource management practices, knowledge management, organizational innovation, organizational culture, and organizational performance between SHRM and innovate performance, investigating how firms preserve and continuously refine human capital and value to sustain competitive advantage and long-term success. Human Capital, which, could be as a determinant to understanding SHRM, employment practices and innovation results [136] since it represents the connections that exist between people and their shared values and behavioural norms to influence the innovative performance firms. These enable and encourage social cooperation and the generation of new knowledge for innovative processes.

Second, in our survey, we will send questionnaires to CEO, finance managers and HR managers. We could also send questionnaire to employees in future research.

Furthermore, we select 250 European technology-based firms as our sample because we think the development of CE is more advanced and developed in Europe. However, we cannot doubt that technology-based firms exist in other countries, China and Japan. Thus, future studies could use a cross-cultural approach to test the hypothesized model.

Besides, research about which, whether, how, and when about different types of employee-experienced SHRM systems jointly influence the innovative organization performance, would be necessary for determinate the human capital organisation influence.

Whereas, our research uses a cross-sectional design, in the future, we plan to use a longitudinal design to ovoid common variance. Although longitudinal studies are more informative about the nature of the relationship between CE and firm performance, they can be challenging to execute because of their resource intensiveness and complex nature and also because of the low follow-up response rate (between two time periods of data collection) caused by the usual time lag between introduction, implementation and impact of CE and firm performance and only allow researchers to suggest a possible causal inference.

Appendix A

This appendix presents the scales selected for this research and that we used them in the fieldwork questionnaire. A Likert-type seven-point scale (1 "totally disagree", 7 "totally agree") will be used.

Internal Innovations **[76]**

In the last three years

1. Invested heavily in cutting-edge product-oriented R&D
2. Invested heavily in cutting-edge process technology-oriented R&D.
3. Introduced many new products or services.
4. Pioneered the development of breakthrough innovations.

Venturing **[32]**
In the last three years

1. The organizational has simulated new demands on the existing products/services in the currents markets through aggressive advertising and marketing.
2. The organization has broadened the business lines in the current industries.
3. The organization has pursued new businesses in new industries that are related to the current business.
4. The organization has entered new businesses by offering new lines and products/services.

Strategic Renewal **[130]**
In the last three years

1. The organization has reorganized units and divisions to increase organizational innovation.
2. The organization has coordinated activities among units to enhance organizational innovation.
3. The organization has adopted flexible organizational structures to increase innovation.

4. The organization has trained and encouraged the employees to be creative and innovative.

Organizational Performance [130]

Comparing to your competitors in the same business area, what is your firm's performance in the last three years?

1. Organizational performance measured by return on assets (economic profitability or ROA).
2. Organizational performance measured by return on equity (financial profitability or ROE).
3. Organizational performance measured by return on sales (percentage of profits over billing volume or ROS).
4. Organization's market share in its main products and markets.
5. Growth of sales in its main products and markets.

Strategic Human Resource Management (SHRM) [29]
How does SHRM performance in your organization?

1. HR strategies are effectively integrated with this organization's strategy
2. HR practices are integrated to be consistent with each other
3. HR strategy will become a more important influence in this organization's strategy in the future
4. This organization matches the characteristics of managers to the strategic plan of the organization
5. This organization modifies the compensation systems to encourage managers to achieve long-term strategic objectives
6. This organization changes staffing patterns to help implement business or corporate strategies
7. This organization evaluates key personnel based on their potential for carrying out strategic goals
8. Job analyses are based on what the job might entail in the future.
9. HR seen by the senior management as a cost to be minimized or source of value creation.

References

1. Berchicci, L.: Towards an open R&D system: internal R&D investment, external knowledge acquisition and innovative performance. Res. Policy **42**(1), 117–127 (2013)

2. Iturriaga, M., Benaiandres, J., Eizaguirre, I.: Del emprendimiento individual a la capacidad emprendedora de nuestras organizaciones un análisis de los factores que la definen. Boletín de estudios económicos **71**(217), 49–69 (2016)
3. Garcia-Morales, V.J., Ruiz-Moreno, A., Llorens-Montes, F.J.: Effects of technology absorptive capacity and technology proactivity on organizational learning, innovation and performance: an empirical examination. Technol. Anal. Strateg. Manag. **19**(4), 527–558 (2007)
4. Goodale, J.C., Kuratko, D.F., Hornsby, J.S., Covin, J.G.: Operations management and corporate entrepreneurship: the moderating effect of operations control on the antecedents of corporate entrepreneurial activity in relation to innovation performance. J. Oper. Manag. **29**(1), 116–127 (2011)
5. Ireland, R.D., Covin, J.G., Kuratko, D.F.: Conceptualizing corporate entrepreneurship strategy. Entrep. Theory Pract. **33**(1), 19–46 (2009)
6. Kuratko, D.F., Covin, J.G., Garrett, R.P.: Corporate venturing: insights from actual performance. Bus. Horiz. **52**(5), 459–467 (2009)
7. Takeuchia, N.: How Japanese manufacturing firms align their human resource policies with business strategies: testing a contingency performance prediction in a Japanese context. Int. J. Hum. Resour. Manag. **20**(1), 34–56 (2009)
8. Davidson, M.C., McPhail, R., Barry, S.: Hospitality HRM: past, present and the future. Int. J. Contemp. Hosp. Manag. **23**(4), 498–516 (2011)
9. Burgelman, R.A.: Corporate entrepreneurship and strategic management: insights from a process study. Manage. Sci. **29**(12), 1349–1364 (1983)
10. Becker, B., Gerhart, B.: The impact of human resource management on organizational performance: progress and prospects. Acad. Manag. J. **39**(4), 779–801 (1996)
11. Jimenez-Jimenez, D., Sanz-Valle, R.: Innovation and human resource management fit: an empirical study. Int. J. Manpow. **26**(4), 364–381 (2005)
12. Jackson, S.E., Schuler, R.S.: The need for understanding human resource management in the context of organizations and their environment. In: Hagan, J., Cook, K.S. (eds.) Annual Review of Psychology, Palo Alto, CA, vol. 46, pp. 237–264 (1995)
13. Dabic, M., Ortiz-De-Urbina-Criado, M., Romero-Martínez, A.M.: Human resource management in entrepreneurial firms: a literature review. Int. J. Manpow. **32**(1), 14–33 (2011)
14. Alpkan, L., Bulut, C., Gunday, G., Ulusoy, G., Kilic, K.: Organizational support for intrapreneurship and its interaction with human capital to enhance innovative performance. Manag. Decis. **48**(5), 732–755 (2010)
15. Lin, C.H., Peng, C.H., Kao, D.T.: The innovativeness effect of market orientation and learning orientation on business performance. Int. J. Manpow. **29**(8), 752–772 (2008)
16. Boxall, P., Purcell, J., Wright, P.: Human resource management: scope, analysis and significance. In: Boxall, P., Percell, J., Wright, P. (eds.) The Handbook of Human Resource Management, pp. 1–16. Oxford University Press, Oxford (2007)
17. De Saá Pérez, P., Falcón, J.M.G.: The influence of human resource management in savings bank performance. Serv. Ind. J. **24**(2), 51–66 (2004)
18. Kidwell, R.E., Fish, A.J.: High-performance human resource practices in Australian family businesses: preliminary evidence from the wine industry. Int. Entrep. Manag. J. **3**(1), 1–14 (2007)
19. Cook, D.S., Ferris, G.R.: Strategic human resource management and firm effectiveness in industries experiencing decline. Hum. Resour. Manage. **25**(3), 441–457 (1986)
20. Storey, J.: The part played by senior and middle line managers in the management of human resources. In: Storey, J. (ed.) Developments in the Management of Human Resources, pp. 98–109. Blackwell Publishers, Oxford (1992)

21. Bennett, N., Ketchen, D.J., Schultz, E.B.: An examination of factors associated with the integration of human resource management and strategic decision making. Hum. Resour. Manage. **37**(1), 3–16 (1998)
22. Schuler, R.S.: Strategic human resources management: linking the people with the strategic needs of the business. Organ. Dyn. **21**(1), 18–32 (1992)
23. Becker, B.E., Huselid, M.A.: High performance work systems and firm performance: a synthesis of research and managerial applications. In: Ferris, G.R. (ed.) Research in Personnel and Human Resource Management, pp. 53–101. JAI Press, Greenwich (1998)
24. Sheehan, C.: A model for HRM strategic integration. Pers. Rev. **34**(2), 192–209 (2005)
25. Wright, P.M., McMahan, G.C.: Theoretical perspectives for strategic human resource management. J. Manag. **18**(2), 295–320 (1992)
26. Azmi, F.T.: Strategic human resource management and its linkage with HRM effectiveness and organizational performance: evidence from India. Int. J. Hum. Res. Manag. **22**(18), 3888–3912 (2011)
27. Delery, J.E., Doty, D.H.: Modes of theorizing in strategic human resource management: tests of universalistic, contingency, and configurational performance predictions. Acad. Manag. J. **39**(4), 802–835 (1996)
28. Twomey, D.F., Harris, D.L.: From strategy to corporate outcomes: aligning human resource management systems with entrepreneurial intent. Int. J. Commer. Manag. **10**(4), 43–55 (2000)
29. Gurbuz, S., Mert, I.S.: Impact of the strategic human resource management on organizational performance: evidence from Turkey. Int. J. Hum. Resour. Manag. **22**(8), 1803–1822 (2011)
30. Martin-Alcazar, F., Romero-Fernandez, P.M., Sanchez-Gardey, G.: Strategic human resource management: integrating the universalistic, contingent, configurational and contextual perspectives. Int. J. Hum. Resour. Manag. **16**(5), 633–659 (2005)
31. Gomez-Mejia, L.R., Balkin, D.B.: Compensation, Organizational Strategy, and Firm Performance. South-Western, Cincinnati (1992)
32. Arthur, J.B.: Effects of human resource systems on manufacturing performance and turnover. Acad. Manag. J. **37**(3), 670–687 (1994)
33. Huselid, M.A.: The impact of human resource management practices on turnover, productivity, and corporate financial performance. Acad. Manag. J. **38**(3), 635–672 (1995)
34. Martell, K., Carroll, S.J.: How strategic is HRM? Hum. Resour. Manage. **34**(2), 253–267 (1995)
35. Gomez-Mejia, L.R., Balking, D.B., Cardy, R.: Managing Human Resources. Prentice-Hall, New York (1995)
36. Welbourne, T.M., Andrews, A.O.: Predicting the performance of initial public offerings: should human resource management be in the equation? Acad. Manag. J. **39**(4), 891–919 (1996)
37. Hicks-Clarke, D., Iles, P.: Climate for diversity and its effects on career and organisational attitudes and perceptions. Pers. Rev. **29**(3), 324–345 (2000)
38. Sidney, E.: Dragging a ball and chain: moving to equal opportunity in the NHS. Int. J. Sel. Assess. **2**(3), 176–185 (1994)
39. Barber, A.E., Daly, C.L.: Compensation and diversity: new pay for a new workforce? In: Kossek, E.E., Lobel, S.A. (eds.) Managing Diversity: Human Resource Strategies for Transforming the Workplace. Blackwell, Cambridge (1996)
40. Pfeffer, J.: Seven practices of successful organizations. Org Dev & Trng, 6E (Iae), 460 (1998)
41. Skaggs, B.C., Youndt, M.: Strategic positioning, human capital, and performance in service organizations: a customer interaction approach. Strateg. Manag. J. **25**(1), 85–99 (2004)
42. Rodwell, J.J., Teo, S.T.: Strategic HRM in for-profit and non-profit organizations in a knowledge-intensive industry: the same issues predict performance for both types of organization. Public Manag. Rev. **6**(3), 311–331 (2004)

43. MacDuffie, J.P.: Human resource bundles and manufacturing performance: organizational logic and flexible production systems in the world auto industry. Ind. Labor Relat. Rev. **48**, 197–221 (1995)
44. Dyer, L., Reeves, T.: Human resource strategies and firm performance: what do we know and where do we need to go? Int. J. Hum. Resour. Manag. **6**(3), 656–670 (1995)
45. Delaney, J.T., Huselid, M.A.: The impact of human resource management practices on perceptions of organizational performance. Acad. Manag. J. **39**(4), 949–969 (1996)
46. Guthrie, J.P.: High-involvement work practices, turnover, and productivity: evidence from New Zealand. Acad. Manag. J. **44**(1), 180–190 (2001)
47. Ichniowski, C., Shaw, K., Prennushi, G.: The effects of human resource management practices on productivity: a study of steel finishing lines. Am. Econ. Rev. **87**(3), 291–313 (1997)
48. Lengnick-Hall, M.L., Lengnick-Hall, C.A., Andrade, L.S., Drake, B.: Strategic human resource management: the evolution of the field. Hum. Resour. Manag. Rev. **19**(2), 64–85 (2009)
49. Wright, P.M., Dunford, B.B., Snell, S.A.: Human resources and the resource based view of the firm. J. Manag. **27**(6), 701–721 (2001)
50. Barney, J.: Firm resources and sustained competitive advantage. J. Manag. **17**(1), 99–120 (1991)
51. Barney, J.B., Wright, P.M.: On becoming a strategic partner: the role of human resources in gaining competitive advantage. Hum. Resour. Manage. **37**, 31–46 (1997)
52. Lado, A.A., Wilson, M.C.: Human resource systems and sustained competitive advantage: a competency-based perspective. Acad. Manag. Rev. **19**(4), 699–727 (1994)
53. Matusik, S.F., Hill, C.W.: The utilization of contingent work, knowledge creation, and competitive advantage. Acad. Manag. Rev. **23**(4), 680–697 (1998)
54. Fombrun, C.J., Tichy, N.M., Devanna, M.A.: Strategic Human Resource Management. Wiley, New York (1984)
55. Baron, J.N., Kreps, D.N.: Strategic Human Resource Management: Framework for General Managers. Wiley, New York (1999)
56. Teo, S.T., Crawford, J.: Indicators of strategic HRM effectiveness: a case study of an Australian public sector agency during commercialization. Public Pers. Manag. **34**(1), 1–16 (2005)
57. Landström, H., Harirchi, G., Åström, F.: Entrepreneurship: exploring the knowledge base. Res. Policy **41**(7), 1154–1181 (2012)
58. Sharma, P., Chrisman, S.J.J.: Toward a reconciliation of the definitional issues in the field of corporate entrepreneurship. Entrep. Theory Pract. **23**(3), 11–27 (1999)
59. Bygrave, W.D., Hofer, C.W.: Theorizing about entrepreneurship. Entrep. Theory Pract. **16**(2), 13–22 (1991)
60. Morris, M.H., Jones, F.F.: Human resource management practices and corporate entrepreneurship: an empirical assessment from the USA. Int. J. Hum. Resour. Manag. **4**(4), 873–896 (1993)
61. Leyden, D.P., Link, A.N., Siegel, D.S.: A theoretical analysis of the role of social networks in entrepreneurship. Working paper, pp. 13–22 (2013)
62. Portugal, M., Rosa, N., Frias, C.: Shimpeter's (1934) influence on enrepreneurship (and managemen) research. REGEPE **6**(1), 4–39 (2017)
63. Schumpeter, J.A.: The Theory of Economic Development: An Inquiry into Profits, Capital, Credit, Interest, and the Business Cycle. Hardvard University Press, Cambridge (1934)
64. Guth, W.D., Ginsberg, A.: Guest editor's introduction: corporate entrepreneurship. Strateg. Manag. J. **11**, 5–15 (1990)

65. Covin, J., Miles, M.: Corporate entrepreneurship and the pursuit of competitive advantage. In: Zahra, S. (ed.) Corporate Entrepreneurship and Growth, pp. 47–63. Edward Elgar Publishing, Indianapolis (2006)
66. Gartner, W.B.: What are we talking about when we talk about entrepreneurship? J. Bus. Ventur. **5**(1), 15–28 (1990)
67. Gartner, W.B.: "Who is an entrepreneur?" is the wrong question. Am. J. Small Bus. **12**(4), 11–32 (1988)
68. Zahra, S.A.: The virtuous cycle of discovery and creation of entrepreneurial opportunities. Strateg. Entrep. J. **2**(3), 243–257 (2008)
69. Martín-Rojas, R., García-Morales, V., Bolívar-Ramos, M.T.: Influence of technological support, skills and competencies, and learning on corporate entrepreneurship in European technology firms. Technovation **33**(12), 417–430 (2013)
70. Teng, B.S.: Corporate entrepreneurship activities through strategic alliances: a resource-based approach toward competitive advantage. J. Manage. Stud. **44**(1), 119–142 (2007)
71. Zahra, S.A.: Goverance, ownership, and corporate entrepreneurship: the moderating impact of industry technological opportunities. Acad. Manag. J. **39**(6), 1713–1735 (1996)
72. Lumpkin, G.T., Dess, G.G.: Clarifying the entrepreneurial orientation construct and linking it to performance. Acad. Manag. Rev. **21**(1), 135–172 (1996)
73. Covin, J.G., Miles, M.P.: Strategic use of corporate venturing. Entrep. Theory Pract. **31**(2), 183–207 (2007)
74. Miller, D., Friesen, P.H.: Archetypes of strategy formulation. Manage. Sci. **24**(9), 921–933 (1978)
75. Burgers, J.H., Jansen, J.J., Van den Bosch, F.A., Volberda, H.W.: Structural differentiation and corporate venturing: the moderating role of formal and informal integration mechanisms. J. Bus. Ventur. **24**(3), 206–220 (2009)
76. Li, X., Zhao, X.L., Zhang, J.Q.: Corporate entrepreneurship, informal ties, and firm performance: evidence from China. Front. Bus. Res. China **7**(1), 26–58 (2013)
77. Antoncic, B., Hisrich, R.D.: Intrapreneurship: construct refinement and cross-cultural validation. J. Bus. Ventur. **16**(5), 495–527 (2011)
78. Hayton, J.C.: Competing in the new economy: the effect of intellectual capital on corporate entrepreneurship in high-technology new ventures. Res. Dev. Manag. **35**(2), 137–155 (2005)
79. Zahra, S.A.: Environment, corporate entrepreneurship, and financial performance: a taxonomic approach. J. Bus. Ventur. **8**(4), 319–340 (1993)
80. Zahra, S.A., Covin, J.G.: Contextual influences on the corporate entrepreneurship-performance relationship: a longitudinal analysis. J. Bus. Ventur. **10**(1), 43–58 (1995)
81. Martin, R.E., Justis, R.T.: Franchising, liquidity constraints and entry. Appl. Econ. **25**(9), 1269–1277 (1993)
82. Bruton, G.D., Rubanik, Y.: Resources of the firm, Russian high-technology startups, and firm growth. J. Bus. Ventur. **17**(6), 553–576 (2002)
83. West, P., III., DeCastro, J.: The Achilles heel of firm strategy: resource weaknesses and distinctive inadequacies. J. Manage. Stud. **38**(3), 417–442 (2001)
84. Antoncic, B., Prodan, I.: Alliances, corporate technological entrepreneurship and firm performance: testing a model on manufacturing firms. Technovation **28**(5), 257–265 (2008)
85. Fernández, S., García, N., Martín, F., Sánchez, G.: Building ambidextrous organizations through intellectual capital: a proposal for a multilevel model. Intangible Cap. **13**(3), 668–693 (2017). ISSN-e 1697-9818
86. Bierwerth, M., Schwens, C., Isidor, R., Kabst, R.: Corporate entrepreneurship and performance: a meta-analysis. Small Bus. Econ. **45**(2), 255–278 (2015). https://doi.org/10.1007/s11187-015-9629-1
87. Espinosa, M.D.M.B., Suanes, A.M.: Corporate entrepreneurship through joint venture. Int. Entrep. Manag. J. **7**(3), 4 (2011)

88. Zhou, Y., Ying Hong, H., Jun, L.: Internal commitment or external collaboration? The impact of human resource management systems on firm innovation and performance. Hum. Resour. Manage. **52**(2), 263–288 (2013)
89. Beugelsdijk, S.: Strategic human resource practices and product innovation. Egos **29**(6), 821–847 (2007)
90. Chen, C.-L., Jaw, Y.-L.: Building global dynamic capabilities through innovation: a case study of Taiwan's cultural organizations. J. Eng. Tech. Manage. **26**(4), 247–263 (2009)
91. Scarbrough, H.: Knowledge management, HRM and the innovation process. Int. J. Manpow. **24**(5), 501–516 (2003)
92. Sanders, K., Lin, C.H.: Human resource management and innovative behaviour: considering interactive, informal learning activities. In: Human Resource Management, Innovation and Performance, pp. 32–47 (2016)
93. Stinglhamber, F., Vandenberghe, C.: Organizations and supervisors as sources of support and targets of commitment: a longitudinal study. J. Organ. Behav. **24**, 251–270 (2003)
94. Seeck, H., Diehl, M.R.: A literature review on HRM and innovation-taking stock and future direction. Int. J. Hum. Resour. Manag. **28**(6), 913–944 (2017)
95. Bos-Nehles, A., Veenendaal, A.R.: Perceptions of HR practices and innovation work behavior: the moderating effect of an innovative climate. Int. J. Hum. Resour. Manag. **30**(18), 2661–2683 (2019)
96. Wright, P., Nishii, L.: Strategic HRM and organizational behavior: integrating multiple levels of analysis. In: Pawauwe, J., Guest, D., Wright, P. (eds.) HRM and Performance: Achivements and Challenges, pp. 97–110. Wiley-Blackwell, Chichester (2013)
97. Wang, Z., Zang, Z.: Strategic human resources, innovation and entrepreneurship fit: a cross-regional comparative model. Int. J. Manpow. **26**(6), 544–559 (2005)
98. Subramony, M.: A meta-analytic investigation of the relation between HRM bundles and firm performance. Hum. Resour. Manag. **48**, 745–768 (2009)
99. Fu, Y.K.: High-performance human resource practices moderate flight attendants' organizational commitment and organizational citizenship behavior. Soc. Behav. Pers. Int. J. **41**(7), 1195–1208 (2013)
100. Grueso-Hinestroza, M.P., López-Santamaría, M., González, J.L., Salcedo, W., Amaya, M.: Organizational culture artifacts and compassionate human resources practices in a healthcare organization. Asian Soc. Sci. **14**(4), 90–99 (2018)
101. Tejada, J.M., Arias Galicia, F.: Prácticas Organizacionales y el Compromiso de los trabajadores hacia la organización. Enseñanza e Investigación en Psicología, Julio Diciembre 10, 002 (2005)
102. Lau, C.M., Ngo, H.Y.: The HR system, organizational culture, and product innovation. Int. Bus. Rev. **13**(6), 685–703 (2004)
103. Heffernan, M., Harney, B., Cafferkey, K., Dundon, T.: Exploring the relationship between HRM, creativity climate and organizational performance: evidence from Ireland. Paper Presented at the Annual Meeting of the Academy of Management, Chicago (2009)
104. Marino, K.E.: Developing consensus on firm competencies and capabilities. Acad. Manag. Exec. **10**(3), 40–51 (1996)
105. Matsuno, K., Zhu, Z., Rice, M.P.: Innovation process and outcomes for large Japanese firms: roles of entrepreneurial proclivity and customer equity. J. Prod. Innov. Manag. **31**(5), 1106–1124 (2014)
106. James, W.M.: Best HR practices for today's innovation management. Res. Technol. Manag. **45**(1), 57–60 (2002)
107. Marsili, O.: Technological regimes and sources of entrepreneurship. Small Bus. Econ. **19**(3), 217–231 (2002)
108. Wang, D., Chen, S.: Does intellectual capital matter? High-performance work systems and bilateral innovative capabilities. Int. J. Manpow. **34**(8), 861–879 (2013)

109. Fini, R., Grimaldi, R., Marzocchi, G.L., Sobrero, M.: The determinants of corporate entrepreneurial intention within small and newly established firms. Entrep. Theory Pract. **36**(2), 387–414 (2012)
110. Casper, S., Stroz, C.: Bounded careers in creative industries: surprising patterns in video games. Ind. Innov. **24**(3), 213–248 (2017)
111. Collins, C.J., Smith, K.G.: Knowledge exchange and combination: the role of human resource practices in the performance of high-technology firms. Acad. Manag. J. **49**(3), 544–560 (2006)
112. Clausen, T., Korneliussen, T.: The relationship between entrepreneurial orientation and speed to the market: the case of incubator firms in Norway. Technovation **32**(9), 560–567 (2012)
113. Therrien, P., Doloreux, D., Chamberlin, T.: Innovation novelty and (commercial) performance in the service sector: a Canadian firm-level analysis. Technovation **31**(12), 655–665 (2011)
114. Luo, J., Olechowski, A.L., Magee, C.L.: Technology-based design and sustainable economic growth. Technovation, 18 Juliol 2012 (2012). https://doi.org/10.1016/j.technovation.2012.06.005
115. Zampetakis, L.A., Beldekos, P., Moustakis, V.S.: "Day-to-day" entrepreneurship within organisations: the role of trait emotional intelligence and perceived organisational support. Eur. Manag. J. **27**(3), 165–175 (2009)
116. Carayannis, E.G., Popescu, D., Sipp, C., Stewart, M.: Technological learning for entrepreneurial development (TL4ED) in the knowledge economy (KE): case studies and lessons learned. Technovation **26**(4), 419–443 (2006)
117. Katila, R., Ahuja, G.: Something old, something new: a longitudinal study of search behavior and new product introduction. Acad. Manag. J. **45**(6), 1183–1194 (2002)
118. Hill, S.A., Birkinshaw, J.: Strategy–organization configurations in corporate venture units: impact on performance and survival. J. Bus. Ventur. **23**(4), 423–444 (2008)
119. Miller, A., Spann, M.S., Lerner, L.: Competitive advantages in new corporate ventures: the impact of resource sharing and reporting level. J. Bus. Ventur. **6**(5), 335–350 (1991)
120. Kuratko, D.F., Ireland, R.D., Covin, J.G., Hornsby, J.S.: A model of middle-level managers' entrepreneurial behavior. Entrep. Theory Pract. **29**(6), 699–716 (2005)
121. Wang, Y., Poutziouris, P.: Entrepreneurial risk taking: empirical evidence from UK family firms. Int. J. Entrep. Behav. Res. **16**(5), 370–388 (2010)
122. Busenitz, L.W.: Entrepreneurial risk and strategic decision making: it's a matter of perspective. J. Appl. Behav. Sci. **35**(3), 325–340 (1999)
123. Covin, J.G., Slevin, D.P.: A conceptual model of entrepreneurship as firm behavior. Entrep. Theory Pract. **16**(1), 7–25 (1991)
124. Sun, L.Y., Aryee, S., Law, K.S.: High performance human resource practices, citizenship behavior, and organizational performance: a relational perspective. Acad. Manag. J. **50**(3), 558–577 (2007)
125. Doz, Y.L., Kosonen, M.: Embedding strategic agility: a leadership agenda for accelerating business model renewal. Long Range Plan. **43**(2), 370–382 (2010)
126. Eisenhardt, K.M., Martin, J.A.: Dynamic capabilities: what are they? Strateg. Manag. J. **21**(10–11), 1105–1121 (2000)
127. Hamel, G., Valikangas, L.: The quest for resilience. Harv. Bus. Rev. **81**(9), 52–65 (2003)
128. Ulrich, D.: A new mandate for human resources. Harv. Bus. Rev. **76**, 124–135 (1998)
129. Prahalad, C.K., Hamel, G.: The core competence of the corporation. In: Zack, M.H. (ed.) Knowledge and Strategy, pp. 41–62. Butterworth-Heinemann, Boston (1990)
130. Walsh, S., Linton, J.D.: The measurement of technical competencies. J. High Technol. Manage. Res. **13**(1), 63–86 (2002)
131. Wang, K.J., Lestari, Y.D.: Firm competencies on market entry success: evidence from a high-tech industry in an emerging market. J. Bus. Res. **66**(12), 2444–2450 (2013)

132. Russell, R.D.: Developing a process model of intrapreneurial systems: a cognitive mapping approach. Entrep. Theory Pract. **23**(3), 65–84 (1999)
133. Martín-Rojas, R., García-Morales, V.J., Mihi-Ramírez, A.: How can we increase Spanish technology firms' performance? J. Knowl. Manag. **15**(5), 759–778 (2011)
134. Zhang, Z., Wan, D., Jia, M.: Do high-performance human resource practices help corporate entrepreneurship? The mediating role of organizational citizenship behavior. J. High Technol. Manage. Res. **19**(2), 128–138 (2008)
135. Wei, L.Q., Lau, C.M.: High performance work systems and performance: the role of adaptive capability. Hum. Relat. **63**(10), 1487–1511 (2010)
136. Sozen, H.C., Varoglu, D., Yeloglu, H.O., Basim, H.N.: Human or social resources management: which conditions force HR departments to select the right employees for organizational social capital? Eur. Manag. Rev. **13**(1), 1740–1754 (2015)

Procedure for the Selection of International Markets: Application in the Cement Business Group of Cuba

Irene García Rondón[1](✉), Alicia Vitale Alfonso[2], and Zamir Cobas[2]

[1] CEEH, La Habana, Cuba
irene.garcia@ceehabana.com
[2] Universidad de La Habana, La Habana, Cuba
{alicia,zamir}@fec.uh.cu

Abstract. Throughout Cuba's economic history, it has been characterized as a country, totally dependent on the international environment. In recent years, this dependence on the foreign sector has been the biggest brake on growth, with a serious trade deficit in the balance of goods. The present work is part of the line of research of the selection of international markets. "The selection of international markets is an intrinsic decision of international marketing of crucial importance for the company" [1]. Therefore, this decision must be taken through a systemic model that rigorously analyzes and evaluates each option. It is proposed to develop a procedure for the selection of international markets for the products of the Cement Business Group (GECEM), based on the elements of the theory of fuzzy subsets, useful for innovation and decision making, in situations of lack of information and uncertainty.

Keywords: Fuzzy sets · Selection of international markets · Innovation

1 Introduction

During the insertion of companies in foreign markets, there are sometimes situations in which companies have several dilemmas and questions; How to enter the international economy, what are the most effective markets for the company, there is enough information to make an accurate and large-scale decision, how to detect if the choice obtained is effective in the absence of data. In short, there are many unknowns that arise at the time of international marketing practice, in which the uncertainty variable is introduced, and this cannot be measured or treated with the same techniques with which the other variables are treated.

This guaranteed, then, to detect recognized techniques in the academic field, which allowed to work with uncertain and at the same time subjective information. The theory of fuzzy subsets, or also known as fuzzy logic, is then presented as an attempt to apply the human way of thinking in its logic and language to the creation of effective systems to process both qualitative and quantitative information [2–5].

A. M. Gil-Lafuente et al. (Eds.): REDCID 2019, LNNS 388, pp. 144–151, 2022.
https://doi.org/10.1007/978-3-030-93787-4_8

"The selection of international markets is an intrinsic decision of international marketing of crucial importance for the company" [1]. Therefore, this decision must be taken through a systemic model that rigorously analyzes and evaluates each option.

In accordance with the foregoing, this article proposes a procedure for the selection of international markets for the products of the Cement Business Group.

2 Selection of International Markets

Conceptualization International marketing in terms of study method has its origin in the marketing itself, transferred to an international context. The differentiation between marketing that is applied in a single country and international marketing is a topic widely discussed among marketing professionals and academics. However, the basic concepts and instruments of analysis applied coincide in national and international marketing, the difference lies in the perspective from which the problems are analyzed and their complexity. In reality, and due to the greater interconnection between countries, marketing is increasingly international, which implies a progressive disappearance in the specialized literature of the distinction between national and international marketing.

When a company decides to operate internationally, it must make three basic international marketing decisions: the selection of international markets, the selection of entry forms and the design of marketing mix strategies. This article is framed in the first of these decisions. The selection of international markets means the decision by which the company chooses the markets to which it will direct its offers.

The selection of foreign markets is the most important decision of the company when it wishes to expand internationally [6]. It is a decision on the competitive positioning of the company, both geographically and psychologically. The selection of a country commits the company to operate in a specific field, which establishes the foundations for its future international expansion.

A bad decision can be very expensive because it can lead the entity to a resounding failure, which beyond the direct costs that this would entail, could demotivate it in its internationalization process. Likewise, a correct selection will allow directing and adapting the offers to those markets where the company has a higher probability of success and long-term maintenance.

The company must take into account not only which markets to address but also how many. It is advisable to start with one or a few markets and then gradually enter others [7]. By concentrating efforts and committing more resources in a few markets, greater penetration is achieved in them, while the company learns about the management of international operations and measures its risks.

Due to the conditions prevailing in the markets and the need to meet the criteria of experts, it is also recommended to make use of tools of fuzzy subsets theory to make an adequate selection of markets, thus attending both external analysis and the inmate.

3 Procedure for the Selection of International Markets in the Cement Business Group (GECEM)

The Cement Glass Business Group CEMVID, subordinate to the Ministry of Basic Industry, is created by resolution 314/2001 of the Ministry of Economy and Planning, which authorizes the merger of the Union Vidrieras Caribe and the Corporation Cementos Cubanos, where, through Resolution no. 148 dated March 30, 2010 of the Minister of Construction, ordered the transfer of the CEMVID Group, subordinate to the Ministry of Basic Industry, for subordination to the Ministry of Construction, as well as its consequent change of denomination by Grupo Empresarial del Cemento, GECEM.

This article adapts the procedure proposed for the selection of international markets to the group [1]. The procedure proposed is structured in two phases: selection of criteria and market evaluation; these phases merge in the present procedure. Similarly, the steps were reduced in order to increase their understanding by specialists in the business sector, as well as the application time. The criteria were adapted to the products under study, without distinction by filters, which increases the flexibility of the proposal.

Step 1. Identification of the countries under study. This step is of vital importance, because discarding attractive markets can compromise the effectiveness of the whole process, likewise choosing a large number of alternatives, increases the time of application of the procedure and therefore, the cost of it.
Step 2. Identification of the factors to consider in the election. The choice of criteria to consider is a crucial step because the success of the entire process depends on it.
Step 3. Selection of experts. In this step two decisions of vital importance for the process are made, referring to which experts to select and how many.

To select the experts, it is proposed to prepare an initial list of possible people to meet the requirements to be considered experts in the field, after consulting their approval to participate. In order to determine their level of competence, the use of the experts' competence coefficient is proposed (K).
Step 4. Determination of the scales of the variables. In the selection of international markets, the use of two variables is proposed: favorable and important. Under the first variable, the behavior of each of the previously selected criteria will be evaluated in the countries under study. In the same way, since not all the criteria evaluated have the same importance in the selection of the markets, it is necessary to carry out a weighting of them, which will be carried out through the variable importance. For this, it is proposed to use the indefinite scale, due to the international acceptance it has [8, 9].
Step 5. Application of expert surveys. Each expert is requested to rate the favorable variable for each of the criteria with respect to each country under study, in which the confidentiality of the criteria of each expert is paramount. As the indefinite scale has been chosen, the expert has the possibility to express his opinion in a range of levels. This gives the possibility of collecting opinions as people think, with their doubts and inaccuracies, without forcing them to decide on a single level of the scale. Also, you are asked to rate the variable importance for each of the criteria that make up the first filter, according to the implementation of a survey.

In the end, there will be (n) confidence intervals for the favorable variable with respect to each criterion in each country under study.

$$\left[A^{(i)}_{1,l,p}, B^{(i)}_{1,l,p}\right] \tag{1}$$

where: $i = 1, 2, \ldots, n$: number of the expert and n the total of these; $l = 1, 2, \ldots, m$: represents the criterion and m the total of these; $p = 1, 2, \ldots, z$: represents the country and z the total of these; subscript 1 corresponds to the first round of the process.

Likewise, n confidence intervals will be obtaines for the importance variable with respect to each criterion:

$$\left[A_{1,l}^{(i)}, B_{1,l}^{(i)}\right] \tag{2}$$

$i = 1, 2, \ldots, n; l = 1, 2, \ldots, m$

where: i: number of the expert and n the total of these; l: represents the criterion and m the total of these, subscript corresponds to the first round of the process.

Step 6. Aggregation of expert opinions. In this step, all valuations for the favorable variable by criteria and by country are computed with the help of an Excel sheet, as well as the valuations of the variable of importance for each criterion. In this phase it is necessary to find the average of the group valuations for each variable, as well as the distances between each valuation and this average.

In this way, the favorable variable for each criterion in each country under study will be obtained:

$$\left[A_{1,l,p}^{(m)}, B_{1,l,p}^{(m)}\right] \tag{3}$$

where: $A_{1,l,p}^{(m)} = \frac{\sum_{i=1}^{n} A_{1,l,p}^{(i)}}{n}, B_{1,l,p}^{(m)} = \frac{\sum_{i=1}^{n} B_{1,l,p}^{(i)}}{n}$.

Likewise, the variable of importance will be obtained for each criterion the interval:

$$\left[A_{1,l}^{(m)}, B_{1,l}^{(m)}\right] \tag{4}$$

where: $A_{1,l}^{(m)} = \frac{\sum_{i=1}^{n} A_{1,l}^{(i)}}{n}, B_{1,l}^{(m)} = \frac{\sum_{i=1}^{n} B_{1,l}^{(i)}}{n}$.

Until now, we have worked with the values of the scale, defined from 1 to 10 for ease of experts. For the calculation of the relative distance of Hamming, all expert valuations are required to be expressed in the interval [0, 1].

For this, it is enough to divide each valuation by the greatest value of the scale, 10 in the case of the indefinite scale.

In this way the intervals will be defined as follows for the favorable variable:

$$\left[A_{1,l,p}^{(i)}/k, B_{1,l,p}^{(i)}/k\right] = \left[a_{1,l,p}^{(i)}, b_{1,l,p}^{(i)}\right] \tag{5}$$

While for the variable importance it will be:

$$\left[A_{1,l}^{(i)}/k, B_{1,l}^{(i)}/k\right] = \left[a_{1,l}^{(i)}, b_{1,l}^{(i)}\right] \tag{6}$$

k being the highest value of the scale applied, and

$$\left[a_{1,l,p}^{(i)}, b_{1,l,p}^{(i)}\right] \subset [0, 1] \text{ y } \left[a_{1,l}^{(i)}, b_{1,l}^{(i)}\right] \subset [0, 1]$$

The average values calculated above are also expressed in the range [0, 1].

$$\left[A_{1,l,p}^{(m)}/k, B_{1,l,p}^{(m)}/k\right] = \left[a_{1,l,p}^{(m)}, b_{1,l,p}^{(m)}\right] \text{y} \left[A_{1,l}^{(m)}/k, B_{1,l}^{(m)}/k\right] = \left[a_{1,l}^{(m)}, b_{1,l}^{(m)}\right] \tag{7}$$

where:

$$\left[a_{1,l,p}^{(m)}, b_{1,l,p}^{(m)}\right] \subset [0, 1]\,\mathrm{y}\left[a_{1,l}^{(m)}, b_{1,l}^{(m)}\right] \subset [0, 1]$$

The relative distance of Hamming between the confidence intervals will be given for the favorable variable for each criterion, country and expert by:

$$d_{1,l,p}^{(i,m)} = \frac{\left|a_{1,l,p}^{(i)} - a_{1,l,p}^{(m)}\right| + \left|b_{1,l,p}^{(i)} - b_{1,l,p}^{(m)}\right|}{2} \quad (8)$$

While for the variable importance for each criterion and expert will be:

$$d_{1,l}^{(i,m)} = \frac{\left|a_{1,l}^{(i)} - a_{1,l}^{(m)}\right| + \left|b_{1,l}^{(i)} - b_{1,l}^{(m)}\right|}{2} \quad (9)$$

Step 7. Repeat steps 5 and 6 until the criteria converge. In this step, the second round of surveys is first carried out, each expert receives information on the averages of the favorable variable intervals for each criterion and country, as well as the average values of the intervals of the variable of importance for each criterion. The distance between your individual responses in the first round and the previous average intervals is also announced. The survey is the same, but each person is free to confirm or vary their previous qualifications. Subsequently, new confidence intervals will be obtained for each variable. Rounds should be repeated until Hamming's distance shows lower values.
Step 8. Development of experts. With the observations of the experts taken to the scale.

[0, 1] the favorable experts $\underset{\sim}{F}$ and importance $\underset{\sim}{I}$ are constructed with the help of an Excel sheet.
Step 9. Calculation of the attractiveness of the countries.

In this step, the expert obtained for each criterion is multiplied first with respect to the favorable variable by the importance expert that evidences the weight given by the experts to each criterion. In this way, the attractive expert will be obtained for each of the criteria of the first filter in each country under study $\underset{\sim}{A_{l,p}}$ which is obtained by multiplying favorably by importance. O sea, $\underset{\sim}{A_{l,p}} = \underset{\sim}{F_{l,p}} \cdot \underset{\sim}{I_l}$.

The attractiveness of a market can be obtained through the distance between its result for each criterion and the ideal market score with respect to them. In the selection of international markets the optimum would be a market that obtained the maximum score in each criterion, expressed in the expert $\underset{\sim}{D_l}$ where

$l = 1, 2, \ldots, 5$; being l: the number of criteria to take in the investigation.

To obtain the distance between the score of each market and the ideal score, we will find the relative distance of Hamming between the mathematical hopes of each expert of each criterion for each country and the mathematical hope of the ideal expert, which is defined for each country as follow:

$$d_{1,l}^{(c,d)} = \frac{\sum_{i=1}^{5}\left|a_{1,l} - d_{1,l}\right| + \left|a_{2,l} - d_{2,l}\right|}{12} \quad (10)$$

Step 10. Hierarchy and selection of countries. Countries are ordered from highest to lowest according to their attractiveness. For this, the distances previously obtained are organized in descending order, obtaining the countries under study ordered according to their attractiveness from highest to lowest.

4 Application of the Proposed Procedure for the Selection of Bulk Cement Markets

To test the procedure, it was decided to apply it to the cement product in bulk, taking as a starting point some selected countries.

Step 1. Identification of the countries under study. Thirteen countries were preselected for the application of the procedure: Venezuela, Mexico, Honduras, Nicaragua, Costa Rica, Haiti, Dominican Republic, Jamaica, Panama, Puerto Rico, Guatemala, Brazil and Barbados. The selection was made according to the criteria of the company's leaders, which was also influenced by the economic situation of the country, which covers the few sources of financing, and the high costs in transportation freight that involves the export of a product.
Step 2. Identification of the factors to consider in the election. Five factors were selected: (i) economic factors, (ii) political factors, (iii) bilateral relations, (iv) belonging to common associations and (v) legal factors. The selection was made according to the bibliography consulted and the criteria of the company leaders.
Step 3. Selection of experts. Seven experts were preselected, to which the competence coefficient was determined, obtaining three with high competition. Taking into account that in the application of these tools the important thing is quality and not quantity (Blanco, 2007) (García I., 2010), as well as the recommendation to only work with experts of high competence in the subject, the Research only with the three experts indicated.
Step 4. Determination of the scales of the variables. The use of an indefinite scale for both variables is decided, because it is the most recommended and used in research, internationally.
Step 5. Application of expert surveys. The questionnaire is applied to the selected experts.
Step 6. Aggregation of expert opinions. With the results obtained, the aggregation of the opinions of the experts is carried out.
Step 7. Repeat steps 5 and 6 until convergence in the criteria. It is decided not to make another round of the method due firstly to the convergence between experts, and secondly for reasons of time.
Step 8. Development of experts. With the results obtained, the experts of the "favorable" variable and the "importance" variable are prepared.
Step 9. Calculation of the attractiveness of the countries. Subsequently, the attractiveness of the countries under study is calculated.
Step 10. Hierarchy and selection of countries. Once Hamming's distance was found between the mathematical hopes of the attractive experts for each country and the ideal of the market, these distances were ordered in descending order, obtaining the level of attractiveness of each country under study. This result is reflected in Table 1.

Table 1. Countries ordered according to their level of attractiveness.

Countries	Value
República Dominicana	0,37
Panamá	0,41
Jamaica	0,44
México	0,48
Costa Rica	0,50
Barbados	0,51
Haití	0,52
Honduras	0,54
Venezuela	0,54
Brasil	0,56
Nicaragua	0,56
Guatemala	0,63
Puerto Rico	0,77

Source: Own elaboration based on the results obtained.

Once the analysis was completed, the experts determined the priority level of each market. The result of this analysis showed a range of countries that should be highly considered for the export of cement in bulk: Dominican Republic, Panama, Jamaica and Mexico, highlighting the Dominican Republic as the most attractive market. These market selections are the most attractive for Cuba given the current conditions of international markets, in which, in addition, appropriate strategic and tactical marketing plans that allow them to take advantage of them must be developed. The rest of the countries are discarded in this process, but they must be considered in future decisions, as market conditions are not static, and a country that is not attractive today, may be within a while.

5 Conclusions

The procedure for the selection of applied markets allows to select the markets with greater rigor from the opinions of the experts, supplying the scarcity of information and uncertainty existing in the sector, explores several markets in order to identify those most attractive for exports.

The application of the procedure to the products of the Cement Business Group (GECEM), taking as a sample 13 countries of Latin America and the Caribbean, allowed to select the most attractive markets for Cuba given the current conditions of the international markets, from the expert opinions, being easily operable in the Microsoft Excel application.

References

1. García-Rondón, I.: Procedimiento para la selección de mercados internacionales de los servicios de gestión ambiental cubanos. Tesis Doctoral, Universidad de la Habana (2010)
2. Blanco, B.: Procedimiento para la evaluación de riesgos empresariales de operación con métodos de las matemáticas borrosas. Tesis Doctoral, Universidad de la Habana (2010)
3. Merigó, J.M.: Fuzzy decision making using immediate probabilities. Comput. Ind. Eng. **58**(4), 651–657 (2010)
4. Blanco-Mesa, F., Gil-Lafuente, A.M., Merigó, J.M.: Aggregation operators for decisión-making in entrepreneurship: an application in sport business. Technol. Econ. Dev. Econ. **24**(2), 335–357 (2018)
5. Alfaro-Garcia, V., Merigó, J.M., Plata-Pérez, L., Alfaro-Calderon, G.G., Gil-Lafuente, A.M.: Induced and logarithmic distances with multi-region aggregation operators. Technol. Econ. Dev. Econ. **25**(4), 664–692 (2019)
6. Galván, I.: La Formación de Estrategias de selección de mercados exteriores. Tesis Doctoral, Universidad de Palmas de Gran Canaria (2003)
7. Cerviño, J.: Marketing international. Nuevas perspectivas para un mercado globalizado. Ediciones Pirámide, España (2006)
8. Merigó, J.M., Gil-Lafuente, A.M., Yager, R.R.: An overview of fuzzy research with bibliometric indicators. Appl. Soft Comput. **27**, 420–433 (2015)
9. Gil-Aluja, J., Gil-Lafuente, A.M., Merigó, J.M.: Using homogeneous groupings in portfolio management. Experts Syst. Appl. **38**(9), 10950–10958 (2011)

Customer Satisfaction at the Roc Presidente Hotel

Thais Zamora Molina(✉), José Manuel Pozo Rodríguez, and Omar Marrero Salgueiro

University of Havana, 10400 Havana, Cuba
{thais,Josem}@fec.uh.cu

Abstract. Quality management is fundamental for every company, organization or business. At present, there is profound technological homogenization as a result of the internationalization of development, which makes quality in products or services what distinguishes then from one another, separates them and positions them within the market. That is why it should be used as tool for the organization, taking advantage of it as a resource that generates advantages. This work covers quality management focused on the perception of it in the hotel Roc Presidente services. Concerning its main weaknesses, from two points of view, internal and external clients, which allows to obtain conclusions that contribute to the improvement of all the services offered. With this image that this installation projects in the market and deepens its development is deepened.

Keywords: Quality · Customers · Company · Techniques · Service

1 Introduction

At present, every company aims to position itself in the market, both nationally and internationally; what is becoming increasingly complicated in the light of the reviving globalization process. Therefore, quality management in its implementation and certification processes has become an indispensable factor to achieve it.

Each company strives to continuously improve and offer customer satisfaction, for them it is necessary to know their main failures through the quantification of the problems detected; being able to make the necessary changes and meet the expectations of customers.

Clients are essential for organizations, so it is necessary to know their expectations and thus be able to find the best tools to meet them. For this, it is necessary to promote teamwork, as well as the provision of all areas of the organization so that the service the client receives is the desired one.

In this work the main problems presented by the Roc Presidente Hotel are clarified and how they influence customer satisfaction, both internal and external. It therefore

Project supported by "Red Iberoamericano para la Competitividad, Innovación y Desarrollo" (REDCID), project number 616RT0515, in "Programa Iberoamericano de Ciencia y Tecnología para el Desarrollo" (CYTED).

A. M. Gil-Lafuente et al. (Eds.): REDCID 2019, LNNS 388, pp. 152–177, 2022.
https://doi.org/10.1007/978-3-030-93787-4_9

focuses on quality management in the entity, providing options for the support of its customers and the satisfaction of their expectations.

Scientific problem: How to structure quality perception studies at the Roc Presidente Hotel, from the perspective of internal and external customers? General Objective: Design a procedure for the study of customer satisfaction at the Roc Presidente Hotel.

2 Materials and Methods

The methods used to find the solution to the problem are the following:

1. Historical-logical analysis, to analyze the evolution of quality management and customer satisfaction. The systemic method allows to base the proposed model, with the decomposition for the analysis and synthesis of all the elements and dimensions of quality management. The modeling method is used to structure the proposed procedure.
2. Empirical method, through measurement, to check the problem situation and measure the level of overall satisfaction of the organization's customers. Mathematical methods are used, through the Theory of Forgotten Effects.

The procedures used are: analysis and synthesis, abstraction, and induction-deduction. The techniques of surveys, documentary analysis and the Theory of Forgotten Effects are used.

2.1 The Roc Hotels Hotel Chain and the Gran Caribe Hotel Group

Roc Hotels is a family hotel chain founded in 1981 in Palma de Mallorca, Spain, which is mainly dedicated to exporting the holiday offer product. It is a competitive offer, always in first-class tourist areas such as Menorca, Mallorca, Almeria, the Costa del Sol, among others in Havana, Varadero and Cayo Santa Lucia.

As a brand, they implement in each of their establishments their own standard of tourist quality and effective hotel management systems, oriented to the satisfaction of all customers. The chain is committed to a basic and efficient structure, based on the professionalism of its components, creating a solid business support with an agile response all aimed at the customer as the main being of any tourism industry.

The Hotel Group Gran Caribe is a commercial company of Cuban nationality, incorporated in Havana on August 1, 1994 according to Cuban laws. The Group is created as a hotel entity awarded to the Ministry of Tourism, under the following name Gran Caribe: Sociedad Anónima, with legal personality. It currently has 12123 rooms distributed in 45 facilities, located in the most important tourist centers of the country.

2.2 Characterization of the Roc Presidente Hotel

The Presidente Hotel was founded in the second decade of the last century, it was the first skyscraper built in Havana, located in a very privileged area of Vedado, very close to the Paseo de los Presidentes and the Malecón of Havana. Although the entity operates under the ownership of the Gran Caribe Hotel Group, it is in a Contract of Administration or Hotel Management with the Spanish chain Roc Hotels, who in addition to offering their experience in the international market, are responsible for executing all the marketing actions of the hotel.

The hotel has 158 comfortable rooms that have all modern comforts and almost all have sea views. On the 2nd and 10th floor rooms are reserved for special and select people.

The stage stands out for its links with Cuban art and culture, in its capacity as host for relevant international events and festivals, both ballet and other modalities. As for its markets, hotel managers mention Germany and rising countries in South America (Fig. 1).

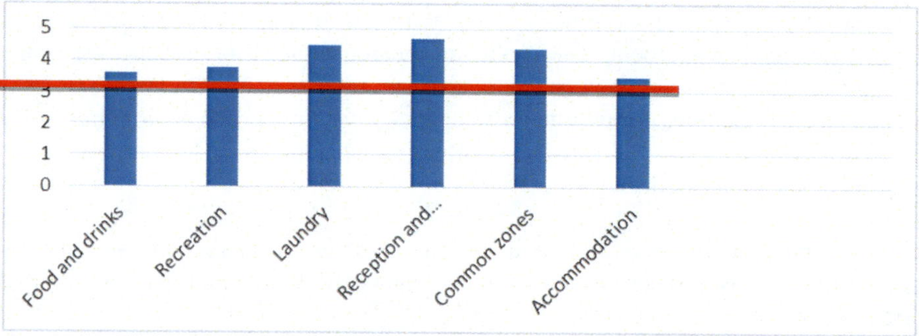

Fig. 1. Evaluation of the attributes of the Hotel.

Taking into account the data offered in the graph, the attributes that the clients have given the lowest score according to their satisfaction with them have been Food and Beverage, Recreation and Accommodation. The study to be developed focuses on the analysis of these three aspects, as well as on the satisfaction that the organization's internal clients present.

3 Procedures to Measure Customer Satisfaction at the Roc Presidente Hotel

The proposed procedure is based on the following principles:

1. Continuous quality improvement: it is a procedure that can be perfected taking into account the Deming cycle (Plan. Do. Verify. Act).
2. Learning and creativity: it creates skills in terms of knowledge, values, communication, cooperation and establishes an environment that fosters the development of quality initiatives in the framework of the quality management system.
3. Customer orientation: quality is seen from the customer approach.
4. Leadership leadership: support and participation of senior management in terms of: policy, objectives, mission and vision.

5. Staff participation: it is done through work areas, improvement teams, quality coites, etc.

The proposed procedure to measure customer satisfaction in the installation consists of three stages, each stage pursues its particular objectives that contribute to the fulfillment of the general objectives of quality in the organization (Fig. 2).

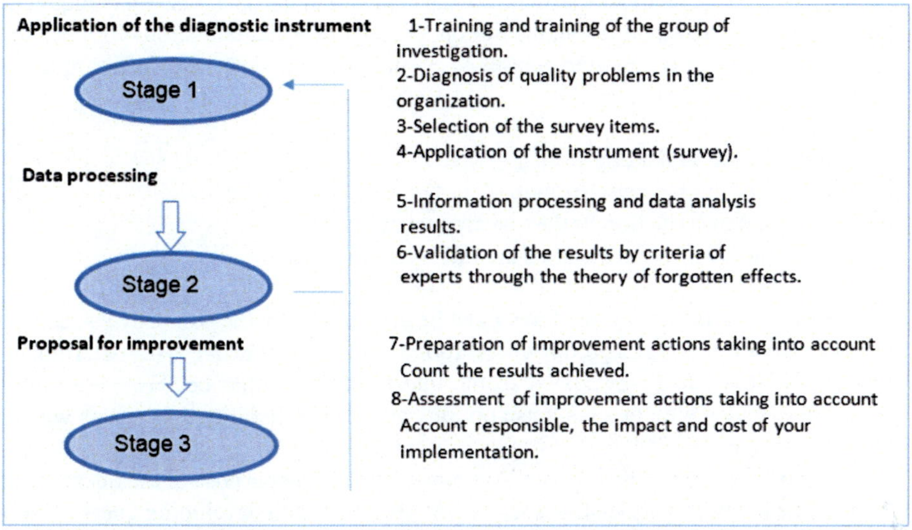

Fig. 2. Procedure to measure customer satisfaction at the Roc Presidente Hotel

Next, each of the stages and steps that make up the proposed procedure are described:

Stage 1: Application of the diagnostic instrument.

Identify the aspects that are affecting the low satisfaction of customers and form the diagnostic instrument.

Step 1: Training and training of the research group.

The research group is formed, composed of workers of the entity, who have previous knowledge on the subject. This research group is trained for it through meetings with the professors in charge of this research, in which they are explained how to carry out all the stages of the procedure.

Step 2: Diagnosis of quality problems in the organization.

The need to conduct a study on quality, more specifically on customer satisfaction, is analyzed based on the situation presented by the entity with respect to the services provided. The Ishikawa Diagram technique is used to state the fundamental aspects that determine low customer satisfaction in that service that was mostly unsatisfied when applying to the Pareto Diagram.

Step 3: Selection of the survey items.

The questionnaire, as a measuring instrument, allows to know the current level of customer satisfaction. For its processing, descriptive statistics are used in this type of research: it provides methods that describe the behavior of the data. These techniques deal with the collection, organization, tabulation, presentation and reduction of information.

Hayes (1999) proposes four phases to prepare the customer satisfaction questionnaires:

1. Determination of the questions.
2. Selection of the response format.
3. Writing the introduction to the questionnaire.
4. Determination of the content of the final questionnaire.

The selection of the survey items must be carried out by the company's specialists, who are part of the research group, who, through a brainstorm, arrive at the determination of what the items to be evaluated in the questionnaire would be. From a statistical perspective, scales with two response options have less reliability than scales with five response options.

The introduction should be brief. You must explain the purpose of the questionnaire and provide the necessary instructions to complete it. When developing questionnaires, it is important to ensure that the data obtained from them reflects valid and reliable information. To estimate the reliability of the questionnaire it is recommended to apply the Cronbach Alpha (determined by J.L Cronbach), which provides a coefficient with a range between zero and one. If the coefficient is above 0.6, it can be noted that the reliability of the questionnaire is good, and as you get closer to the unit, it will be better, until 1 is excellent.

Step 4: Application of the instrument (surveys)

The samples of internal clients and external clients to be surveyed are calculated. The calculation of the sample size is carried out taking into account the behavior of the population, whereby the type of sampling may vary from one Entity to another, since it is related to the type of information it has. The most commonly used probabilistic samples are:

- Simple Random Sampling M.A.S
- Random Unrestricted Sampling M.I.A
- Random Stratified Sampling M.A.E

Sample sizes calculated, items selected and questionnaires developed, both for internal and external clients, both questionnaires with different approaches; in which some of the items may not match since they will be applied to two different samples of customers, the surveys are applied.

Stage 2: Data Processing

Process the data obtained after the application of the diagnostic instrument and quantitatively reduce the study variables in order to influence the client's satisfaction with less resources.

Step 5: Information processing and analysis of the results

After collecting the information, both univariate and multivariate techniques are applied; for this, the variables are created, in the SPSS statistical package, with which it is going to work, taking into account the measurement scale that must be applied with each variable, after that the collected data is introduced and the procedures are applied techniques.

From the univariate statistics, a relationship can be calculated, both of position and central tendency (average, mode, median) and dispersion (variance, standard deviation, coefficient of variation). To conduct a deeper study, bivariate statistics are used, which allows analyzing the relationship between two variables taken in the same.

From the univariate statistics, a relationship can be calculated, both of position and central tendency (average, mode, median) and dispersion (variance, standard deviation, coefficient of variation). To conduct a more in-depth study, bivariate statistics are used, which allows the relationship between two variables taken at the same time to be analyzed, specifically through the Chi-Square test.

By working simultaneously with more than two variables, we have the methods of multivariate analysis, Principal Component Analysis and Multidimensional Scaling, which enrich the results and conclusions obtained by the responses observed in a survey. In this study of customer satisfaction, the statistical technique will be the Analysis by Main Components.

Step 6: Validation of the results by expert criteria through the Theory of Forgotten Effects

After applying the statistical techniques and evaluating the results, these are validated through expert criteria and through the Theory of Forgotten Effects, taking into account the results produced by external clients.

First, the experts to be surveyed are identified, to whom a questionnaire is designed and applied to know their level of expertise. Experts who are internal clients of the organization, as well as academics should be surveyed to triangulate the analysis performed.

The expert analysis is carried out using the experts' competence coefficient, which is obtained from two coefficients: Knowledge coefficient (Kc) and argumentation coefficient (Ka). To obtain Kc, experts are asked to evaluate their level of knowledge about the subject on a growing scale from 1 to 10, where value one means that the consulate considers not having any competence on the subject and value ten means which considers having a maximum domain over it (Garcia R 2010).

With the above score, Kc is obtained through the following formula:

$$Kc = n/10$$

Where n is the range selected by the expert.

To obtain Ka, experts are instructed to assess their ability to argue in high, medium or low, taking into account a group of aspects which will be weighted and adapted to the type of organization under study and are shown in the form of table.

The answers of the experts are contrasted with the comparison values of that table and the coefficient is calculated through the formula

$$ka = \sum_{n=1}^{6} ni$$

Where neither is the value corresponding to the source of argument.

Finally, the competence coefficient (K) for each expert is obtained through the expression:

$$K = 0.5\,(Kc + Ka)$$

The competition scale is made up of the following ranges;

$$K \geq 0.75\,(Kc + Ka)\ \text{High competition coefficient}$$
$$0.5 \leq K \leq 0.75\ \text{Average competition coefficient}$$
$$K < 0.5\ \text{Low competition coefficient}$$

After selecting the experts with high competence, the questionnaire is designed and applied based on the theory of Forgotten Effects, with the aim of validating the grouping by components carried out in the previous step.

The theory of Forgotten Effects is established in 1988 by professors Kaufman and Gil Aluja. This theory takes into account the incidence or chance relationships and allows obtaining all direct and indirect relationships, recovering what is called "forgotten effects" (Kaufman. A and Gil Aluja. J 1988).

The membership function concerns the interval [0, 1], with the highest incidence ratio being the closest to the valuation at 1 and will be weaker when the value that corresponds to it is closer to 0. A chained scale is used, which is shown in the following Table 1.

Stage 3: Proposal for improvement actions

Prepare the improvement actions that must be carried out in the company with the aim of raising the level of satisfaction of the service provided, assessing its impact, cost and responsibility.

Step 7: Development of the best actions taking into account the results achieved.

Table 1. Semantic correspondence of the endecaria scale.

Value	Correspondence
0	No incident
0,1	Virtually no incident
0,2	Almost without incident
0,3	Very weak incidence
0,4	Weak incidence
0,5	Average incidence
0,6	Considerable incidence
0,7	0.7 Enough incidence
0,8	0.8 Strong incidence
0,9	0.9 Very strong incidence
1	1 The highest incidence

The actions that improve the service provided and increase customer satisfaction are elaborated. Which are prepared taking into account the groupings of the variables according to the main components and stating in each of them the person responsible for carrying out their implementation and the date on which it will be carried out.

Step 8: Assessment of improvement actions taking into account the person responsible, the impact and the cost of their implementation.

The administration undertakes to apply the improvement actions and to repeat the study carried out in a given period of time, in order to compare the results and be able to evaluate whether the improvement actions were effective.

4 Results of the Application of the Procedure to Measure Customer Satisfaction at the Hotel Presidente

- Training and training of the research group of the hotel under study.

To begin with this first stage of the procedure developed to manage the quality of the services of the Roc Presidente Hotel, a research group formed by workers of the center was created. Which were trained on each of the stages of the procedure to be implemented, as well as the importance of it for the entity.

The group is made up of:

1. Management Specialist Hotel Roc Presidente
2. Deputy Director of the Department of Human Resources
3. Specialist of the Human Resources Department

4. Director General
5. Assistant Director General
6. Head of Reception

- Diagnosis of quality problems in the organization.

The main quality problems that the Hotel Roc Presidente is currently facing are studied. It was observed that the attributes that present less satisfaction are: food and drink, recreation and accommodation.

- Selection of survey items, application of the instrument and analysis of the purpose of the questionnaire.

Two questionnaires were applied as a survey. One of them is intended for external clients (guests) where all those points that are of interest for the management of the hotel and another questionnaire for internal clients (workers) are measured, with questions regarding the quality of the hotel and its position before your workplace.

To know the total number of external customers to be surveyed, the sample is calculated for the Roc Presidente Hotel. Probabilistic sampling is used since the probability corresponding to each subject of the population is known in advance, therefore, all elements of the population have the possibility of being chosen in the sample.

In addition, simple random sampling without replacement is used, since the populations under study are finite and the objective is that each extracted element discards the next extraction. The value of the sample error with which it is going to work is 8%, since according to (Bolea et al. 2007) it is advisable that the sample error be less than 10%, the most typical error values being 5% and 8%.

In this regard, during the month of February of the year 2019, 140 external clients were surveyed. The population was 2000 customers as it is the historical average that the hotel has calculated for the month of February. For internal clients, it was concluded to survey 100 workers since the total number of workers in the company is 121 and between holidays, free days and leave, the active workforce is 100 workers.

Questionnaire reliability analysis

For this calculation, Cronbach's alpha is used, which perceives the level at which the different questions are correlated with each other, to what is called internal consistency. When the values of this coefficient are closer to 1 the internal consistency is greater, although if they are between 0.6 and 0.8 it can be said that this is good.

Cronbach's Alpha is calculated for internal and external customers through the output of the IBM SPSS statistics program. The results are shown in the following table (Table 2).

From the results obtained from this study, reliability is considered that the questionnaires applied to customers, have a high equivalence, homogeneity and internal consistency for all items.

- Results of the survey applied to internal clients

This diagnosis allows to know the factors that affect the satisfaction of the workers of the entity and therefore to the quality of the product.

Table 2. Results of the Conbach Alpha for external and internal clients.

Customers	Alfa de Cronbach
External	,620
Internal	,820

- Frequency analysis. Working environment.

The work environment allows to know the state of the company in terms of organizational aspects, environment culture, among others that influence its performance within the organization.

The survey yielded the following results related to the work environment of the Roc Presidente Hotel:

1. In the entity, there is dissatisfaction on the part of the workers with their job, since 16% of the internal clients declare themselves very dissatisfied with the conditions of their job, 3% declare themselves unsatisfied, 32% are indifferent, 34% satisfied and 15% very satisfied.
2. Internal clients face the work with great motivation and express it as follows: 20% declare themselves very satisfied with their motivation towards work and 55% satisfied, while 14% are indifferent to this aspect and only 5% are little satisfied and 6% very little satisfied.
3. Internal clients in general feel satisfied with their work performance and 61% express it. 17% are very satisfied with their work performance, only 4% are not very satisfied and 18% of them declare it indifferent.
4. In the entity, internal clients are not part of the decision-making process and they demonstrate this. 22% declare themselves very dissatisfied and 21% unsatisfied, with the freedom to make decisions they have, while for 29% they are indifferent. 27% are satisfied and only 1% declare themselves very satisfied.
5. Regarding the mood in the different work areas, 56% declare themselves satisfied and 17% very satisfied, while 8% are not very satisfied and 19% are indifferent.

The following figure shows the main reasons that threaten the levels of satisfaction of workers through the average of each item, demonstrating that the motivation at work is not adequate and also the non-participation of internal customers in the Decision making, are the main causes of discontent against the entity (Fig. 3).

- Results of the survey regarding the importance and assessment of internal customers.

Knowing the level of satisfaction of external customers in terms of quality at the Roc Presidente Hotel is one of the bases of this research, but another point of reference is the importance of these for internal customers and how they value the satisfaction of the external customer expectations, as for the same items.

Fig. 3. Main reasons that undermine workers' satisfaction levels.

1. In the entity the results showed that the importance of cleaning the rooms is indifferent to 3% of internal customers, while for 14% it is important and for 78% this element is very important. At the same time internal customers think of the satisfaction of external customers with respect to this factor as follows: 7% are indifferent, 53% describe it as satisfactory and 30% as very satisfactory. According to the comparison between the means of importance and satisfaction granted by internal clients, this aspect should be strengthened.
2. 25% of internal customers rate the comfort of the rooms as important, while 75% rate it as very important. For 21% of them it is indifferent in the satisfaction of external customers in terms of room comfort, while 42% believe it is satisfactory and 37% that it is very satisfactory. According to the comparison between the means of importance and satisfaction for internal customers, this aspect should be strengthened.
3. Complementary services within the rooms are very important for 52% of internal customers and important for 47% of them, while only 1% consider it indifferent. At the same time they consider that external customers find this aspect satisfactory by dividing as follows: 29% very satisfactory, 57% satisfactory and 14% indifferent. According to the comparison between the means of importance and satisfaction for internal customers, this aspect should be strengthened.
4. The security of the rooms is very important for internal customers as evidenced by 78% of the surveys, 14% consider it important, and 8% is indifferent. Internal customers consider that the majority of external customers find the security of the room satisfactory, granting 49% and 43% consider it very satisfactory, while only 4% consider it unsatisfactory and 4% indifferent. According to the comparison between the means of importance and satisfaction for internal customers, this aspect should be strengthened.
5. In the area of food and beverages, the variety of the offer is very important for 48% of internal customers, important for 32% of them, while 3% considers it unimportant, 7% very little important and 9% is indifferent to this aspect. While they think that 17% of external customers rate this aspect as very satisfactory, 53% as satisfactory, 17% are indifferent, 4% consider it unsatisfactory and 9%

very unsatisfactory. According to the comparison of the means of importance and satisfaction for internal customers, this aspect should be strengthened.

6. The acquisition schedule of these gastronomic offers is very important for 58% of internal customers, 34% also consider it important, 7% are indifferent to this element and only 1% consider it very unimportant. They also consider that 27% of external customers are very satisfied with the acquisition schedule of these offers, 56% satisfactory, 13% are indifferent, and only 3% think they feel little satisfaction and 1% very little satisfaction. According to the comparison of the means of importance and satisfaction for internal customers, this aspect should be strengthened.
7. The presence of employees is considered very important for 62%, important for 24%, unimportant for 3% and indifferent for 11%. Likewise, 60% believe that external customers value the presence of employees as highly satisfactory, 25% satisfactory, only 3% find it unsatisfactory and 12% remain indifferent to this aspect. According to the comparison of the means of importance and satisfaction for internal customers, this aspect should be strengthened.
8. Recreational options are representative within a hotel, since this is usually frequented by vacationers, who seek to have fun, but this facility wants to compensate for their lack of variety in these offers with personalized attention and other offers that make their stay more enjoyable. external customers In this sense, 24% of internal customers consider the variety of offers very important, 32% important, 12% unimportant and 22% remain indifferent. They also believe that 9% of external customers find the variety of this offer very satisfactory, 34% satisfactory, while 36% consider it unsatisfactory, 7% very unsatisfactory, and 14% remain indifferent in The incidence of this aspect. According to the comparison of the means of importance and satisfaction for internal customers, this aspect should be strengthened.
9. If the variety of recreation options is important, if it is the time at which they can be accessed, therefore 32% of internal customers consider it very important, 60% important, 7% It remains indifferent and only 1% declares it unimportant. For internal customers, 21% of external customers consider them very satisfactory, 47% satisfactory, 12% indifferent, and 14% and 6% unsatisfied and dissatisfied respectively. According to the comparison of the means of importance and satisfaction for internal customers, this aspect should be strengthened.
10. The conditions of the leisure or recreation areas are very important for 57% of internal clients, important for 27% of them, unimportant for 3% and indifferent for 13% of them. According to their criteria, 15% of external customers feel very satisfied with these conditions, 71% are satisfied, 12% remain indifferent to these conditions and only 1% in each case are not satisfied or very dissatisfied. According to the comparison of the means of importance and satisfaction for internal customers, this aspect should be strengthened.

Table 3 shows a summary of the above.

Analysis of main components of the results of the questionnaire applied to internal clients.

Table 3. Comparison between the average importance and satisfaction of internal customers.

Attributes	Importance	Satisfaction	Assessment
Room cleaning	4,75	4,13	Strengthen
Room comfort	4,75	4,16	Strengthen
Complementary services	4,51	4,15	Strengthen
Room security	4,70	4,31	Strengthen
Variety of offer	4,12	3,65	Strengthen
Acquisition schedule	4,48	4,05	Strengthen
Presence of employees	4,45	4,42	Strengthen
Professionalism	4,47	4,46	Keep
Variety of recreation	3,88	3,02	Strengthen
Schedule of recreation	4,23	3,63	Strengthen
Conditions of the areas	4,38	3,98	Strengthen

A factor analysis attempts to identify underlying factors or variables that explain the correlations established between the set of observed variables. It is necessary to analyze the different attributes studied in a multivariate way.

- KMO test

The Kaiser-Meyer-Olkin test is a measure of sample adequacy that contrasts whether the partial correlations between the variables are small. For the results of this test to be acceptable, very good or excellent, they must be above 0.7. The results for internal customers are shown in Table 4.

Table 4. KMO test for the attributes of the questionnaire to internal clients.

Appearance	KMO	Qualification
Importance level	0.647	Not acceptable
Satisfaction level	0.652	Not acceptable

The test results for both types of attributes do not reach acceptable values for this type of analysis. In the same way it happens with the accumulated percentage of explained variance, which does not reach the required 80%.

- Bartlett's sphericity test

Bartlett's test of sphericity contrasts whether the correlation matrix is an identity matrix. The results are shown in the following table (Table 5).

Table 5. Bartlett's sphericity test.

Appearance	Value
Importance level	0
Satisfaction level	0

Although the KMO test does not obtain acceptable values, the Bartlett Test does, which determines that the correlation matrix does not match the identity matrix, this means that there is a correlation between the variables. Once these results are obtained, the phase of extracting new factors continues, so that the grouping of original variables in the new components is obtained (Table 6).

Table 6. Contribution of the variables in the model taking into account their importance and valuation (communalities).

Attributes	Initial	Extraction importance	Extraction satisfaction
Cleaning	1,000	,690	,642
Comfort	1,000	,736	,763
Complementary services	1,000	,707	,581
Room security	1,000	,705	,775
Variety of offer	1,000	,797	,696
Acquisition schedule	1,000	,688	,818
Presence of employees	1,000	,797	,778
Professionalism	1,000	,845	,533
Variety of recreation	1,000	,510	,813
Schedule of recreation	1,000	,364	,676
Conditions of the areas	1,000	,697	,510

This analysis yields as a result of the attributes of importance for internal customers, the most relevant is the professionalism of the employees, followed by the variety of the gastronomic products offer and the presence of the employees. It can be said that the rest of the variables are of high relevance for the model, except for the variety of recreation offers and the recreation schedule. This means that more than 50% of the variables are explained through the model.

In the same way, the valuation attributes in the model have an important importance since they mostly exceed 50% of the contribution, except the recreation time. Among the valuation attributes that most contribute to the model are the professionalism of the employees and in the same way they also contribute the acquisition schedule of the gastronomic offers and the presence of the employees.

Taking into account the total variance explained, it was possible to determine the number of factors in which the original attributes were grouped. This method aims to obtain a minimum of components that describe the total explained variability, hence, in reducing the dimensions, it must be met that the new number of variables is lower than the previous one. The values derived from this analysis can then be verified, in the following table (Table 7):

Table 7. Number of factors and total variability value explained by type of attribute.

Appearance	Number of factors	% of variance
Importance	3	68.52%
Satisfaction	4	67.11%

These results show that, for the attributes of importance the rotation was reduced to three main factors and for the valuation to four main factors; which are a linear combination of the rest of the attributes. In both cases the factors explain the total variation in more than 50%.

- Main components for the attributes of the questionnaire applied to internal customers

Next, the main components are analyzed from the model matrix, obtained as a result of the rotation for each attribute of importance and valuation. The elements of the matrix represent the correlation coefficient between each attribute and the components. The nomenclature used for each factor corresponds to the attributes included in them, which is shown in Table 8 and Table 9.

The attributes found in A1 are those related to the overall satisfaction that customers obtain. They are fundamental for the provision of the service and for the person who receives it. The conditions of the facilities, A2, are those factors that are visible and perceived by the customers at the time of obtaining the product and interaction with the staff. The complementary services are those that are complementing the attractiveness of the products but are not directly reflected in the A3.

The main components for the assessment aspect shown in Table 9 are shown below.

The attributes found in A1 are those related to the basic requirements of the service. They are those that confer an aggregate, to provide a better service. The variety of services includes those factors that offer diversity, are tangible factors, and offer options for customers. The conditions of the services are those factors visible and perceived by the client upon receiving their services. The safety factor of the room allows the client to have confidence in the hotel.

- Diagnosis of the survey results for external clients

The questionnaire aimed at external customers was applied to 140 customers who visited the facility in February 2019 to learn about the factors that affect the quality of service and customer satisfaction. The importance of this study lies in the fact that

Table 8. Grouping of the main components for the importance conferred by internal clients to each of the attributes.

Variable	Correlation	Component	Name
Complementary services	0.675	1	General satisfaction
Variety of offer	0.743		
Acquisition schedule	0.668		
Presence of employees	0.879		
Professionalism of employees	0.917		
Cleaning	0.808	2	Room conditions
Comfort	0.820		
Acquisition schedule	0.558		
Condition of recreation areas	0.832		
Room security	0.815	3	Complementary services
Variety recreation	0.593		

Table 9. Grouping of the main components for valuation that internal clients confer to each of the attributes of the questionnaire.

Variable	Correlation	Component	Name
Acquisition schedule	0,787	1	Attractiveness of services
Presence of employees	0,797		
Professionalism of employees	0,606		
Schedule of recreation	0,611		
Condition of recreation areas	0,719		
Variety of offer	0,758	2	Variety of services
Variety recreation	0,823		
Cleaning	0,665	3	Conditions of services
Comfort	0,884		
Complementary services	0,609		
Room security	0,871	4	Room security

a satisfied customer ignores the competition and usually acquires the service again; communicating to other potential clients their positive experiences.

For the analysis of the results, descriptive statistics were taken into account, which was used to describe the data through the frequency distribution. In addition, we worked with inferential statistics through the mean and the median, in order to generalize the data obtained from the sample.

The following results were obtained from the imposition of surveys on external clients:

1. The cleanliness of the rooms is unimportant for 7%, indifferent for 4.2% and is important and very important for 49 and 44.1% respectively. Likewise, 23.1% are dissatisfied, 14.7% indifferent, and 45% and 41% satisfied and very satisfied respectively. In a comparison of the means, it can be established that the majority of customers think that the cleanliness of the rooms is important in the quality of the installation and at the same time most of them feel satisfied about it, so the conditions of this aspect are They must strengthen.
2. The comfort in the rooms is indifferent for 2.8%, important for 44.8% and very important for 50.3%. Regarding the level of satisfaction, 33.6% consider it indifferent, 37.1% satisfied and 27.3% very satisfied. In a comparison of the averages it can be established that most of the clients think that comfort in the rooms is very important and at the same time most of them feel satisfied with this aspect, so the conditions with respect to it in The entity must be strengthened.
3. Complementary services within the rooms are very unimportant for 6.3%, unimportant for 21.7%, indifferent for 43.4% and important for 22.4% of external customers. In terms of satisfaction, 18.7% are dissatisfied, 24.5% indifferent, 35% satisfied and 20.3% very satisfied. Comparing their means, it can be seen that this element is indifferent in terms of importance for most customers and at the same time they feel satisfied with its quality, so that the condition of this element must be maintained.
4. 11.2% of customers consider the security of the rooms important and 86.7% very important. They feel satisfied and very satisfied 45.5 and 42% respectively, while 10.5% declare themselves indifferent. Comparing their means, it can be seen that it is very important for most respondents and most of them feel satisfied, so this aspect should be strengthened.
5. Regarding the importance of the variety in the gastronomic offer offered by the facility, 49% is indifferent, 36.4% considers it important and 12.6% very important. 56% feel satisfied with the gastronomic offer, 37.1% consider themselves satisfied and 16.8 declare themselves indifferent. When comparing their stockings, it can be seen that customers give medium importance to the variety in the offer, while they are satisfied with that offer, so this aspect must be maintained.
6. 80.4% of customers consider the acquisition schedule of gastronomic offers important and 16.8% very important, while 7% remain indifferent. 59.4% feel satisfied with this schedule and 38.5, very satisfied. When comparing their stockings, it can be seen that the majority of external customers think that this element is important and they feel satisfied with it, so their conditions must be maintained.
7. 67.8% consider the presence of employees important, 26.6% consider it very important and 3.5% remain indifferent. 59.4% of customers feel satisfied and 44.8% very satisfied. When comparing their averages, it can be seen that the presence of employees is considered important for the majority of customers and at the same time they feel satisfied with it, by which this element can be maintained.
8. 36.4% of customers consider the professionalism of employees important, 31.5% very important and 30.1% indifferent. At the same time 30.8% feel satisfied with this service, 66.4% very satisfied and 7% consider themselves indifferent. When

comparing their stockings, it can be seen that the professionalism of the employees is important for the majority of the clients and at the same time they feel satisfied with this aspect, so their conditions must be maintained.

9. 53% consider the variety of recreation very important, 38.5% consider it important, 7% unimportant and 21.7% are indifferent. 84.6% feel satisfied with this aspect, and 12.6% feel very satisfied, while 7% declare themselves indifferent. When comparing their stockings, it can be appreciated that the variety of recreation is important for clients and at the same time they feel satisfied with it, so they must strengthen the conditions in this aspect.
10. The recreation schedule is very important for 43.4% of clients, important for 37.8% and indifferent for 16.8%. 49% are satisfied, 48.3% very satisfied and 7% indifferent. When comparing their averages, it can be appreciated that the recreation schedule is important for the clients and at the same time they feel satisfied with said schedule, therefore, these conditions must be maintained.
11. 40.6% of customers consider the conditions of recreation areas very important, 50% important and 7% indifferent. 42% of customers feel very satisfied with these conditions, 39.9% satisfied, and 16.1% are indifferent. When comparing their averages, it can be seen that the conditions of the areas are important for the majority of the clients and at the same time they feel satisfied with these conditions, therefore, this element must be strengthened.

The following table shows a synthesis of the analysis previously performed (Table 10).

Table 10. Comparison between the median satisfaction and importance of the attributes of the questionnaire applied to external clients

Attributes	Importance	Satisfaction	Assessment
Room cleaning	4,39	3,67	Strengthen
Room comfort	4,49	3,94	Strengthen
Complementary services	2,96	3,59	Keep
Room security	4,89	4,32	Strengthen
Variety of offer	3,63	4,11	Keep
Acquisition schedule	4,16	4,39	Keep
Presence of employees	4,24	4,46	Keep
Professionalism	4,01	4,67	Keep
Variety of recreation	4,14	4,12	Strengthen
Schedule of recreation	4,27	4,49	Keep
Conditions of the areas	4,34	4,26	Strengthen

- Customer loyalty index.

You need to know what your hotel insurance customers are. For the analysis, the results of the questions were crossed: how do you feel about the service received in general? Would you choose us again? And yes, would you recommend our products?

The analysis was performed through a contingency table obtained from the output of the statistic SPSS (Table 11).

Table 11. Cross tabulation.

I would return to our entity			recommendations of our services: yes	recommendations of our services: no	Total
yes	Overall satisfaction with the service received	indifferent	15	2	17
		satisfied	54	3	57
		very satisfied	59	6	65
	Total		128	11	139
no	Overall satisfaction with the service received	satisfied		1	1
	Total			1	1
Total	Overall satisfaction with the service received	indifferent	15	2	17
		satisfied	54	4	58
		very satisfied	59	6	65
	Total		128	12	140

For secure customers, the following responses are taken into account (Fig. 4):

42.14% of the customers surveyed are classified as safe customers.
38.57% are classified as favorable customers.
10.71% are classified as vulnerable customers.

The company must take measures to improve its loyalty index since only 42.14% is considered safe customers.

- Analysis of satisfaction means of external and internal customers

This analysis is established by comparing the means of satisfaction issued by internal and external customers. In the following table it can be observed that the cleanliness of

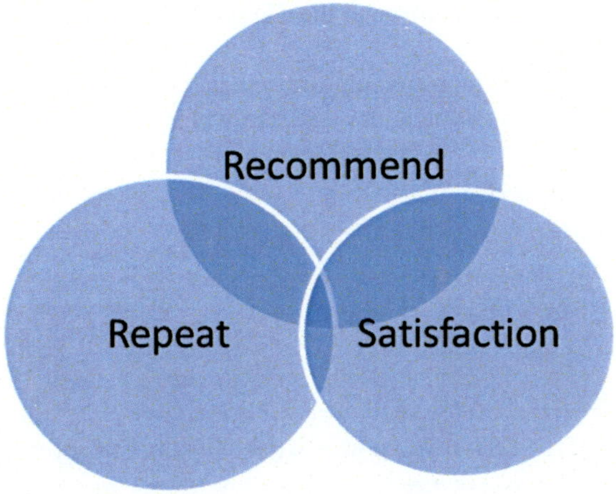

Customer loyalty index

Fig. 4. Customer loyalty

the rooms, the comfort of the rooms and the complementary services in them should be strengthened, while the rest of the aspects must maintain their conditions (Table 12).

Table 12. Comparison between the average satisfaction of external and internal customers.

Attributes	ICS	ECS	Assessment
Room cleaning	4,13	3,67	Strengthen
Room comfort	4,16	3,94	Strengthen
Complementary services	4,15	3,59	Strengthen
Room security	4,31	4,32	Keep
Variety of offer	3,65	4,11	Keep
Acquisition schedule	4,05	4,39	Keep
Presence of employees	4,42	4,46	Keep
Professionalism	4,46	4,67	Keep
Variety of recreation	3,02	4,12	Keep
Schedule of recreation	3,63	4,49	Keep
Conditions of the areas	3,98	4,26	Keep

In general, internal clients agree with the answers given by external clients, which demonstrates that they succeed in estimating the satisfaction that external clients perceive with the service provided by them, indicating them as affected aspects the cleanliness of the room, the comfort of the room and complementary services, with the aim of strengthening them.

- Validation by expert criteria through the Theory of Forgotten Effects (TEO)

Six experts on quality management are selected, four of them academics and two specialists from the organization under study, to which the questionnaire that appears in it is applied (Fig. 5).

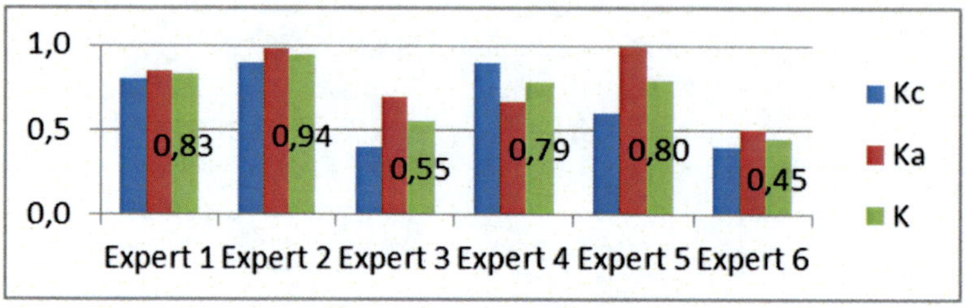

Fig. 5. Expertise coefficient of experts

When processing the data, it was obtained that of the total of respondents, four have a high competence coefficient. Two of the total respondents are eliminated from the analysis because they have competence coefficients on the medium and low subject.

The questionnaire is applied to experts with a high coefficient of competence, in order to analyze whether there are hidden relationships between the variables included in the study. As effects are taken the four main components thrown by the study of satisfaction that internal customers and as causes the 11 attributes. First, the direct incidences between the causes and the effects were sought, then the incidences of the causes on themselves and finally the effects are crossed.

In the incidence cause cause table, of the total of 44 incidence, 14 stood out as very strong, of which 16, 21, 25, 28, 58 and 41, have the highest incidence in all attributes. The incidence table between different causes showed that, of 110 non-unit incidents, they were determined as strong 52. The effect incidence table shows how effect 1 has a strong incidence on effect 2, 3 and 4, effect 2 has a very strong impact on 3, and effect 4 has a strong impact on 1.

With the aim of analyzing the causes of forgotten effects, when applying this theory, the program showed that there are no forgotten effects (Table 13).

Figure 6 shows how the cleanliness of the room has a very strong incidence of 0.9 on the basic requirements of the service. It also has a strong impact on the conditions of the services and also has a strong impact on the security of the room.

Figure 7 shows how the variety of the offer has a very strong impact on the variety of services, while the influence of other variables indirectly makes the variety of the offer have a very strong impact on the schedule of the service. recreation, which in turn has the greatest impact on the variety of services.

Figure 8 shows how the acquisition schedule has the highest impact on the basic requirements of the service. It also strongly influences the variety of services and also the security of the room.

Table 13. Table of forgotten effects.

	E_1	E_2	E_3	E_4
C_1	0	0	0	0
C_2	0	0	0	0
C_3	0	0	0	0
C_4	0	0	0	0
C_5	0	0	0	0
C_6	0	0	0	0
C_7	0	0	0	0
C_8	0	0	0	0
C_9	0	0	0	0
C_{10}	0	0	0	0
C_{11}	0	0	0	0

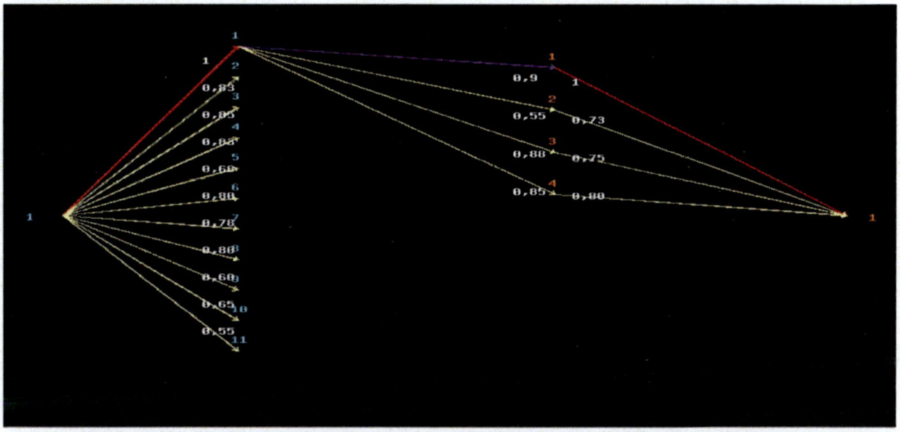

Fig. 6. Variation of incidence (forgotten effect) between the cause cleaning the room and the effect basic service requirements

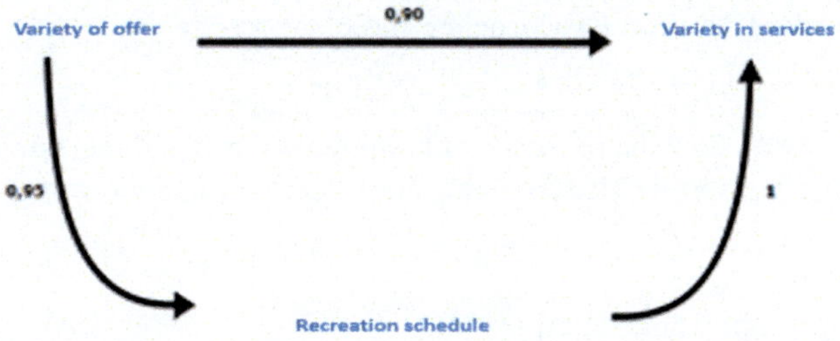

Fig. 7. Graph of the variation of incidence (forgotten effect) between the cause Variety of the offer and the effect Variety in the services

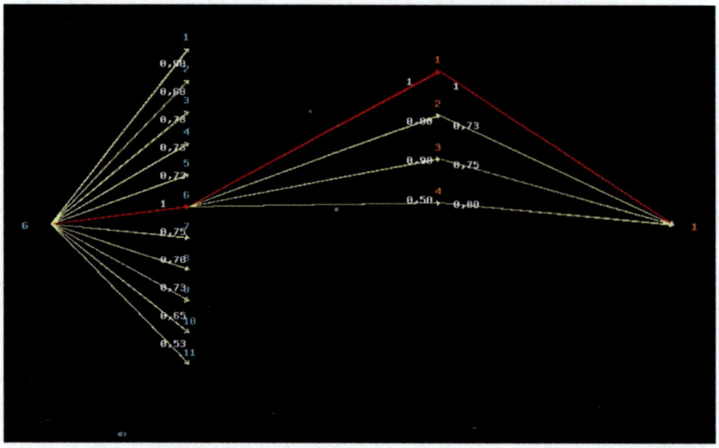

Fig. 8. Graph of the variation of incidence (forgotten effect) between the cause Acquisition schedule and the effect Basic service requirements

Figure 9 shows how the professionalism of the employees has the highest impact on the basic requirements of the service. It also strongly influences the variety of services and the conditions of services.

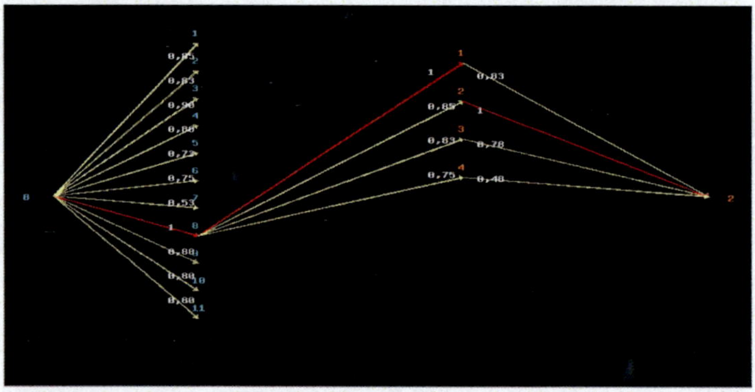

Fig. 9. Graph of the variation of incidence (forgotten effect) between the cause Professionalism of the employees and the effect Basic requirements of the service

Figure 10 shows how the presence of employees has the highest impact on the basic requirements of the service. It also has a considerable impact on the conditions of the services.

- Proposals for improvement actions

Arriving at the last stage of the procedure and taking into account all the previous analysis in conjunction with the Ishikawa diagram. This causal diagram is the graphical

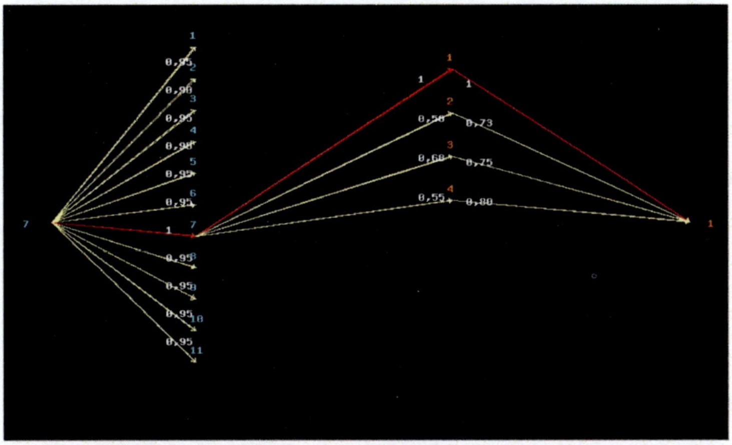

Fig. 10. Graph of the variation of incidence (forgotten effect) between the cause Presence of the employees and the effect Basic requirements of the service.

representation of the multiple cause-effect relationships between the various variables involved in a process. Therefore, we propose proposals for improvement actions in the institution based on the results of said analysis.

Room Security:

- Deterioration of the rooms: The rooms have different standards, which are not met in all cases. Effects can be observed especially in the bathrooms. For this, it is proposed to allocate part of the annual budget that the hotel consists of, through a well-justified investment plan, to gradually restore (by floors), all the rooms.
- Detriment to complementary services in the rooms: In some rooms you can find old models of televisions, mini-bars and safe deposit boxes, some of which are in malfunction. It is proposed to renew all these articles, if not possible, replace at least those that have difficulties in operation.

Conditions of services:

- Malfunction of the elevators: The elevators of the hotel are small and old, making it difficult for customers to travel. For this, the expansion and modernization of the same is proposed.
- Lack of service ladder: This aspect directly affects the service, forcing workers to travel along the same ladder as customers. For this, it is proposed to install an elevator only for the service, since this does not affect the structure of the hotel.

Variety of services:

- Product instability: Due to the current situation facing the country, it is difficult to keep a product on offer. For this, a rethinking of the position before the suppliers is proposed, demanding the respect of the agreed deadlines; In addition to a conscious preparation of orders to these suppliers.
- Low quality of the inputs: A fundamental part of the service are the inputs used for them, which sometimes do not meet the established requirements, so it is proposed to have greater rigor in terms of the entry of these products to the hotel, and demand from the suppliers the established quality.
- Time limitations: Customers are affected because the hotel does not have a 24-h gastronomic offer, which forces them to purchase it outside the hotel, sometimes late at night. To this end, it is proposed to establish an area of the hotel that maintains a light letter throughout the day capable of satisfying the needs of the clients.

Attractive services:

- It does not have a night center: The hotel does not have a soundproof area, therefore, the actions are carried out on the hotel's external terrace, this means that it has a restricted schedule since they cause discomfort to multi-family residential buildings that They border the hotel. For this purpose, the adaptation of an area of the hotel to this type of activities is proposed, which complies with the pertinent regulations to support a varied recreational offer with a longer schedule.
- Little variety of recreation offers during the day: The hotel does not present options for those clients who choose to enjoy a vacation on the premises, since during the day the only option is the pool and this does not have any recreation staff. For this purpose, the hiring of animation staff is proposed, and at the same time the adaptation of an area near the pool for their work.
- Insufficient provision of tools for recreation: The offer for renting books, board games, etc. is not in the facility, so it is proposed to the facility to purchase this type of product to offer them to customers.

Nonconformity of workers

- Lack of freedom of decision: Workers claim that they cannot make decisions concerning their job, which directly affects customer service. To this end, it is proposed to hold staff meetings and trainings that provide workers with the necessary tools to solve the problems that the client may present, in addition to considering their criteria in the formation of new measures, either in the form of surveys or of staff meetings.
- Inefficient working conditions: Many work teams are obsolete, sometimes they do not have all the cleaning tools, the lack of a service ladder, among other things they slow down and encrude the work. For this, it is proposed to modernize both equipment and tools so that they have the best conditions for the correct performance of their work.

5 Conclusions

Once the investigation was completed, they reached the following conclusions:

1. Through the procedure developed, it was possible to assess the importance and satisfaction of external customers with the attributes that describe and characterize the service. What could be corroborated with the surveys applied to the experts.
2. The surveys also showed, through the comparison of the means, between the results obtained in the items evaluated by internal clients and external clients, that the aspect of the professionalism of the workers must be maintained, while the rest of the aspects evaluated should be strengthened.
3. When making a comparison between the Satisfaction of the attributes raised; by internal customers from the perspective of external customers and the real ones, it can be seen that the differences are minimal, and when assessing overall satisfaction, the two parties turn out to be similar.
4. The application of the Theory of Forgotten Effects, allows to see the level of relations between dependent and independent variables. Demonstrating that influencing a single component can improve various aspects.
5. The main improvement actions are aimed at improving the structure and organization of the hotel, as well as incorporating activities into the management.

References

Bolea, L.G., Calvo, M.Á., Zapata, M.Á.: Guía para la medición directa de la satisfacción de los clientes. Instituto Andaluz de Tecnología, Andalucía (2007)

Blanco-Mesa, F., Gil-Lafuente, A.M., Merigó, J.M.: Subjective stakeholder dynamics relationships treatment: a methodological approach using fuzzy decision making. Comput. Math. Organ. Theory **24**(4), 441–472 (2018)

Blanco-Mesa, F., Merigó, J.M., Gil-Lafuente, A.M.: Fuzzy decision making: a bibliometric-based review. J. Intell. Fuzzy Syst. **32**(3), 2033–2050 (2017)

García, R.: Procedimiento para la selección de los mercados internacionales de los servicios de gestión medioambiental cubanos. Tesis Doctoral, Facultad de Economía, Universidad de La Habana, Cuba (2010)

Hayes, B.E.: Cómo medir la satisfacción del cliente, desarrollo y utilización de cuestionarios. Gestión 2000. 2da Edición, España (1999)

Kaufmann, A., Gil Aluja, J.: Modelo para la investigación de efectos olvidados. Miladoiro, España (1988)

Menéndez, J.R.: Evolución del concepto de Calidad. Revista Española de Transplante **10**(3), 169–175 (2010)

Mulet-Forteza, C., Horrach, P., Socias, A. Merigo, J.M.: The forgotten effects: an application in the social economy of companies of the Balearic Islands, Econ. Comput. Econ. Cybern. Stud. Res. **52**(3), 147–160 (2018)

Palacios, J.L.: Estrategias de ponderación de la respuestas en encuestas de satisfacción de usuarios de servicios. Metodologías De Encuestas **4**(2), 175–193 (2002)

Stefano, N.M., Casarotto, N., Pentiado, L.: Medición de la calidad percibida en organización de servicio utilizando índice potencial de aumento de la satisfacción del consumidor o modelo de 5 gaps. Sistemas Gestión **5**(1), 17–31 (2010)

The Internationalization in the Learning of the Accounting Profession: Use of the Kahoot as a Teaching Tool for Learning

M. Pilar Curós Vilá[1](✉), Elisa Isabel Cano Montero[2], Julián Chamizo González[3], and Jordi Martí Pidelaserra[1]

[1] University of Barcelona, 08034 Barcelona, Spain
{pilar.curos,2marti2}@ub.edu
[2] University of Castilla- La Mancha, 45016 Talavera de la Reina, Spain
Elisaisabel.cano@uclm.es
[3] Autonomous University of Madrid, 28049 Madrid, Spain
Julian.chamizo@uam.es

Abstract. Business organizations, as well as society, evolve based on various inputs, whether due to technological advances, cultural changes or the incidence of the environmental situation. In the case of the accounting profession, it is usual to face legislation, documentation, reports, ratios, databases, etc. in other languages. Our students must be familiar with the different environments that allow them to consult any type of information, such as balance sheets, income statements, among other national and international companies. Information and Communication Technologies (ICTs) have provided an opportunity as a means of innovation and motivation in university education. Specifically, gamification in learning contributes to improving attention, concentration, complex thinking, strategic planning or problem solving. An example of this is the use of Kahoot. This work gathers the experience of applying gamification in the classroom to students from different backgrounds, to improve teaching-learning, specifically focusing on the use of kahoot as a learning tool in accounting for national and international students. The objective that we set ourselves is, first of all, to characterize the students who participate in different accounting classes at the University of Barcelona, and if there are significant differences in participation in the accounting subject between national and international students and, on the other hand, to identify if there are differences by origin of country and gender, and finally show whether the use of kahoot produces a difference in the learning outcomes between students of different origins and gender.

Keywords: Internationalization · Teaching learning · Gamification accounting

1 Introduction

Business organizations, as well as society, evolve based on various inputs, whether due to technological advances, cultural changes or the incidence of the environmental situation.

A. M. Gil-Lafuente et al. (Eds.): REDCID 2019, LNNS 388, pp. 178–193, 2022.
https://doi.org/10.1007/978-3-030-93787-4_10

This is reflected in the accounting profession. On the one hand, they have to use more and more sophisticated tools, computer programs, and on the other, be aware of the continuous changes in legislation, as well as access documentation, reports, ratios and databases of an international nature, typical of a globalized market. In this sense, our students must be familiar with the different environments that allow them to consult any type of information.

It is indisputable that due to many circumstances the classroom is increasingly heterogeneous; on the one hand, migration has made it possible for second-generation emigrants to have the opportunity to access the university education system; on the other hand, exchange programs, such as Erasmus (1987), promote the mobility of students at a European level, as well as other programs at the intercontinental level and, finally, non-European students who decide to fully study in Europe.

The Socrates program (1995) advocates European cooperation in all areas of education. This takes different forms: mobility (moving around Europe) and the elaboration of common projects, the creation of European networks (dissemination of ideas and good practices) or the carrying out of comparative studies and analysis.

The type of students has changed and the university must be prepared for this new challenge of global change, in fact, the subjects taught in English are more and more numerous. According to Urquía and del Campo (2016) the students of the Erasmus programs who study abroad understand and develop skills to study and work in a culturally diverse and different environment. Therefore, they perceive that they develop sociocultural competences and feel more international and European in line with Campbell (2011), Bennet et al. (2013) and Leask (2009).

Regardless of the destination country, according to Knight (2004) and Willis (2010) the academic, cultural and integration experiences in the host universities determine the general satisfaction perception of Erasmus students in line with Rienties et al. (2011) The least valuable countries are Austria, Finland, Germany and Sweden, probably due to their high cost of living.

In this framework, Information and Communication Technologies (ICTs) have provided an opportunity as a means of innovation and motivation in university education (Camacho et al. 2016; Urquía et al. 2009, Zabalza 2006). Gamification in learning (Deterding et al. 2011a, b; Kapp 2012; Domínguez et al. 2013), is a medium that has been used at different educational levels, both nationally and internationally. In the case of teaching-learning, students are involved in such a way that they improve attention, concentration, complex thinking, strategic planning or problem solving, sometimes even managing to capture the attention of the students. less aware to learn (Cano et al. 2019; Brehm and Self 1989; Holzinger and Maurer 1999; Higgins et al. 1999 Holzinger et al. 2001; Kettenanurak et al. 2001, Kirriemuir and McFarlane 2004, Kapp 2012).

Kahoot is an educational gamification tool. The use of Kahoot has been previously studied at the university level in degree programs in different areas such as physics, computer engineering or law. Del Cerro (2015) indicates that kahoot increases the class attendance rate, and motivates the student, who perceives it as a game and not as an evaluation system (Fernández et al. 2016).

This article aims to apply gamification in the classroom, as a tool for improving teaching-learning, specifically focusing on the use of kahoot as a learning tool in accounting for national and international students. The objective that we set ourselves is, first of all, to characterize the students who participate in different accounting classes at the University of Barcelona, and if there are significant differences in participation in the accounting subject between national and international students and, on the other hand, to identify if there are differences by origin of country and gender and, finally, to show whether the use of kahoot produces a difference in the learning results in students between students of different origins and gender.

For this we have asked ourselves the following questions:

Q1 Do students participate equally by degrees, gender and place of origin?
Q2: Is there a difference in the results obtained in the kahoot based on gender, degree and provenance?
Q3 Does the use of kahoot affect in the achievement of academic results?

This research includes a study on the use of gamification in higher education, specifically in accounting, as well as its influence and that of other factors on the learning outcomes of this subject according to the origin of the students. The study is carried out in the degrees of ADE (Business Administration and Management), DER (Business Laws) + ADE, ECO (Economics) of the University of Barcelona, and TUR (Tourism) of the Barcelona School of Tourism in the 2016/2017 academic year. Through the students' mobile phone and the Kahoot application with its corresponding questionnaire, the accounting contents are evaluated. The results show that there are gender differences by degree and origin.

This article is divided into the following parts: The present introduction where the bibliographic review is included, then the methodology and main results, to finish with the conclusions.

2 Methodology

The study is carried out in the academic year 2016–2017 on a total of 308 students enrolled in the accounting subject in the degrees of Business Administration and Management (ADE), double degree ADE + Law (DADE), Economics (ECO), from the University of Barcelona and the degree in Tourism (TUR), at the University School of Tourism in Barcelona.

The kahoots are carried out before the first partial exam, with 20 theoretical questions and another 10 practice questions on accounting entries, in order to know the degree of understanding of the subject explained. The participation by groups according to the origin is that presented in Table 1.

To carry out the Kahoot, a questionnaire is made, each question has four possible answers associated, only one of them being correct. The questions are projected in the classroom and the student individually through their computer, Smartphone or Tablet, have to choose the color where the correct answer is from the four colors that come out, red, yellow, blue and green and adjusting to the time marked by the teacher, specifically

Table 1. Identification and participation of the groups

Kahoot Y/N	Group				
	ADE	DADE	ECO	TUR	Total
Yes	20	12	16	18	66
Practice[a]		7	1		8
Theory[b]	7	8	12	2	29
No[c]	67	31	74	33	205
Total	**94**	**58**	**103**	**53**	**308**

[a]Students who only participated in the practical part.
[b]Students who obly participated in the theoretical part.
[c]Students who either did not participate in the game or could not be identified as player name) of the game.

20 seconds. The teacher is the one who manages the computer screen to the rhythm of the questions, this is important because he can stop to comment on the answer. Each question shows the respective winner and the points are accumulated to offer a final ranking as if it were a competition. It shows the first three in the classroom, in turn allows the teacher to have a database of the platform with the ratings reached by the players, which can be downloaded in the spreadsheet as evaluation support.

When determining the score obtained in the Kahoot test for each student, the count of the correct answers was chosen, instead of the score provided by the application, since it included additional points depending on the speed of response and of the number of correct questions followed by each student/player.

Given that some of the Research Questions (RQ) refer to the gender of the students and given the impossibility of accessing the personal data of the students, a correspondence table has been drawn up between the player's name and the list of final scores of the students enrolled in each group. Despite the fact that the students had been instructed so that their identification as a player would allow them to relate it to their real identity, in some cases this was not the case and they were considered as students who had not played Kahoot.

All the students who appeared on the list of scores were attributed the gender when it could be easily inferred from the name, sometimes resorting to the personal knowledge of their teacher, but in some cases especially of students with non-EU origins, it could not be inferring gender and therefore not assigned.

All the students were classified according to:

- From the group of belonging (Degree in which they were enrolled),
- From their participation in both Kahoot tests (Yes), in none (No), only in theory (Theory), or only in practice (Practice)
- From the score obtained in the kahoot tests (0–10, No Kahoot)
- Of the score obtained in the course (0–10)
- Gender (Male, Female, Undetermined)

- Whether the course score obtained was equal to or higher than the average for their group (0 = No; 1 = Yes)
- Whether the pass rate of your group was equal to or higher than the average of the entire sample (0 = No; 1 = Yes).
- Origin
- In relation to the origin, the classification depends on:
- From the continent of origin of the students (Africa, America, Asia, Europe and Unknown)
- From the nation of origin of the students (Germany, Andorra, Belgium, Bolivia, Brazil, Colombia, Cuba, Unknown, Ecuador, Spain, France, The Gambia, Georgia, Equatorial Guinea, Honduras, Italy, Morocco, Mexico, Pakistan, Paraguay (Peru, Portugal, Dominican Republic, Romania, Russia, Salvador, Ukraine, Venezuela, China)
- Of Spanish or Foreign nationality
- Of the Race or Ethnicity attributable to each student (Far Easterner, Arab/North African, Caucasian, Mestizo (from Latin America), Mulatto, Sub-Saharan, or Unknown)

With all the data thus related, a database was created in a spreadsheet, in which the students were anonymized by creating a unique identifier for each one, made up of an order number and grade.

All the students who appeared on the list of grades were attributed the gender when it could be easily inferred from the name, sometimes resorting to the personal knowledge of their teacher, but in some cases, especially of students with non-EU origins, it could not be possible to infer the gender and therefore not to assign it. The same was done with regard to race, nationality and origin.

Once the database was formed, it was analyzed and studied, and the results described in Sect. 3 have been found.

3 Results

Given that some of the Research Questions (RQ) refer to the gender, race or origin of the students and given the impossibility of accessing the personal data of the students, a correspondence table has been prepared between the student's name/player and the list of final grades of the students enrolled in each group. Despite the fact that the students had been instructed so that their identification as a player would allow them to relate it to their real identity, in some cases this was not the case and they were considered as students who had not played Kahoot.

The questions asked were:

Q1 Do students participate equally by degrees, gender and place of origin?
Q2: Is there a difference in the results obtained in the kahoot based on gender, degree and provenance?
Q3 Does the use of kahoot affect in the achievement of academic results?

3.1 About Students

Participation by gender has been analyzed (Table 2), based on the name of the player, we observed that of the students who make up the sample, 53.6% were male, 45.5% were female and 0.97% gender could not be determined. By groups in ADE, men represented 62% and women 38%, in DADE the majority portion was women with 67%, for only 31% of men with 1.72% pending to determine their gender. In the degree in Economics, men predominate with 74% for 26% of women, finally in the degree in Tourism, the most abundant gender is that of women with 72% for 25% of men and 3.77% pending to determine.

Table 2. Gender distribution of students

Gender	Group				
	ADE	DADE	ECO	TUR	Total
Male	61,70%	31,03%	73,79%	24,53%	53,57%
Female	38,30%	67,24%	26,21%	71,70%	45,45%
Pending	0,00%	1,72%	0,00%	3,77%	0,97%
Total	**100,00%**	**100,00%**	**100,00%**	**100,00%**	**100,00%**

Looking at this table, it follows that in all the groups the students are mostly of Spanish nationality, with the degrees in ADE and ECO being the most international with respectively 22.34% and 20.39% and the degree in TUR the least with 3.77% non-Spanish students (Table 3).

Table 3. Nationality distribution of students

Group					
Spanish/Foreigner	ADE	DADE	ECO	TUR	Total
Spanish	74,47%	82,76%	77,67%	96,23%	80,84%
Foreigner	22,34%	17,24%	20,39%	3,77%	17,53%
Unknown	3,19%	0,00%	1,94%	0,00%	1,62%
Total	**100,00%**	**100,00%**	**100,00%**	**100,00%**	**100,00%**

If we consider the attributed race of the students (Table 4) it is observed that the majority group is Caucasian with 88% followed by mestizos with 8.77%.

If we take into account the origin of the students we can see (Table 5) that the majority origin is European, followed by the American, with no representative from Oceania.

Table 4. Race distribution of students

Race	Group				
	ADE	DADE	ECO	TUR	Total
Caucasian	84,04%	87,93%	87,38%	96,23%	87,99%
Mestizo (from Latin America)	10,64%	8,62%	9,71%	3,77%	8,77%
Far East	0,00%	1,72%	1,94%	0,00%	0,97%
Unknown	3,19%	0,00%	0,00%	0,00%	0,97%
Arab/North African	2,13%	0,00%	0,00%	0,00%	0,65%
Sub Saharian	0,00%	0,00%	0,97%	0,00%	0,32%
Mulatto	0,00%	1,72%	0,00%	0,00%	0,32%
Total	**100,00%**	**100,00%**	**100,00%**	**100,00%**	**100,00%**

Table 5. Continent distribution of students

Continent	Group				
	ADE	DADE	ECO	TUR	Total
Europe	81,91%	87,93%	83,50%	98,11%	86,36%
America	12,77%	6,90%	9,71%	1,89%	8,77%
Asia	1,06%	1,72%	3,88%	0,00%	1,95%
Unknown	3,19%	0,00%	1,94%	0,00%	1,62%
Africa	1,06%	3,45%	0,97%	0,00%	1,30%
Total	**100,00%**	**100,00%**	**100,00%**	**100,00%**	**100,00%**

Table 6. Participation by gender and group in kahoot

Kahoot Y/N	Gender	Group				Total
		ADE	DADE	ECO	TUR	
Yes	Male	75,0%	25,0%	68,8%	11,1%	47,0%
	Female	25,0%	75,0%	31,3%	88,9%	53,0%
Practice	Male		71,4%	100,0%		75,0%
	Female		28,6%	0,0%		25,0%
Theory	Male	71,4%	0,0%	66,7%	100,0%	51,7%
	Female	28,6%	100,0%	33,3%	0,0%	48,3%
No	Male	56,7%	32,3%	75,7%	27,3%	55,1%
	Female	43,3%	64,5%	24,3%	66,7%	43,4%
	Pending	0,0%	3,2%	0,0%	6,1%	1,5%

Table 7. Group players

ADE		DADE		ECO		TUR	
%	Nº	%	Nº	%	Nº	%	Nº
28,72%	27	46,55%	27	28,16%	29	37,74%	20
71,28%	67	53,45%	31	71,84%	74	62,26%	33
100%	**94**	**100%**	**58**	**100%**	**103**	**100%**	**53**

3.2 About the Players

From the reading of Table 6, the Q1 can be answered since, it is found that on average among the players women predominate with 53% while in non-players men represent 55.1%., Men predominate among ADE players (75%) and among non-players (56.7%), in DADE women predominate among players (75%) and among non-players (64.5%), in ECO men predominate both Among the players (68.8%) and among the non-players (75.7%), finally in TUR, women who play Kahoot (88.9%) and those who don't play (66.7%) are mostly women.

Participation by degrees ranges from 15.53% of players to the two tests (Theory and Practice) of ECO to 33.96% of TUR.

If we add the continent of origin and group all the players who have participated in the practical, theoretical, or both, under the label of yes and non-players under no. (Table 7, Table 8).

Table 8. Players by continent

Continent	Yes		No	
	Nº	%	Nº	%
Africa	1	25,00%	3	75,00%
America	5	18,52%	22	81,48%
Asia	4	66,67%	2	33,33%
Unknown		0,00%	5	100,00%
Europe	93	34,96%	173	65,04%
Total	**103**	**33,44%**	**205**	**66,56%**

We can see that except in Asians who play 67%, the rest do not exceed 35% of players compared to the total number of students. Likewise, by race, the Far Easterners and Sub-Saharans all played and the Arabs in 50%, among the non-players, the only mulatto of the students did not play, the Caucasians did not play in 66% and the mestizos did not participate in 81% (Table 9).

Table 9. Players by race

Race	Yes		No	
Far Easterner	3	100,00%		0,00%
Arabe	1	50,00%	1	50,00%
Caucasian	93	34,32%	178	65,68%
Unknown		0,00%	3	100,00%
Mestizo	5	18,52%	22	81,48%
Mulatto		0,00%	1	100,00%
Sub-Saharan	1	100,00%		0,00%
Total	**103**	**33,44%**	**205**	**66,56%**

Considering the Spanish nationality or not of the students, of these were players as shown in Table 10, those of unknown nationality, 100% did not play Kahoot, 34.5% of Spaniards were players and while foreigners 31.5% played.

Table 10. Players by nationality

Nationality	Yes		No	
	Nº	%	Nº	%
Unknown		0,00%	5	100,00%
Spanish	86	34,54%	163	65,46%
Foreigner	17	31,48%	37	68,52%
Total	**103**	**33,44%**	**205**	**66,56%**

The answer to Q1 is therefore no that they do not participate and do not play the same by degrees, by gender, nor by continent, nationality or race.

Likewise, from Table 1 it appears that the average participation of students in the practical test was 24.03% and in practice 30.84%, the maximum participation was reached in TUR with 33.96% and the minimum in ECO with 16.5%. In Theory, the average participation was 30.84% with a maximum of 37.74% in TUR and a minimum of 27.18 in ECO. Therefore, also the answer to Q1, is no, and there is no similar participation by groups in the theory and practice tests.

If we take into account the final grade for the course and participation in some or all of the Kahoot tests (Table 4), we can highlight that the average final mark of the 66 students from all groups who have participated in both tests is higher as a whole and within their group with respect to the 205 who have not participated in any of them. Likewise, with the 21 students who have participated only in the theoretical part of the Kahoot, the same thing happens with respect to those who have not participated in any.

3.3 On the Results Obtained by the Kahoot Players

If we analyze the results obtained in the Kahoot test, considering exclusively players who have participated in one of the two tests (Theory/Practice), giving the result zero out of ten to those who have not done it.

Table 11. Results in Kahoot by gender and group

Gender	ADE	DADE	ECO	TUR	Total
Male	5,75	4,84	4,81	4,44	5,15
Female	5,57	4,05	4,31	5,56	4,78
Total	**5,70**	**4,29**	**4,66**	**5,34**	**4,97**

It is observed (Table 11) that the average score is slightly higher in men than in women, something that is repeated in all groups except TUR.

Table 12. Results in Kahoot by continent and group

Continent	ADE	DADE	ECO	TUR	Total
America	7,08	5,13			6,30
Europe	5,57	4,24	4,74	5,34	4,94
Asia	4,75	3,75	4,25		4,25
Africa			3,25		3,25
Total	**5,70**	**4,29**	**4,66**	**5,34**	**4,97**

Table 12 shows that the best scores in Kahoot are obtained by students from the American continent (6.30), followed by Europeans (4.94), Asians (4.25) and finally Africans (3.25).

The results in Kahoot in terms of the players' race (Table 13), show us similar results to the origin by continents. Being the Mestizos the ones that obtain the highest average mark in the Kahoot (6.3.) Followed by the Caucasians (4.94), Arabs/North Africans (4.75), Far Easterners (4.08) and finally the Sub-Saharans (3.25).

The best score per group corresponds to the Mestizos who study ADE (7.08) followed by the Caucasians who study ADE (5.57).

The worst results correspond to Sub-Saharans who study ECO (3.25) preceded by Far Easterners who study DADE (3.75).

Regarding the nationality of the players (Table 14), it is the Germans who study DADE those who obtain the best results (8.25), followed by the Peruvians who study ADE (7.08), the Romanians who study ECO (7.00). The worst results correspond to the Georgians who study ADE (3.00), preceded by hearing the Equatoguineans who study ECO (3.25) and the Brazilians who studied DADE (4.00).

Table 13. Results in Kahoot by race and group

Race	ADE	DADE	ECO	TUR	Total
Mestizo	7,08	5,13			6,30
Caucasian	5,57	4,24	4,74	5,34	4,94
Arab/ North Africa	4,75				4,75
Far Eastern		3,75	4,25		4,08
Sub-Saharan			3,25		3,25
Total	**5,70**	**4,29**	**4,66**	**5,34**	**4,97**

Table 14. Results of Kahoot by nationality and group

Nationality	ADE	DADE	ECO	TUR	Total
Germany		8,25			8,25
Peru	7,08				7,08
Romania			7,00		7,00
France			6,75		6,75
Italy			6,50		6,50
Equator		6,25			6,25
Spain	5,68	4,07	4,48	5,34	4,87
Pakistan	4,75				4,75
China		3,75	4,25		4,08
Brazil		4,00			4,00
Equatorial Guinea			3,25		3,25
Georgia	3,00				3,00
Total	**5,70**	**4,29**	**4,66**	**5,34**	**4,97**

Based on the data shown in Table 11, Table 12, Table 13 and Table 14, we can answer affirmatively to Q2, since in fact differences in the results obtained in Kahoot can be seen depending on gender, race, continent and nationality, as well as the group.

3.4 On the Effect of Play on Learning Outcomes

In relation to the findings and results obtained, as observed in Graph 1. The descending green points represent the final grade of the course obtained by the students who have participated in the Kahoot test, the black points represent the border between the pass and the failure, this the rating of 5 out of 10, consequently representing the red dots the result of the Kahoot test.

As can be seen in Table 15 of the 96 students who have played Kahoot in some way 92 of them (95.8%) have passed the course and 84 (87.5%) have had a better grade in the course than in the Kahoot, which graphically can be seen in Fig. 1.

Table 15. Relation between playing and improving/passing

	Improve		Pass	
Yes	84	87,5%	92	95,8%
No	12	12,5%	4	4,2%
	96	100,0%	96	100,0%

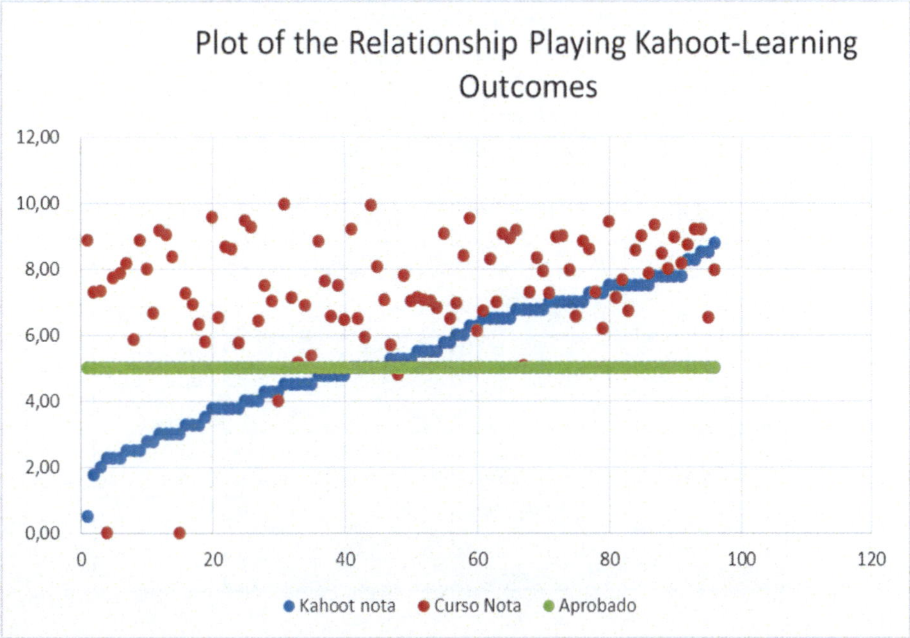

Fig. 1. Relationship between playing Kahoot and learning outcomes. Blue points: Kahoot scores. Orange points: Course scores. Grey points: Pass score.

3.5 Effect of Other Factors on Learning Outcomes

Standardized Grade			
Rhoofe Spearman	Pass	Correlation coefficient	,209**
		Sig. (bilateral)	0,000
		N	308
	Play	Correlation coefficient	,461**
		Sig. (bilateral)	0,000
		N	308
	Gender	Correlation coefficient	-,169**
		Sig. (bilateral)	0,003
		N	308
	Spanish	Correlation coefficient	0,048322862
		Sig. (bilateral)	0,398
		N	308
	Continent	Correlation coefficient	-0,070043363
		Sig. (bilateral)	0,220
		N	308
	Race	Correlation coefficient	-0,049552834
		Sig. (bilateral)	0,386
		N	308
	Nationality	Correlation coefficient	-0,076876837
		Sig. (bilateral)	0,178
		N	308
	Kahoot nota	Correlation coefficien	,461**
		Sig. (bilateral)	0,000
		N	308
	Standardized Grade	Correlation coefficient	1
		Sig. (bilateral)	
		N	

4 Conclusions

The use of the gamification tool allows us to make quick studies on different aspects, in addition to providing an evaluation component of the subject's own learning. The analysis of the findings and results obtained has made it possible to answer the research questions asked and to leave future lines of research open.

When analyzing participation by gender, it has been observed that this depends more on the degree studied by the student than on the use of Kahoot in the teaching-learning process, since in degrees where there is a prevalent gender by number, this is usually the one that participates the most in the gamification. It is also observed that the use of the Kahoot influences the improvement in the learning results of all the groups, since there are differences by gender, in all the groups it happens that the group of students who have participated in the gamification obtain better results (from course) than those who have not.

In relation to learning outcomes, it should be added that carrying out tests of this type have a triple task, on the one hand towards teaching and the teacher, in that it facilitates continuous evaluation, since its execution is fast and immediate correction This allows it to be carried out once each content block has been completed, and to be certain that the planned objectives have been achieved. On the other, towards the student, related to motivation, in that he/she seeks success and recognition of the group through being the best in the game and therefore in learning. And finally, in relation to obtaining the learning results, since to obtain good results in the game they must keep the subject up-to-date, this also reflects on the final grades of the subject.

Therefore, we can say that the gamification experience with Kahoot facilitates the student's feedback and monitoring procedures in an easy and real-time way, that is, it allows the direct connection between student and teacher in addition to providing indicators for the evaluation of the Learning Outcomes (LO). This direct connection with the evolution of student learning, and the ease of its registration, make it possible to collect the observed needs and the results obtained, facilitating the development of more effective learning activities. Students are motivated, so involved that the game leads them to learning without realizing it. This motivational effect does not occur with other similar applications such as Plickers or Socrative, which although it allows the immediacy of evaluation and feedback, does not include the competitive playful element that makes it so interesting for students.

As future lines of research, the possibility of categorizing the questions raised in relation to the topics or units in which they are integrated is considered, by their importance, to verify if the success or failure of certain questions considered key has or not influence on the grade obtained in the gamification itself and in the course. For them, it is expected to have a larger sample of students to include different gamification applications with different tools and to show if they contribute differently to obtaining the learning outcomes.

References

Brehm, J.W., Self, E.A.: The intensity of motivation. Annu. Rev. Psychol. **40**, 109–131 (1989)

Camacho Miñano, M.M., Urquía Grande, E., Rivero Menéndez, M.J., Pascual Ezama, D.: Recursos multimedia para el aprendizaje de Contabilidad Financiera en los grados bilingües. Educación XX1 **19**(1), 63–89 (2016). https://doi.org/10.5944/educxx1.15578

Cano, E.I., Chamizo, J., Valiente, C.: Los resultados de aprendizaje como indicador de mejora continua y herramienta de gobernanza. Los estudios de Contabilidad y Finanzas en la London South Bank University (Reino Unido) Revista de Formación e Innovación Educativa Universitaria **7**(1), 59–74 (2014). En http://webs.uvigo.es/refiedu/

Cano, E., Martín, T., et al.: Experiencia de coordinación docente con entornos virtuales en el Grado de Administración y Dirección de Empresas en la Facultad de Ciencias Sociales de Talavera de la Reina., Revista de estudios económicos y Empresariales. no. 22, Plasencia (Cáceres), Ed. Centro Universitario de Plasencia, pp 137–147 (2010). ISSN 0212-7237

Chamizo, J., Cano, E., et al.: La plataforma virtual como instrumento de coordinación y comunicación, el nexo entre profesorado y estudiantes. El caso del Grado en Administración y Dirección de Empresa en la Facultad de Ciencias Sociales de Talavera. Revista de Formación e Innovación Educativa Universitaria **6**(1) (2013)

Curós, P., et al.: Gamificación en la universidad. Aplicación en la asignatura de contabilidad. In: XXIV congreso Internacional de aprendizaje, 19–21 July 2017. Honolulu, USA (2017)

Del Cerro, G.: Aprender jugando, resolviendo: diseñando experiencias positivas de aprendizaje. XII Jornadas Internacionales de Innovación Universitaria Educar para transformar: Aprendizaje experiencial (2015)

De-Marcos, L., Domínguez, A., Saenz-de-Navarrete, J., Pagés, C.: An empirical study comparing gamification and social networking on e-learning. Comput. Educ. **75**, 82–91 (2014). https://doi.org/10.1016/j.compedu.2014.01.012

Deterding, D., Sicart, M., Nacke, L., O'Hara, K., Dixon, D.: Gamification: using game design elements in non-gaming contexts. In: CHI 2011 Extended Abstracts on Human Factors in Computing Systems, pp. 2425–2428 (2011a). https://doi.org/10.1145/1979742.1979575

Deterding, S., Dixon, D., Khaled, R., Nacke, L.: From game design elements to gamefulness: defining gamification. In: Proceedings of the 15th International Academic MindTrek Conference: Envisioning Future Media Environments, pp. 9–15. ACM (2011b). https://doi.org/10.1145/2181037.2181040

Domínguez, A., Saenz-de-Navarrete, J., De-Marcos, L., Fernández-Sanz, L., Pagés, C, Martínez-Herráiz, J.: Gamifying learning experiences: practical implications and outcomes. Education. 380–392 (2013). https://doi.org/10.1016/j.compedu.2012.12.020

González, X.A., Hevia, I.: El practicum de la licenciatura de pedagogía: estudio empírico desde la perspectiva del alumnado. Rev. Educ. **354**, 209–236 (2011)

Grifoll, J., Hopbach, A., Kekäläinen, H., Lugano, N., Rozsnyai, H., Shopov, T.: Quality procedures in the European Higher Education Area and Beyond – Visions for the Future. Third ENQA survey (2012)

Fernández, A., Olmos, J., Alegre, J.: Valor pedagógico del repositorio común de conocimientos para cursos de dirección de empresas. Revista d'innovació Educativa **16** (2016)

Hamari, J., Koivisto, J.: Social motivations to use gamification: an empirical study of gamifying exercise. In: Proceedings of the 21st European Conference on Information Systems. Utrecht, Netherlands, 5–8 June 2013

Hamari, J., Koivisto, J., Sarsa, H.: Does gamification work?-A literature review of empirical studies on gamification. In: System Sciences (HICSS) Proceedings: 2014 47th Hawaii International Conference on IEEE, pp. 3025–3034 (2014). https://doi.org/10.1109/HICSS.2014.377

Hanus, M.D., Fox, J.: Assessing the effects of gamification in the classroom: a longitudinal study on motivation, satisfaction, effort, and grades. Comput. Educ. **80**, 152–161 (2015). https://doi.org/10.1016/j.compedu.2014.08.019

Herzberg, F.: Work and the Nature of Man, World Publishing co

Herzberg, F., Mausner, B., Snyderman, B.: The Motivation to Work. Wiley, New York (1959). (1966)

Higgins, E., Grant, H., Shah, J.: Self-Regulation and quality of life: emotional and nonemotional life experiences. In: Kahneman, D., Diener, E., Schwarz, N. (eds.) Well-being: The foundations of hedonic psychology, pp. 244–266. Russell Sage Foundation, Nueva York (1999)

Hirsh-Pasek, K., Zosh, J.M., Golinkoff, R.M., Gray, J.H., Robb, M.B., Kaufman, J.: Putting education in educational apps: lessons from the science of learning. Psychol. Sci. Public Interest **16**(1), 3–34 (2015). https://doi.org/10.1177/1529100615569721

Holzinger, A., Maurer, H.: Incidental learning, motivation and the tamagotchi eVect: VR-Friends, chances for new ways of learning with computers. In: Proceedings of the Computer Assisted Learning, CAL 99, p. 70. London (1999)

Holzinger, A., Pichler, A., Almer, W., Maurer, H.: TRIANGLE: a multi-media test-bed for examining incidental learning, motivation and the tamagotchi-eVect within a game-show like computer based learning module. In: Proceedings of the Educational Multimedia, Hypermedia and Telecommunication 2001, pp. 766–771. Tampere, Finland (2001)

Jere-Folotiya, J., et al.: The effect of using a mobile literacy game to improve literacy levels of grade one students in Zambian schools. Educ. Tech. Res. Dev. **62**(4), 417–436 (2014). https://doi.org/10.1007/s11423-014-9342-9

Kapp, K.M.: The Gamification of Learning and Instruction: Game-based Methods and Strategies for Training and Education. Pfeiffer. Impreso, San Francisco (2012). https://doi.org/10.1145/2207270.2211316

Ke, F.: Designing and integrating purposeful learning in game play: a systematic review. Educ. Tech. Res. Dev. **64**(2), 219–244 (2015). https://doi.org/10.1007/s11423-015-9418-1

Kettenanurak, V., Ramamurthy, K., Haseman, W.: User attitude as a mediator of learning performance improvement in an interactive multimedia environment: an empirical investigation of the degree of interactivity and learning styles. Int. J. Hum. Comput. Stud. **54**(4), 541–583 (2001). https://doi.org/10.1006/ijhc.2001.0457

Kirriemuir, J., McFarlane, A.: Literature review in games and learning (2004). http://www.futurelab.org.uk/download/pdfs/research/lit_reviews/Games_Review

Pechenkina, E, Laurence, D., Oates, G., Eldridge, D., Hunter, D.: Using a gamified mobile app to increase student engagement, retention and academic achievement. Int. J. Educ. Technol. High. Educ. (2017). https://doi.org/10.1186/s41239-017-0069-7

Rienties, B., Grohnert, T., Kommers, P., Niemantsverdriet, S., Nijhuis, J.: Academic and Social Integration of International and Local Students at Five Business Schools, a Cross-Institutional Comparison. In: Van den Bossche, P., Gijselaers, W., Milter, R. (eds.) Building Learning Experiences in a Changing World. Advances in Business Education and Training, vol. 3, pp. 121–137. Springer, Dordrecht (2011). https://doi.org/10.1007/978-94-007-0802-0_8

Sánchez-Martín, J., Cañada-Cañada, F., Dávila-Acedo, M.A.: Just a game? Gamifying a general science class at university Collaborative and competitive work implications. Think. Skills Creat. **26**(2017), 51–59 (2017). https://doi.org/10.1016/j.tsc.2017.05.003

Urquía Grande, E., Muñoz Colomina, C.I., Cano Montero, E.I.: Generating knowledge in management accounting for the EHEA: using a simulation to learn about the balanced scorecard. Int. J. Crit. Account. **4**(5/6), 542–571 (2012). https://doi.org/10.1504/IJCA.2012.051464

Urquía Grande, E., Del Campo, C.: Erasmus programme effects: a Spanish case. Papeles de Europa **29**(2), 94–110 (2016)

Zabalza, M.A.: Competencias docentes del profesorado universitario. Narcea, Madrid (2006). ISBN: 84-277-1399-1

An Input of Fuzzy Logic to Sustainable Tourism: A Case from Brazil

Luciano Barcellos de Paula[1,2](✉), Anna Maria Gil-Lafuente[3], and Daniela Fantoni Alvares[4]

[1] CENTRUM Cátolica Graduate Business School, Jirón Daniel Alomía Robles 125, Urbanización Los Álamos de Monterrico, Santiago de Surco, 15023 Lima, Perú
lbarcellosdepaula@pucp.edu.pe
[2] Pontificia Universidad Católica del Perú, Av. Universitaria 1801, San Miguel 15088, Perú
[3] University of Barcelona, Av. Diagonal 690, 8034 Barcelona, Spain
[4] Federal Institute of Education, Belo Horizonte, Minas Gerais, Brazil

Abstract. Tourism has experienced continuous expansion and diversification to become one of the fastest-growing economic sectors in the world. However, researches affirm the complexity of this sector and the fact that sustainable development depends on various topics that are not correctly identified by policymakers. In this context, this manuscript aims to reflect on the effects of tourism and propose alternatives that can be sustainable. Results showed gaps in knowledge about the subject and, through a case study analysis in Brazil, identified the forgotten effects of tourism activity that can impact sustainable development. Also, it presented the Forgotten Effects Theory to manage uncertainty and assist in decision-making.

Keywords: Tourism · Sustainable tourism · Decision-making · Fuzzy logic · Forgotten effects theory

1 Introduction

Tourism is an important economic sector that represents 10% of the global Gross Domestic Product (GDP), and 30% of service exports. Researches also indicate that tourism is the third-largest export category in the world after chemicals and fuels [1].

However, there is a growing societal concern for sustainability, which increases the sector's responsibility for sustainable development (SD). Tourism has the potential to contribute, directly or indirectly, to the Sustainable Development Goals (SDGs) [2]. Furthermore, studies affirm the complexity of this sector, e.g., the relationships between sustainable development, tourism, and localities [3] and that sustainable development depend on various topics that are not correctly identified by policymakers. In this context, this manuscript aims to reflect on the effects of tourism and propose alternatives that can be sustainable. The research undertaken can be classified as applied, with the explanatory purpose and combined approach (quantitative-qualitative), through modeling and simulation, and a case study [4].

A. M. Gil-Lafuente et al. (Eds.): REDCID 2019, LNNS 388, pp. 194–204, 2022.
https://doi.org/10.1007/978-3-030-93787-4_11

Due to the complexity involved in sustainable tourism, this paper uses the Forgotten Effects Theory [5]. This algorithm has proven to be a handy tool in sustainable development issues in enterprises [6, 7]. This paper provides a novel contribution by applying the algorithm to the concept of sustainable tourism. In terms of results, knowledge gaps have identified, and the forgotten effects of tourism activity on sustainable development have been exposed through the case analysis in Brazil. The main contribution of the manuscript is to reflect on the complexity of the tourism sector and present an algorithm that can cope with uncertainty and contribute to sustainable development. The most significant limitation concerns the number of small and medium-sized enterprises (SMEs) that participated in the analysis. The paper is structured as follows: Sect. 2 introduces the methodology. Section 3 explains the results. Section 4 discusses the obtained results. Section 5 provides conclusions and supported by references.

2 Methodology

This section is organized into three parts. First, it presents a summary of the sustainable tourism (ST) literature, considering the most cited papers from Web of Science (WoS) on the subject. Secondly, it describes a case analysis of ST in Brazil. Thirdly, it explains an algorithm of fuzzy logic that can be used in ST.

2.1 Sustainable Tourism

Research using the specific term "ST" began just two decades ago [8]. As a reference, "ST may be regarded most basically as the application of the sustainable development idea to the tourism sector" [9]. In other words, tourism activity is developed sustainably without compromising future generations. ST includes many topics, such as heritage management [10], environmental concern [11], respect for local culture and gastronomy [12], and must be linked to sustainable development [13].

However, tourism can make both a positive and negative contribution to the environment and communities [3]. Scientific studies suggest that the global environmental consequences of tourism occur locally, but they are added to global phenomena [14]. For example, CO_2 emissions from tourism have a significant impact on the environment. It is estimated that approximately 40% of this impact comes from air transport, 32% from car transport, and 21% from accommodation [15]. Other research identified five areas of impacts stemming from tourism: "land cover and land-use change, energy use, biotic exchange and extinction of wild species, exchange, and dispersal of diseases, and changes in perception and understanding of the environment through travel" [14].

The tourism sector is not yet close to sustainability, and "the main driver of improvement is regulation rather than market measures" [8]. Governance faces several obstacles, mainly at the political level, and it can be challenging to influence the private sector. Social theory can help to build a strategic relationship between stakeholders [9]. Sustainability is a matter of both local and global responsibilities, with globalization presenting an enormous challenge for political and economic actors to place much greater emphasis on human relations and ethics in tourism [3].

2.2 Sustainable Tourism Certification Program

This Section presents a case analysis on ST in Brazil based on the implementation of the Sustainable Tourism Certification (STC) Program [16] in Brazil. This program was a pioneer on the subject and has served as the basis for the ISO 21401:2018 standard. This program aimed to improve the quality and competitiveness of SMEs in tourism by stimulating their best performance in environmental, socio-cultural and economic areas, through the adoption of standards, a sustainability management system and pursuit of certification.

In Brazil, this program was applied to 365 accommodation establishments in 19 different tourist destinations from 2004 to 2007. The object of this study is the implementation of this STC program in the "Serra do Cipó," a tourist region located in the province of Minas Gerais, Brazil. The STC program lasted seven months in "Serra do Cipó" in 2006 and included the participation of 20 accommodation establishments. Due to confidentiality reasons, the names of the SMEs and owners are not disclosed. The program used the Brazilian technical standard (NBR 15401 - Sustainability Management System - Accommodation facilities), based on the Sustainable Tourism Principles (STP), as a reference. The concept of the standard is established on a Sustainability Management System (SMS) and contemplates environmental, socio-cultural, and economic dimensions.

The positive results of the program, combined with the importance of ST for society, were decisive for creating a new version of the NBR 15401 standard in 2014 when the process began by actively participating in the ISO Tourism Committee. In 2016, the ST working group comprising 60 specialists from 26 different countries was created. As results were approved of ISO 21401:2018 – Tourism and related services — Sustainability management system for accommodation establishments — Requirements [17]. The standard is aligned with Agenda 2030, and its implementation will help organizations achieve the SDGs, primarily by managing the main aspects and impacts of sustainability-related to the business.

2.3 Forgotten Effects Theory

The approach about the Forgotten Effects Theory [5] starts with the existence of a direct incidence relationship; that is, an uncertain cause-effect matrix defined by two sets of elements: $A = \{a_i/i = 1, 2, \ldots, n\}$ which act as causes; $B = \{b_j/j = 1, 2, \ldots, m\}$ which act as effects and a causality relationship $\tilde{M}$ defined by the $n \times m$ dimension matrix: $\left[\tilde{M}\right] = \{\mu_{a_i b_j} \in [0, 1]/i = 1, 2, \ldots, n; j = 1, 2, \ldots, m\}$ being $\mu(a_i, b_j)$ of the values the characteristic function of belonging of each one of the elements of the matrix $\tilde{M}$ (formed by the rows corresponding to the elements of the set - causes - and the columns corresponding to the elements of the set - effects).

It can be said, then, that the matrix $\tilde{M}$ is composed of the estimates made around all the effects that the elements of set A exert on the elements of set B. The more significant this incidence ratio is, the higher the valuation assigned to each of the elements of the matrix. In this case, since it is assumed that the characteristic function of belonging had to belong to the interval [0, 1], it is understood that the higher the incidence ratio, the closer to 1 the assigned valuation is. Conversely, the weaker a causal relationship between two elements is considered, the closer the corresponding valuation is to 0. It should be emphasized that this initial matrix $\tilde{M}$ is elaborated from direct cause-effect relationships; that is, from the first generation.

The objective of this study is based on obtaining a new matrix of incidences that not only reflects the direct causal relationships, but also those that, although not evident, exist and are sometimes fundamental for the appreciation of phenomena. It is necessary to establish the devices that determine that different causes may have effects on themselves and, at the same time, take into account that specific effects may also give rise to incidences on themselves. For this reason, it is necessary to build two new incident relationships, which include the possible effects derived from relating both causes to each other and effects on each other. These two auxiliary matrices are square matrices that are expressed as follows: $\left[\tilde{A}\right] = \left\{\mu_{a_i a_j} \in [0, 1]/i, j = 1, 2, \ldots, n\right\}$ and $\left[\tilde{B}\right] = \left\{\mu_{b_i b_j} \in [0, 1]/i, j = 1, 2, \ldots, m\right\}$. Matrix $\left[\tilde{A}\right]$ lists the incidence relations that can occur between each of the elements that act as causes and matrix $\left[\tilde{B}\right]$ contains the incidence relations that can occur between each of the elements that work as effects. Both $\left[\tilde{A}\right]$ and $\left[\tilde{B}\right]$ coincide in the fact that both are reflective matrices, that is to say: $\mu_{a_i a_j} = 1 \forall_{i=1,2,\ldots,n}$ and $\mu_{b_i b_j} = 1 \forall_{j=1,2,\ldots,m}$.

This means that an element, whether cause or effect, affects itself with the greatest presumption. Neither $\left[\tilde{A}\right]$ nor $\left[\tilde{B}\right]$ are symmetrical matrices, there is at least some pair of subscripts i, j so: $\mu_{a_i a_j} \neq \mu_{a_j a_i} and \mu_{b_i b_j} \neq \mu_{b_j b_i}$.

Once the matrices $\left[\tilde{M}\right]$, $\left[\tilde{A}\right]$ and $\left[\tilde{B}\right]$ have been constructed, direct and indirect incidences are established; that is, incidences in which, at the same time, some cause or effect intervenes. For this the max-min composition of the three matrices is used: $\left[\tilde{A}\right] \circ \left[\tilde{M}\right] \circ \left[\tilde{B}\right] = \left[\tilde{M}^*\right]$.

The order in the composition must always allow the number of elements in the row of the first matrix to coincide with the number of items in the column of the second matrix. The result obtained is a new $\left[\tilde{M}^*\right]$ matrix that collects the incidences between second-generation causes and effects, or the initial causal relationships affected by the possible interposed incidence of some cause or effect. In this sense, the matx is:

$$[\tilde{M}^*] = \begin{array}{cccccc} \Gamma & b_1 & b_2 & \cdots & b_m \\ a_1 & \mu^*_{a_1 b_1} & \mu^*_{a_1 b_2} & \cdots & \mu^*_{a_1 b_m} \\ a_2 & \mu^*_{a_2 b_1} & \mu^*_{a_2 b_2} & \cdots & \mu^*_{a_2 b_m} \\ \vdots & \vdots & \vdots & \vdots & \vdots \\ a_n & \mu^*_{a_n b_1} & \mu^*_{a_n b_2} & \cdots & \mu^*_{a_n b_m} \end{array}$$

From this new matrix $\left[\tilde{M}^*\right]$, the difference between the second-generation effects matrix and the direct incidences matrix will allow for the determination of the degree to which some causality relationships have been forgotten or ignored: $\left[\tilde{F}\right] = \left[\tilde{M}^*\right] - \left[\tilde{M}\right]$.

$$[\tilde{F}] = \begin{array}{c|ccc} \tilde{F} & b_1 & \cdots & b_m \\ a_1 & \mu^*_{a_1b_1} - \mu_{a_1b_1} & \cdots & \mu^*_{a_1b_m} - \mu_{a_1b_m} \\ a_2 & \mu^*_{a_2b_1} - \mu_{a_1b_1} & \cdots & \mu^*_{a_2b_m} - \mu_{a_2b_m} \\ \vdots & \vdots & \vdots & \vdots \\ a_n & \mu^*_{a_nb_1} - \mu_{a_nb_1} & \cdots & \mu^*_{a_nb_m} - \mu_{a_nb_m} \end{array}$$

Using the degree of forgetfulness of some incident, it is possible to determine the element (cause or effect) that acts as a link. To do this, the steps made from the max-min composition of the matrices must be followed (see Fig. 1).

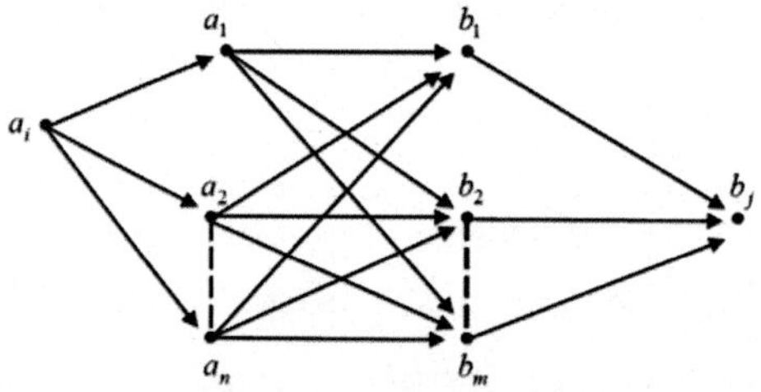

Fig. 1. The max-min composition of the matrices.

Finally, the higher the corresponding value between an element a_i and an element b_j of the characteristic function of belonging to the matrix $\left[\tilde{F}\right]$, the higher the degree of oblivion between a_i and b_j they are produced in the initial relationship and incidence. This means that the implications derived from incidents that are not considered or taken into account in their proper intensity may give rise to erroneous or, at the very least, poorly valued actions.

3 Application of Forgotten Effects Theory

In the section, we present an application of the Forgotten Effects Theory in the STC Program case study. The data was processed using the FuzzyLog[1]© software. It should be noted that this is a conceptual study based on modeling and simulation.

To apply the Forgotten Effects Theory, it was necessary to select a set of external and other elements related to ST. The criterion used to determine the external elements is based on the behaviors observed in a western economy and the application of the model to the tourism sector. The sustainability requirements listed in NBR 15401 are used as

[1] Anna María Gil-Lafuente.

Table 1. Causes (External elements)

1. Educational level
2. The purchasing power of the population
3. Natural resources
4. Cultural resources
5. Security level
6. Political stability
7. Social stability
8. The climate of the country
9. Energy price
10. Communication infrastructure
11. Legislation
12. Level of corruption
13. Economic stability
14. Transport infrastructure
15. Immigration/Emigration Relationship
16. Stakeholder engagement
17. Health services
18. Natural disasters
19. Institutional support and partnerships
20. Currency volatility

a reference to determine the elements that can act as effects. The algorithm was applied as follows.

First, based on the group of experts' validation, the set of C elements was considered, these are supposed to be external to strict business control and act as causes that may have an impact on sustainability [18]. Table 1 presents the external elements.

Secondly, search terms comprising the set of E elements are considered, which represents the dimensions (environmental, social-cultural, and economic) and act as effects and can have an impact on business sustainability. Table 2 shows the sustainability requirements.

Thirdly, a group of experts assessed the causes and effects of sustainability in companies. Semantic correspondence was used for 11 values, from 0 to 1, with the help of the level of truth in the notion of incidence, where the value 0 means no incidence, and the value 1 means the highest incidence.

In the "Direct Incident Matrix" $[\tilde{M}]$, the cause-effect relationships are shown in different degrees that are produced between the elements of set C (causes) and the elements of set E (effects) (see Fig. 2).

Table 2. Effects (Sustainability requirements)

1. Preparation and Assistance to Environmental Emergencies
2. Natural Areas, Flora, and Fauna
3. On-site Construction Architecture and Impacts
4. Landscaping
5. Emissions, Effluents and Solid Waste
6. Energy Efficiency
7. Conservation and Management of Water Use
8. Selection and Use of Inputs
9. Local Communities
10. Work and Income
11. Working Conditions
12. Cultural Aspects
13. Health and Education
14. Traditional Populations
15. Economic Viability of the Enterprise
16. Quality and Customer Satisfaction
17. Customer Safety and Health and Safety at Work

However, this initial matrix $[\tilde{M}]$ is elaborated from the opinion expressed by experts and represents direct cause-effect relationships, that is, first-generation. The objective is to obtain a new matrix of incidences that reflects not only the direct causal relationships but also those that, although they not evident, exist and are sometimes fundamental for the appreciation of the phenomena.

It is necessary to establish the devices that make it possible for different causes to have effects on themselves and, at the same time, take into account that certain effects may also affect themselves. For this reason, it is necessary to construct two relations of further incidences that include the possible effects derived from relating causes to each other, on the one hand, and effects to each other, on the other. To achieve these results, the experts' opinions are again requested to assess the actual incidences between the causes, establishing a square matrix $[\tilde{A}]$, and the matrix $[\tilde{B}]$ with the actual incidences among the effects. Once the matrices $[\tilde{M}]$, $[\tilde{A}]$ and $[\tilde{B}]$ have been constructed, direct and indirect incidences are established; that is to say, incidences in which, at the same time, some cause or effect intervenes. For this purpose, the max-min composition of the three matrices is carried out: $\left[\tilde{A}\right] \circ \left[\tilde{M}\right] \circ \left[\tilde{B}\right] = \left[\tilde{M}^*\right]$.

The result obtained is a new matrix $\left[\tilde{M}^*\right]$, cumulative effects matrix which collects the incidences between second-generation causes and effects, or the initial causal relationships affected by the possible interposed incidence of some cause or effect.

↱	E_1	E_2	E_3	E_4	E_5	E_6	E_7	E_8	E_9	E_{10}	E_{11}	E_{12}	E_{13}	E_{14}	E_{15}	E_{16}	E_{17}
C_1	0,5	0	0,5	0,5	0,9	0,9	0,8	0,7	0	0,7	0,8	0	0,7	0	1	0,7	0,6
C_2	0,6	0	0,7	0	0,5	0,8	0,5	0,4	0	0,7	0,6	0,5	0,4	0	0,6	0,2	0,5
C_3	0,8	1	0,6	0,6	0	0,8	0,9	0	0	0,3	0,2	0,6	0,6	0	0,3	0,4	0,3
C_4	0	0	0	0	0,4	0,5	0,4	0,5	0,6	0,3	0,6	1	0,3	1	0,3	0,4	0,3
C_5	0,2	0	0	0	0	0	0	0	0,9	0,5	0,6	0	0,5	0	1	0,7	0,5
C_6	0,5	0	0	0	0,5	0	0,2	0,1	0,7	0,8	0,4	0	0,6	0	0,5	0,4	0,2
C_7	0	0	0	0	0,2	0,1	0,2	0,1	1	0,5	0,4	0,5	0,4	0,8	0,5	0,5	0,4
C_8	0,8	1	0	0	0	0,8	0,7	0,5	0	0,3	0,4	0	0,3	0	0,4	0,4	0,2
C_9	0	0	0	0	0	1	0,9	0,3	0	0,5	0,4	0	0,2	0	0,5	0,2	0
C_{10}	0,8	0	0	0	0	0	0,2	0,3	0,6	0,5	0,4	0	0,3	0	0,5	0,7	0,2
C_{11}	0	0,6	0	0	0,7	0,6	0,8	0	0	0,5	0,6	0	0,5	0	0,4	0,3	0,2
C_{12}	0	0	0	0	0	0	0,2	0,5	0,4	0,6	0,7	0	0,5	0	0,5	0,4	0,3
C_{13}	0,8	0	0	0	0,7	0,8	0,7	0,6	0,6	0,6	0,4	0,5	0,7	0	0,8	0,4	0,2
C_{14}	0	0	0	0	0,5	0	0	0	0,3	0,6	0,3	0,4	0,5	0,3	0,6	0,7	0,3
C_{15}	0	0	0	0	0	0	0,4	0,6	0,7	0,5	0,3	0,4	0,6	0,4	0,3	0,4	0,2
C_{16}	0,8	0	0	0	0,6	0	0,5	0	0,6	0,5	0,4	0	0,4	0	0,5	0,6	0,4
C_{17}	0,7	0	0	0	0,2	0	0,8	0,2	0,7	0,2	0,6	0,5	1	0,5	0,2	0,5	0,6
C_{18}	0,6	0,9	0,9	0,8	0	0	0,9	0	0,8	0,9	0,1	0	0,2	0,8	0,7	0,6	0,6
C_{19}	0,8	0,6	0	0	0,5	0,7	0,4	0,5	0,7	0,7	0,6	0,7	0,7	0,6	0,6	0,5	0,5
C_{20}	0	0	0	0	0	0,5	0,5	0,5	0,5	0,6	0	0	0	0	0,7	0,5	0,5

Fig. 2. Direct incident matrix.

Finally, the difference between the cumulative effects' matrix and the direct incidences matrix allows for the determination of the degree to which some causality relationships have been forgotten. The forgotten effects matrix is then obtained $\left[\tilde{F}\right] = \left[\tilde{M}^*\right] - \left[\tilde{M}\right]$.

4 Results

The results presented demonstrate that the cause-to-effect relationships that were initially rated 0 (i.e., no incidence) in the direct incidence matrix, showed a great incidence relationship of 0.9 at by the end of the forgotten effects matrix, with which it had been forgotten to consider a critical incidence (see Fig. 3). Significant degrees are highlighted that reveal some forgotten effects that can influence the sustainable development of tourism.

To show the elements that contributed the most to the indirect effects, the cause-effect relationships were analyzed in the matrix of forgotten effects. First, we present the incidence variation between Education level and Natural Areas, Flora, and Fauna (incidence C1, E2) (see Fig. 4).

This incidence ratio indicates that although an initial estimate of 0 was established in the Education level incidence on Natural Areas, Flora, and Fauna, in reality, this ratio increases to 0.9 given that there is an interposed element (Emissions, Effluents and Solid Waste) that potentiates and accumulates effects in the causality relationship.

	E_1	E_2	E_3	E_4	E_5	E_6	E_7	E_8	E_9	E_{10}	E_{11}	E_{12}	E_{13}	E_{14}	E_{15}	E_{16}	E_{17}
C_1	0,2	0,9	0,2	0,3	0	0	0,1	0,1	0,8	0	0	0,6	0	0,6	0	0,7	0,2
C_2	0,2	0,8	0	0,7	0,2	0	0,3	0,4	0,7	0	0,1	0,1	0,3	0,6	0,2	0,3	0,2
C_3	0	0	0,3	0,2	0,7	0	0	0,8	0,8	0,6	0,5	0,2	0	0,8	0,4	0,1	0,3
C_4	0,5	0,6	0,6	0,6	0,2	0,1	0,2	0,1	0,3	0,3	0	0	0,3	0	0,3	0,1	0,3
C_5	0,4	0,6	0,6	0,6	0,6	0,9	0,8	0,8	0	0,1	0,1	0,8	0,1	0,8	0	0,7	0,1
C_6	0,3	0,8	0,6	0,7	0,2	0,8	0,6	0,7	0	0	0,3	0,7	0,1	0,7	0,3	0,1	0,5
C_7	0,7	0,7	0,7	0,7	0,5	0,8	0,6	0,7	0	0,3	0,3	0,3	0,3	0	0,4	0	0,3
C_8	0	0	0,6	0,6	0,7	0,1	0,2	0,2	0,8	0,2	0,2	0,6	0,3	0,6	0,1	0,1	0,4
C_9	0,7	0,8	0,7	0,7	0,7	0	0	0,4	0,8	0,2	0,2	0,6	0,4	0,6	0,1	0,3	0,6
C_{10}	0	0,8	0,6	0,6	0,6	0,6	0,4	0,3	0	0,1	0,2	0,6	0,3	0,6	0,1	0,7	0,4
C_{11}	0,6	0,2	0,6	0,7	0	0,1	0	0,7	0,8	0,2	0,1	0,7	0,1	0,7	0,1	0,2	0,5
C_{12}	0,7	0,7	0,6	0,7	0,7	0,7	0,5	0,2	0,3	0,2	0	0,7	0,2	0,7	0,2	0,1	0,4
C_{13}	0	0,8	0,7	0,7	0	0	0,1	0,2	0,2	0,1	0,3	0,3	0	0,8	0	0,1	0,5
C_{14}	0,5	0,5	0,5	0,5	0	0,6	0,6	0,6	0,3	0	0,3	0,1	0	0,2	0	0,7	0,2
C_{15}	0,7	0,7	0,6	0,6	0,7	0,6	0,3	0,1	0,2	0,1	0,4	0,4	0,1	0,4	0,2	0,1	0,5
C_{16}	0	0,8	0,6	0,6	0	0,7	0,2	0,7	0,1	0,3	0,3	0,7	0,3	0,7	0,1	0,8	0,3
C_{17}	0	0,8	0,4	0,5	0,5	0,6	0	0,5	0,1	0,4	0,1	0,2	0	0,2	0,2	0	0,2
C_{18}	0,2	0	0	0	0,7	0,9	0	0,8	0	0	0,6	0,8	0,4	0	0	0,8	0
C_{19}	0	0,2	0,6	0,6	0,1	0	0,3	0,2	0	0	0,1	0	0	0,1	0	0	0,2
C_{20}	0,8	0,8	0,7	0,7	0,7	0,3	0,3	0,3	0,2	0,1	0,7	0,6	0,7	0,6	0,1	0	0,2

Fig. 3. Forgotten effects matrix $\left[\tilde{F}\right]$.

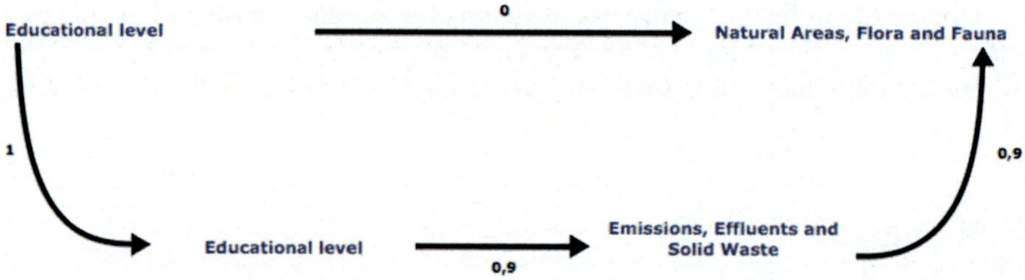

Fig. 4. Incidence variation (forgotten effect) between the Education level cause and the Natural Areas, Flora, and Fauna effect.

Secondly, we display the incidence variation between Natural disasters and Energy Efficiency (incidence C18, E6) (see Fig. 5).

This incidence ratio demonstrates that, although initially an estimate of 0 was established in the Natural disasters incidence on Energy Efficiency, in reality, this ratio increases to 1 given that there is an interposed element (Energy price) potentiating and accumulating effects in the causality relationship.

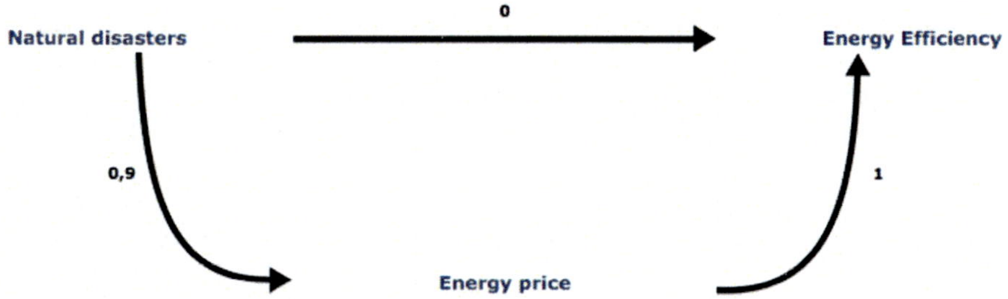

Fig. 5. Incidence variation (forgotten effect) between the Natural disasters cause and the Energy Efficiency effect.

5 Discussion

Social theory can create a strategic relationship among stakeholders [9], and the results of this research show that this relationship generated a positive impact. Likewise, the results provide evidence to support the statement [14] that case studies can help us to assess the interactions among the various dimensions of tourism, sustainable development, and governance at specific junctures.

It is important to note that accommodation establishments are responsible for most of the direct land disturbances related to tourism [13] and for 21% of the sector's CO2 emissions [15]; this reinforces the contribution of this paper in addressing ST in this type of tourism business. The study also provides evidence to support the assertion [3] that sustainability is a matter of both local and global responsibilities. The creation of the ISO 21401:2018 strengthens the tourism sector's concern for sustainable development [13], it also demonstrates the contribution of the pilot project in sustainable tourism that has now become an international standard [17]. The main contribution of the paper is to reflect on the complexity of the tourism sector [3] and to present an algorithm that can address uncertainty [6, 7] and contribute to sustainable development [19].

6 Conclusions

In this paper, the studied subject is broad and complex, as there are multiple factors that directly and indirectly influence the achievement of sustainable tourism. A case analysis of sustainable tourism in Brazil has conducted applying the forgotten effects methodology, which has made it possible to identify some elements that are not readily observable and may impact sustainable development. Also, an algorithm presented to manage uncertainty and facilitate decision-making in sustainable tourism.

It is essential to recognize the limitations of this research in terms of the relatively small number of SMEs that participated in the analysis and the focus of the study on one country (Brazil), which means that the results cannot be generalized.

This work's contribution consists of providing a causality model to the study of sustainable tourism because it is a useful tool in objective selection processes. It has shown that the combination of the different elements that form a direct or indirect part of the business context decisively affect decision-making. Failure to consider forgotten or indirect causal relationships can lead to irreversible errors.

This article brings a significant input that will support future lines of research in the field of incidence matrices and the application of the methodology of recovery of forgotten effects on the sustainable development of tourism. Future research could also include studies in other countries with companies implementing ISO 21401:2018.

Acknowledgments. The authors wish to thank the Royal Academy of Economic and Financial Sciences, University of Barcelona, Spain, and CENTRUM Católica Graduate Business School, Peru.

References

1. World Tourism Organization 2017 International Tourism Results: the highest in seven years. UNWTO World Tourism Barometer, vol. 16, pp. 1–7 (2018)
2. World Tourism Organization International Tourism Highlights, 2019 Edition; World Tourism Organization (UNWTO): Madrid (2019)
3. Saarinen, J.: Traditions of sustainability in tourism studies. Ann. Tour. Res. **33**, 1121–1140 (2006)
4. Will, M., Bertrand, J., Fransoo, J.C.: Operations management research methodologies using quantitative modeling. Int. J. Oper. Prod. Manag. **22**, 241–264 (2002)
5. Kaufmann, A., Gil-Aluja, J.: Modelos para la investigación de efectos olvidados; Editorial Milladoiro: Vigo (1988). ISBN 84-404-3657-2
6. Vizuete-Luciano, E., Gil-Lafuente, A.M., García González, A., Boria-Reverter, S.: Forgotten effects of corporate social and environmental responsibility. Kybernetes **42**, 736–753 (2013)
7. Mulet-Forteza, C., Horrach, P., Socias, A., Merigó, J.M.: The forgotten effects: an application in the social economy of companies of the Balearic Islands. Econ. Comput. Econ. Cybernet. Stud. Res. **52**(3), 147–160 (2018)
8. Buckley, R.: Sustainable tourism: research and reality. Ann. Tour. Res. **39**, 528–546 (2012)
9. Bramwell, B.: Governance, the state and sustainable tourism: a political economy approach. J. Sustain. Tour. **19**, 459–477 (2011)
10. Moscardo, G.: Mindful visitors: heritage and tourism. Ann. Tour. Res. **23**, 376–397 (1996)
11. Santos-Lacueva, R., Clavé, S.A., Saladié, Ò.: The vulnerability of coastal tourism destinations to climate change: the usefulness of policy analysis. Sustainability (Switzerland) **9** (2017)
12. Sims, R.: Food, place and authenticity: local food and the sustainable tourism experience. J. Sustain. Tour. **17**, 321–336 (2009)
13. Hunter, C.: Sustainable tourism as an adaptive paradigm. Ann. Tour. Res. **24**, 850–867 (1997)
14. Gössling, S.: Global environmental consequences of tourism. Glob. Environ. Chang. **12**, 283–302 (2002)
15. Scott, D., Peeters, P., Gössling, S.: Can tourism deliver its aspirational greenhouse gas emission reduction targets? J. Sustain. Tour. **18**(3), 393–408 (2010)
16. Instituto de Hospitalidade PCTS – Programa de Certificação de Turismo Sustentável (2006)
17. Garrido, A.: Gestão da Sustentabilidade (In portuguese), Homepage. https://alexandregarrido.blog/category/gestao-da-sustentabilidade/. Accessed 01 Nov 2019
18. Barcellos de Paula, L., Gil Lafuente, A.: Una aplicación de la metodología de los efectos olvidados: los factores que contribuyen al crecimiento sostenible de la empresa. Cuadernos Del CIMBAGE **12**, 23–52 (2010)
19. Hansmann, R., Mieg, H.A., Frischknecht, P.: Principal sustainability components: empirical analysis of synergies between the three pillars of sustainability. Int. J. Sustain. Dev. World Ecol. **19**, 451–459 (2012)

A Bibliometric Study of Key Journals in Corporate Social Responsibility

Verónica Pizarro[1(✉)], José M. Merigó[1,2], Leslier Valenzuela[3], and Sebastián Aciares[1]

[1] Department of Management Control and Information Systems, School of Economics and Business, University of Chile, Avenue Diagonal Paraguay 257, 8330015 Santiago, Chile
{vpizarro,jmerigo}@fen.uchile.cl

[2] School of Information, Systems and Modelling, Faculty of Engineering and Information Technology, University of Technology Sydney, 81 Broadway, Ultimo, NSW 2007, Australia

[3] Department of Business Administration, School of Economics and Business, University of Chile, Avenue Diagonal Paraguay 257, 8330015 Santiago, Chile

Abstract. Corporate social responsibility (CSR) has acquired great relevance in the last decades, due to the growth in the social trend of giving greater emphasis to social problems and those activities carried out by companies and corporations. The purpose of this study aims to investigate and present the evolution of academic research and the publications on this topic, between 1990 and 2014. To accomplish this, the research used the Web of Science digital platform, whose database is usually considered the most important and complete for bibliometric research of this kind. Results show a strong increase in research and the study of CSR for the last years.

Keywords: Corporate Social Responsibility · Bibliometrics · Journals · Web of Science · VOS viewer

1 Introduction

Corporate Social Responsibility (CSR) is defined as the active and volunteer contribution towards the economic, environmental and social improvement by companies, corporations or firms to the society, in general looking for improving their competitive situation and evaluation. In some way, this is observed as the interest that firms keep with their environment and the concern in the care and development for this. Given that, the research on this topic is growing exponentially for the last years, even in areas and in journals where this has not been sufficiently taken into account.

Bibliometrics is a good method to analyze the research conducted in this topic and also its growth. Defined as the field of research studying quantitatively the bibliographic material [1], bibliometrics is a tool which became more and more popular in order to analyze the fields of study. There are several bibliometric studies in a wide range of areas, such as business and economics [2], management [3], finance [4], economics [5, 6] and decision making [7].

A. M. Gil-Lafuente et al. (Eds.): REDCID 2019, LNNS 388, pp. 205–216, 2022.
https://doi.org/10.1007/978-3-030-93787-4_12

Many of these studies have focused on the evaluation of journals [8, 9]. Merigó and Núñez [10] present a general overview of health journals. Mingers et al. [11] measures the quality of business and management journals with the h-index [12] and Podsakoff et al. [13] makes an evaluation of their quality in the eighties and nineties. Simsek et al. [14] measures the impact as a diffusion process focusing on the citation structure.

Some other studies have focused on the area of business ethics and corporate social responsibility. Chan et al. [15, 16] provides a general overview of business ethics research. They also develop a deep analysis by using threshold citation analysis [17]. In corporate social responsibility, Aguinis and Glavas [18] present a general overview of the main issues affecting the field. Lu & Liu [19] develop the main paths of the discipline between 1970 and 2011. Danilovic et al. [20] explores the diffusion and dynamics of corporate social responsibility identifying the leading topics and research areas and De Bakker et al. [21] develops a bibliometric analysis also considering the leading topics in the field during the last thirty years. However, there is no previous study that has developed a deep analysis of the leading journals in the field identifying the key citation trends.

The aim of this research is to evidence the constant growth in the study of corporate social responsibility all over the world, as well as providing a general view about what currently is this field of research in society, studying the journals that publish on this matter, their publications and the amount of citations they get, in order to measure their level of influence, all of this through bibliometric indicators. For this purpose, database obtained from Web of Science was used, seeking that these indicators can be as objective as possible. These data were used to construct different tables where journals are ordered according to their level of publication according to the topic and number of citations; also by rank of time and the structure of citations. All this leads to investigate the level of evolution that the research on corporate social responsibility has achieved and also a progressive increase in the coming years is expected to be evidenced.

Along with this, some graphics are presented and a detailed description about the analysis on the growth perceived by the research on this topic, comparing it with the increase in research, in general with all other topics published in journals that can be found on the Science website.

A detailed description is presented below on the search method and analysis used to get the tables with the results showed at a later point, the results. Hereafter, the results obtained in tables and graphics are analyzed, providing a general overview of the growth in corporate social responsibility as research topic and their main exponents in terms of publications. Finally, the article ends by concluding about the results, conjointly with a general reflection of them, in order to make it clear what has been achieved through this study.

2 Methodology

Bibliometrics is the field of research that studies bibliographic material quantitatively [1], providing, this way, a general perspective on a field of research according to a wide range of indicators [22].

The following study uses indicators which include the number of publications, total number of citations and h-index [12], seeking as objective, to provide a general view on the research [23, 24] conducted in the field of corporate social responsibility.

This work focuses on specialized journals on corporate social responsibility and those who publish on this topic, considering those publications in a rank of 25 years between 1990 and 2014, according to Web Science database, considered as one of the most complete and relevant academic database for bibliometric research.

With the aim of using these indicators, conjointly with a time gap/, another series of filter in the search for the Web Science database is established, only looking for in its Core Collection. In the database the term "Corporate Social Responsibility" is sought in the "Topic" category, specifying the temporary space from 1990 to 2014. Next is specified that the search is only for articles, reviews, letters and notes in the selection of types of documents. Hereafter it is filtered by research area, selected by Business & Economics, Social Sciences Other Topics, Environmental Sciences Ecology, Engineering, Public Administration, Government Law, Operation Research – Management Science, Public Environmental Occupational Health & International Relations. Given those filters, at the time of the research, conducted between August and December 2015, the Web of Science database yielded 4151 articles, with which the research was made.

The work also uses the VOS viewer software [25] to visualize the leading citation trends in the field by using bibliographic coupling [26, 27], co-citation [28, 29] and citation [30].

3 Results of the Bibliometric Review

This section shows the results of the research. Firstly, the study analyzes the evolution of CSR publications during the 25 years period, from 1990 to 2014. Afterwards, the number of publications of journals is analyzed in gaps of five years, dividing the 25 years period in five segments.

3.1 Publication and Citation Structure of CSR

Table 1 presents the number of publications that have reached a certain threshold between 1990 and 2014. The analysis considers thresholds of citations of 200, 100, 50, 20, 10, 5 and 1. The total number of citations obtained by the paper each year since it was published is also included.

As we can see in the previous table, most of the most mentioned papers were published between 2003 and 2008, starting to decrease marginally within the last year. However, even with such reduction, the number of publications increases almost every year, but every time with a minor difference through the years.

One reason to explain the low level of citations for the last five years is that these publications still need more time to consolidate and obtain a stronger level of citations. In general, only a 0.99% of papers have received over 200 citations, and the 51.53% less than five citations. It also important to emphasize that almost the 18% of publications had not been cited until the study was conducted.

Table 1. Citation structure in CSR research

Year	>200	>100	>50	>10	>1	Articles per year	Cites per year
1990	0	0	0	3	4	4	75
1991	1	2	4	9	14	14	1271
1992	1	1	2	6	9	10	407
1993	0	0	3	7	10	11	301
1994	1	4	4	13	19	19	909
1995	0	1	1	7	12	12	289
1996	2	3	4	10	20	21	1131
1997	5	6	8	13	23	23	3479
1998	1	4	7	16	21	22	1017
1999	2	3	6	12	18	19	1030
2000	1	3	9	20	24	25	1528
2001	3	7	12	25	32	32	3411
2002	2	4	8	23	29	29	1443
2003	4	10	19	43	55	58	4472
2004	4	12	21	69	92	94	4140
2005	2	6	20	73	93	96	3389
2006	3	25	44	88	112	117	6838
2007	6	14	37	118	166	180	6472
2008	3	101	26	172	263	276	7025
2009	0	3	24	218	395	424	6636
2010	0	2	15	182	389	412	5163
2011	0	2	6	143	439	493	4603
2012	0	1	1	115	430	501	3370
2013	0	0	0	34	446	596	1721
2014	0	0	0	3	305	663	649
Total	41	214	281	1422	3420	4151	70769
Percentage	0,99%	5,16%	6,77%	34,26%	82,39%	100,00%	

3.2 Ranking of Most Productive Journals in CSR

In Table 2, the top 30 Journals with more publications on CSR during the last 25 years are analyzed, firstly ordered by the amount of publications and, as second criteria, the number of citations about the topic.

In this Table, h-index is exposed for each of the journals in the table, according to the search conducted on CSR. Also we have the TC and TP columns, which detail the amount

Table 2. Leading journals in CSR

R	Journal	TP	H	TC	C/P	%CSR	> 50
1	J business ethics	1057	56	16928	16,02	19,62%	68
2	Corporate social respon env manage	131	18	1244	9,50	66,16%	1
3	J cleaner production	97	19	1591	16,40	2,81%	8
4	Business ethics - a Eur review	76	11	437	5,75	36,89%	0
5	Business and society	75	13	537	7,16	36,76%	0
6	J business research	63	12	811	12,87	1,86%	5
7	Public relations review	62	10	526	8,48	3,25%	1
8	Business strategy and the environment	56	14	492	8,79	24,78%	0
9	Business ethics quarterly	52	18	1086	20,88	9,12%	4
10	Management decision	47	12	432	9,19	5,58%	0
11	Amfiteatru economic	46	7	129	2,80	8,88%	0
12	Corporate governance - an int Review	44	16	646	14,68	5,47%	1
13	Resources policy	35	11	362	10,34	3,96%	1
14	African J business management	35	5	91	2,60	1,78%	0
15	Int J hospitality management	31	9	306	9,87	3,66%	2
16	Strategic management journal	30	18	2965	98,83	1,67%	11
17	J management studies	29	18	1373	47,34	1,75%	12
18	Ecological economics	29	9	303	10,45	0,67%	0
19	International J production economics	27	10	487	18,04	0,54%	1
20	California management review	26	14	860	33,08	3,28%	7
21	Academy management review	25	20	5294	211,76	1,61%	18
22	Sustainable development	23	9	213	9,26	4,80%	1
23	J academy marketing science	22	16	1245	56,59	2,48%	7
24	J management	22	16	983	44,68	1,74%	8
25	Supply chain management - an int J	22	12	380	17,27	3,92%	1
26	J marketing	21	13	2227	106,05	1,81%	11
27	Academy management perspectives	21	8	477	22,71	4,98%	4
28	Asian business & management	21	4	46	2,19	9,21%	0
29	Eur J marketing	19	5	232	12,21	2,18%	1
30	Accounting auditing & accountab. J	19	7	129	6,79	7,31%	0

Abbreviations: R = Rank; H, TC, TP, C/P = H-index, total citations, total publications and cites per paper in corporate social responsibility; %CSR = Percentage of articles of corporate social responsibility in the journal; >50 = Citation threshold of 50 citations.

of citations and publications with regard to the topic under research, and conjointly create the TC/TP column, that creates a value dividing the number of citations per publication.

In this table it can be seen consequently that, those journals with more publications on the topic are most cited. However, if we compare the TC/TP index of citations and publications on the field treated, the order in the table would vary, as there are some journals with a very good average, besides the Journal of Business Ethics, for example like the Journal of Cleaner Production, the Business Ethics Quarterly or the Corporate Governance – An International Review. Nevertheless, there are some journals that show an excessively high average, due to the few publications on the topic but with many citations, such as the Strategic Management Journal and the Journal of Management Studies, among others, ranked in the top 20 in the table.

3.3 Mapping the Results with VOS Viewer Software

In order to analyze deeper the results of the previous sections, let us develop a graphical visualization of the bibliographic material by using VOS viewer software [25]. Note that this software is freely available at: http://www.vosviewer.com/. In this study, the focus is on journals. Therefore, the analysis presents co-citation, bibliographic coupling and citation of journals. However, it is also possible to use it for other variables including authors, universities and countries [31]. In order to visualize the results, Fig. 1 presents the co-citations of journals of the articles published in Web of Science Core Collection between 1990 and 2014 about corporate social responsibility. The figure uses a threshold of three hundred citations and the one hundred most significant co-citation connections.

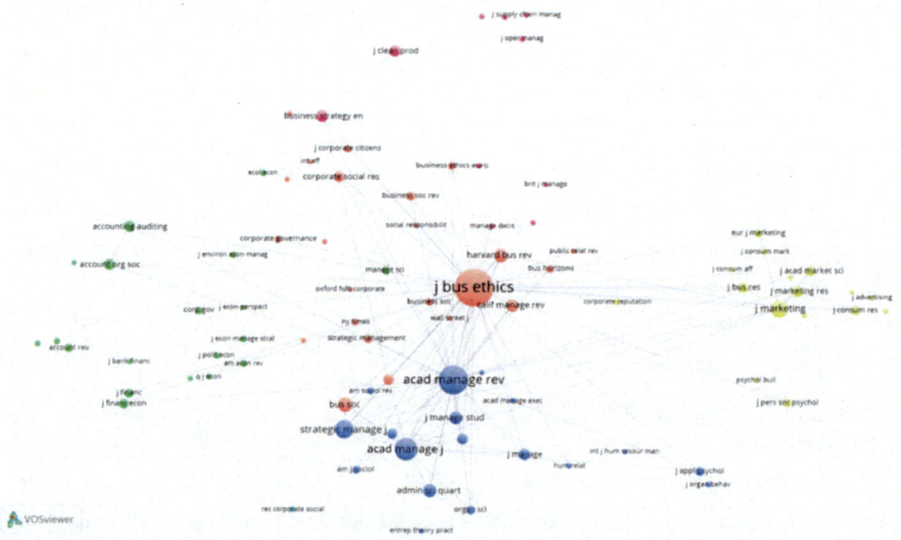

Fig. 1. Co-citation analysis of journals

The Journal of Business Ethics is the journal with the highest influence in the field. However, it is worth noting that some general management journals also achieve a

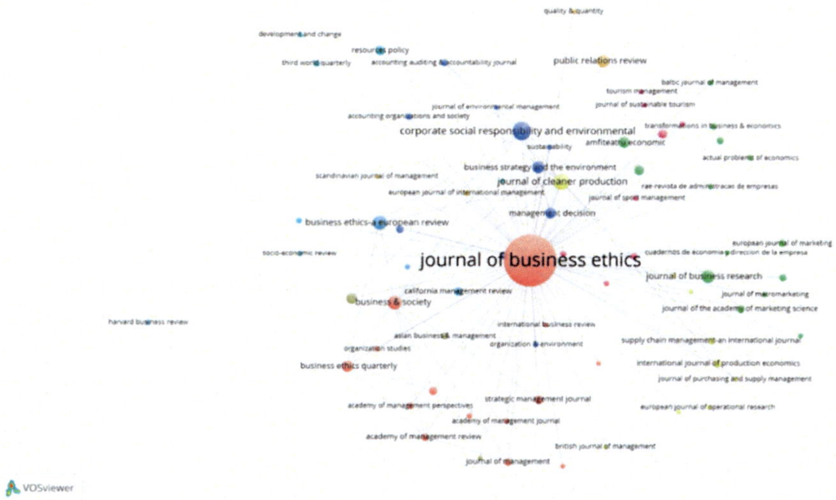

Fig. 2. Bibliographic coupling of journals

significant position including the Academy of Management Review and the Academy of Management Journal. Journals from a wide range of fields of business and management appear in the graph including marketing, finance and psychology.

Next, let us look into bibliographic coupling between journals. That is, when two journals cite the same third journal. Figure 2 presents the results by using a threshold of ten documents and one hundred connections.

The size of the circles indicates the number of publications and the network connections the bibliographic coupling links. The Journal of Business Ethics is by far the journal that publishes more on corporate social responsibility as it has been shown in Table 2. Moreover, it has the widest bibliographic coupling links. Corporate Social Responsibility and Environmental Management, and some other business ethics and general business journals obtain remarkable results in this area.

Another interesting issue is to consider citation connections between journals. That is, the number of times two journals cite each other. Figure 3 presents the results with a threshold of ten and one hundred connections.

The results have strong similarity with bibliographic coupling, especially because it also considers the number of documents published. The Journal of Business Ethics is clearly the leading journal in the field although Corporate Social Responsibility and Environmental Management is more focused on the topic but much smaller.

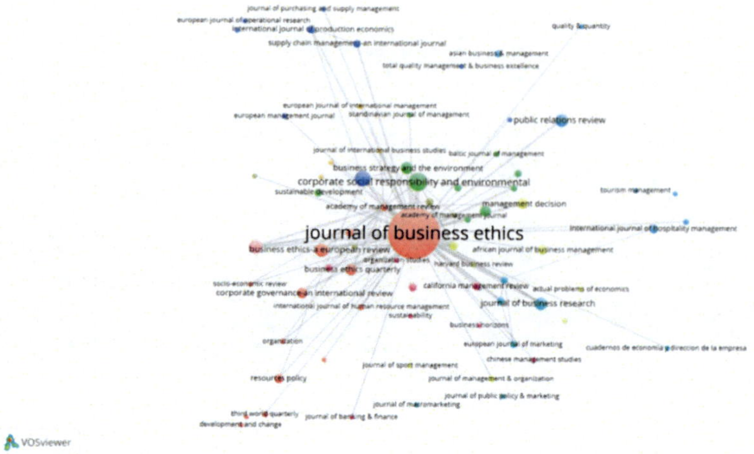

Fig. 3. Citation analysis of journals

To understand better the bibliographic connections, let us classify the information in periods of time in order to visualize the temporal evolution. Figures 4–7 present the co-citation results. Note that Fig. 4 considers documents between 1990 and 1999 with a threshold of ten citations. Figure 5–7 considers periods of five years: 2000–2004, 2005–2009, and 2010–2014. Figure 5 has a threshold of five, Fig. 6 of fifty and Fig. 7 of two hundred. All of the figures visualize the one hundred most significant connections.

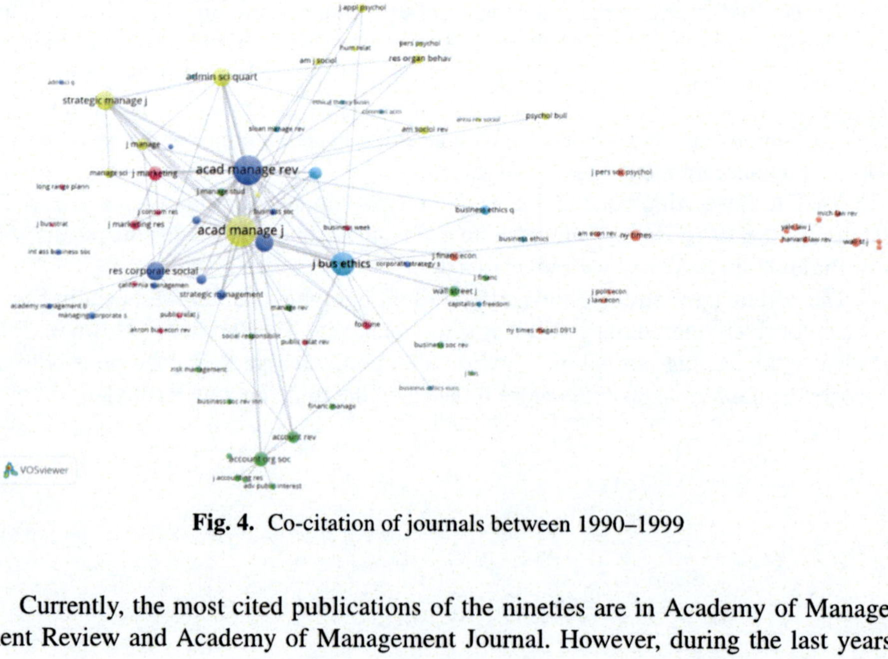

Fig. 4. Co-citation of journals between 1990–1999

Currently, the most cited publications of the nineties are in Academy of Management Review and Academy of Management Journal. However, during the last years,

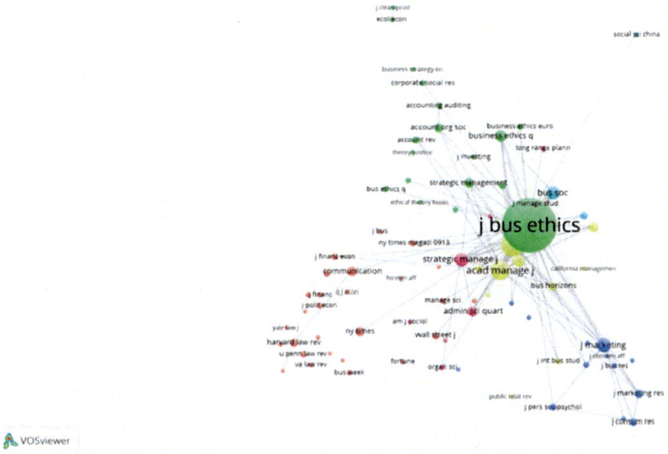

Fig. 5. Co-citation of journals between 2000–2004

the publications in the Journal of Business Ethics are obtaining more citations. This is particularly clear between 2000 and 2004.

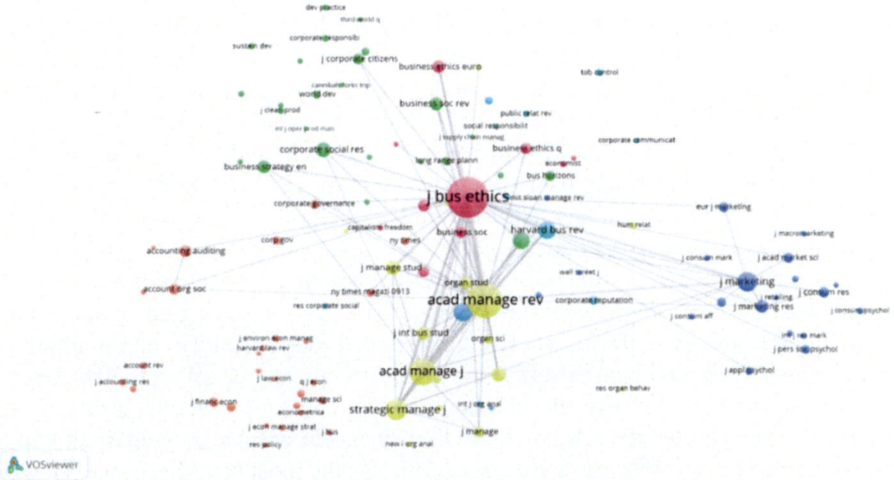

Fig. 6. Co-citation of journals between 2005–2009

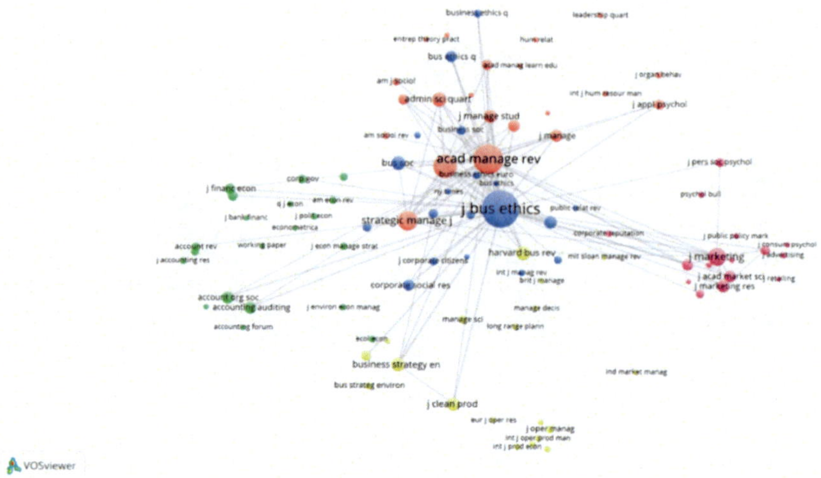

Fig. 7. Co-citation of journals between 2010–2014

4 Conclusions

This study provides a general view on CSR, its research and the influence that has made to the business world, between 1990 and 2014, by using bibliometric tools for measurement. According to these measures, it is clear that the research on this field increased exponentially during this period, but with a slight deceleration for the last years, it can be foreseen that in the future both influence and relevance will increase due to the global contingency that CSR provokes inside profit and nonprofit organizations. Given this, it is almost mandatory for journals keep researching and publishing on this topic, which in comparison to other fields of research, has not got many material. Therefore there is a long road ahead.

It is noteworthy the Journal of Business Ethics among all the journals studied, as this is the journal with more publications to date, and keeps publishing on social responsibility. On the other side, the journal Corporate Social Responsibility and Environmental Management is consolidating itself as the main specialized journal on this matter.

In order to deepen in the results, the study has also developed a graphical visualization of the bibliographic material by using VOS viewer. The results are consistent with the tables identifying the Journal of Business Ethics as the most significant one. However, here the work also identifies the journals highly cited by articles in this field including the Academy of Management Review, the Academy of Management Journal and the Strategic Management Journal. Moreover, the graphical visualization also identifies the most cited documents in the field and the most common keywords.

However, and in conclusion, as previously mentioned, the material is scarce on this matter and it becomes necessary conducting more specialized studies, in order to indicate the course of research on social responsibility. Future research should consider studies in various regions and specific journals as well as the significant influence among them, and also the influence of authors publishing and researching on this topic.

References

1. Broadus, R.N.: Toward a definition of "Bibliometrics." Scientometrics **12**, 373–379 (1987). https://doi.org/10.1007/BF02016680
2. Merigó, J.M., Rocafort, A., Aznar-Alarcón, J.P.: A bibliometric overview of business and economics research. J. Bus. Econ. Manage. **17**, 397–413 (2016)
3. Aguinis, H., Suárez-González, I., Lannelongue, G., Joo, H.: Scholarly impact revisited. Acad. Manage. Perspect. **26**, 105–132 (2012)
4. Alexander, J.C., Jr., Mabry, R.H.: Relative significance of journals, authors, and articles cited in financial research. J. Finan. **49**, 697–712 (1994)
5. Bonilla, C., Merigó, J.M., Torres-Abad, C.: Economics in Latin America: a bibliometric analysis. Scientometrics **105**, 1239–1252 (2015). https://doi.org/10.1007/s11192-015-1747-7
6. Coupé, T.: Revealed performances: worldwide rankings of economists and economics departments, 1990–2000. J. Eur. Econ. Assoc. **1**, 1309–1345 (2003)
7. Yu, D., Li, D.F., Merigó, J.M., Fang, L.: Mapping development of linguistic decision making studies. J. Intell. Fuzzy Syst. **30**, 2727–2736 (2016)
8. Tur-Porcar, A., Mas-Tur, A., Merigó, J.M., Roig-Tierno, N., Watt, J.: A bibliometric history of the journal of psychology between 1936 and 2015. J. Psychol. **152**, 199–225 (2018)
9. Valenzuela, L., Merigó, J.M., Johnston, W., Nicolás, C., Jaramillo, F.: Thirty years of the journal of business and industrial marketing: a bibliometric analysis. J. Bus. Ind. Mark. **32**, 1–18 (2017)
10. Merigó, J.M., Núñez, A.: Influential journals in health research: a bibliometric study. Glob. Health **12**, 46 (2016)
11. Mingers, J., Macri, F., Petrovici, D.: Using the h-index to measure the quality of journals in the field of business and management. Inf. Process. Manage. **48**, 234–241 (2012)
12. Hirsch, J.E.: An index to quantify an individual's scientific research output. Proc. Natl. Acad. Sci. U.S.A. **102**, 16569–16572 (2005)
13. Podsakoff, P.M., MacKenzie, S.B., Bachrach, D.G., Podsakoff, N.P.: The influence of management journals in the 1980s and 1990s. Strateg. Manage. J. **26**, 473–488 (2005)
14. Simsek, Z., Heavey, C., Jansen, J.J.P.: Journal impact as a diffusion process: a conceptualization and the case of the journal of management studies. J. Manage. Stud. **50**, 1374–1407 (2013)
15. Chan, K.C., Fung, H.G., Yau, J.: Business ethics research: a global perspective. J. Bus. Ethics **95**, 39–53 (2010). https://doi.org/10.1007/s10551-009-0346-z
16. Chan, K.C., Fung, A., Fung, H.G., Yau, J.: A citation analysis of business ethics research: a global perspective. J. Bus. Ethics **136**, 557–573 (2016). https://doi.org/10.1007/s10551-014-2533-9
17. Chan, K.C., Fung, H.G., Yau, J.: Predominant sources and contributors of influential business ethics research: evidence and implications from a threshold citation analysis. Bus. Ethics: Eur. Rev. **22**, 263–276 (2013)
18. Aguinis, H., Glavas, A.: What we know and don't know about corporate social responsibility: a review and research agenda. J. Manage. **38**, 932–968 (2012)
19. Lu, L.Y.Y., Liu, J.S.: The knowledge diffusion paths of corporate social responsibility: from 1970 to 2011. Corp. Soc. Responsib. Environ. Manage. **21**, 113–128 (2014)
20. Danilovic, M., Hensbergen, M., Hoveskog, M., Zadayannaya, L.: Exploring diffusion and dynamics of corporate social responsibility. Corp. Soc. Responsib. Environ. Manage. **22**, 129–141 (2015)
21. De Bakker, F.G.A., Groenewegen, P., Den Hond, F.: A bibliometric analysis of 30 years of research and theory on corporate social responsibility and corporate social performance. Bus. Soc. **44**, 283–317 (2005)

22. Ding, Y., Rousseau, R., Wolfram, D.: Measuring Scholarly Impact: Methods and Practice. Springer, Heidelberg (2014). https://doi.org/10.1007/978-3-319-10377-8
23. Merigó, J.M., Gil-Lafuente, A.M., Yager, R.R.: An overview of fuzzy research with bibliometric indicators. Appl. Soft. Comput. **27**, 420–433 (2015)
24. Merigó, J.M., Mas-Tur, A., Roig-Tierno, N., Ribeiro-Soriano, D.: A bibliometric overview of the journal of business research between 1973 and 2014. J. Bus. Res. **68**, 2645–2653 (2015)
25. Van Eck, N.J., Waltman, L.: Software survey: VOSviewer, a computer program for bibliometric mapping. Scientometrics **84**, 523–538 (2010). https://doi.org/10.1007/s11192-009-0146-3
26. Kessler, M.M.: Bibliographic coupling between scientific papers. Am. Doc. **14**, 10–25 (1963)
27. Martínez-López, F.J., Merigó, J.M., Valenzuela, L., Nicolás, C.: Fifty years of the European journal of marketing: a bibliometric analysis. Eur. J. Mark. **52**, 439–468 (2018)
28. Laengle, S., et al.: Forty years of the European journal of operational research: a bibliometric overview. Eur. J. Oper. Res. **262**, 803–816 (2017)
29. Small, H.: Co-citation in the scientific literature: a new measure of the relationship between two documents. J. Am. Soc. Inform. Sci. **24**, 265–269 (1973)
30. Cancino, C., Merigó, J.M., Coronado, F., Dessouky, Y., Dessouky, M.: Forty years of computers and industrial engineering: a bibliometric analysis. Comput. Ind. Eng. **113**, 614–629 (2017)
31. Merigó, J.M., Cancino, C., Coronado, F., Urbano, D.: Academic research in innovation: a country analysis. Scientometrics **108**, 559–593 (2016). https://doi.org/10.1007/s11192-016-1984-4

A Bibliometric Review of Decision Models in Uncertainty Between 1990 and 2018

Luciano Barcellos de Paula[1,2(✉)], Iván de La Vega[1,2], and Anna Maria Gil-Lafuente[3]

[1] CENTRUM Cátolica Graduate Business School, Urbanización Los Álamos de Monterrico, Jirón Daniel Alomía Robles 125, 15023 Santiago de Surco, Lima, Perú
lbarcellosdepaula@pucp.edu.pe

[2] Pontifica Universidad Católica del Perú, Avenue Universitaria 1801, 15088 San Miguel, Perú

[3] University of Barcelona, Avenue Diagonal 690, 08034 Barcelona, Spain

Abstract. The increase in uncertainty due to frequent changes affects decision making, increasing the complexity of choosing the best decisions. For this reason, the decision-maker is looking for alternatives to reduce risks and meet these new challenges. In this context, the document aims to carry out a bibliometric review of decision models under uncertainty between 1990 and 2018 and to learn about the scientific progress of seventeen peer-validated models. It analyses 21,535 articles in nine bibliometric indicators extracted from the Web of Science database. As main results, a positive trend of publications in the analysed models observed, being the Analytical Hierarchical Process (AHP) the most used one. Also, China presented as a country with more scientific collaborations. There is excellent potential for future lines of research on the subject.

Keywords: Decision-Making · Uncertainty · AHP · Bibliometric · Fuzzy Logic

1 Introduction

Increased uncertainty due to frequent changes affects decision-making, which increases the complexity of choosing the best decisions. The decision-maker seeks alternatives to reduce risks and meet these new challenges.

In this context, academics play an essential role in proposing new solutions. Appropriately, Fuzzy Logic [1] was born to guide the application of decision models that reduce uncertainty and facilitate decision-making. Over time, Fuzzy Logic models have proven effective in addressing the new needs of society. The mathematics of uncertainty [2] is useful in various areas of knowledge [3]. For this reason, it defined as an object of study in this article. As indicated in other studies [4–6], decisions in the real world occur in uncertain environments the consequences of actions are not accurately known. In recent years, bibliometric studies have addressed the issue Fuzzy decision-making [5, 6]. However, there is a knowledge gap to evaluate some decision models in uncertainty. Researches have reinforced that "this discipline has strong potential and expectations for the future are that it will continue to grow" [7].

A. M. Gil-Lafuente et al. (Eds.): REDCID 2019, LNNS 388, pp. 217–225, 2022.
https://doi.org/10.1007/978-3-030-93787-4_13

For these reasons, the article aims to carry out a bibliometric review of decision models in uncertainty through scientific mapping and performance analysis between 1990 and 2018 and to know the scientific progress of seventeen models that were validated by specialists.

Results showed a positive trend of publications in the models analyzed has been revealed, being the AHP the most used. Other findings have shown China as the country with the most scientific collaborations, the growth of research lines linked to sustainable development, and the emergence of journals specialized in decision models. The manuscript's main contribution is to deepen the studies on decision models in uncertainty through bibliometric analysis. The most significant limitation concerns the number of decision models in uncertainty included in the report. The article is structured as follows: Sect. 2 presents the methodology used. Section 3 shows the results of the bibliometric review. Section 4 describes the conclusions, contributions, and future research lines, followed by acknowledgments and references.

2 Methodology

This section presents the methodology used to carry out the bibliometric review that follows an approach combined [8, 9] with the scientific mapping (SM) and performance analysis (PA). According to Cobo et al. [10], the SM displays the structure and dynamic aspects of scientific research. On the other hand, the PA shows the evaluation of the groups of scientists and the impact of their activity on the bibliographic database [10]. According to this methodological line, nine bibliometric indicators will be used: papers retrieved from each keyword, papers by year, cites and countries, keywords analysis; countries collaboration; journal analysis; research areas; most cited articles; study by institutions; and author analysis.

The study begins with the validation of seventeen decision models in uncertainty: Analytic Hierarchical Process; Branch and Bound; Hamming Distance; Intuitionistic Fuzzy Set; Owa Operator; Type-2 Fuzzy Set; Galois Theory; Fuzzy Subset; Adequacy Ratio; Markov Chain; Fuzzy Delphi Method; Hungarian Algorithm; Subjective Preferences; Fuzzy Measure Theory; Forgotten Effect; Theory of Experton; and Type-2 Intuitionistic Fuzzy Set. To achieve this objective, five specialists on the subject participated. The consultation was carried out in September 2019 via e-mail. For reasons of confidentiality, the names of the specialists cannot be disclosed.

Bibliometric indicators used, and all languages considered. For the keywords, all the keywords analyzed one by one to refine the results. Finally, for the authors, the same analysis of the keywords was carried out. Four filters used to guarantee the quality of the results:

- Out of a total of 27,128 papers, 26,465 identified as single articles.
- Twenty-three thousand eight hundred fifty-two documents were only the Science Citation Index, Social Sciences Citation Index, and Art & Humanities Citation Index.
- Twenty-one thousand five hundred thirty-five papers were for the period 1990–2018.
- Twenty-one thousand five hundred thirty-two were only single papers (duplicates eliminated).

The process of data extraction was carried out on October 23, 2019, using the Web of Science (WoS) database. The WoS used because this database allows a more detailed citation analysis than others, such as Scopus. Also, it is the database collection with the broadest coverage of structured information in terms of the time range. The data extraction followed three steps. Firstly, a scraping process performed using Python [11]. The results stored in Excel format. Secondly, R software [12] used tables and graphs. Thirdly, the VosViewer software [13] was used to produce the figure of connections and obtain the complete list of authors. Finally, the analysis of 21,535 elements in nine bibliometric indicators carried out. The results showed in the next section.

3 Results of the Bibliometric Review

This section shows the results of the bibliometric review. The outcomes will be presented through figures with the respective analysis, and in the following order: papers retrieved from each keyword; papers by year, cites and countries; keyword analysis; countries collaboration; journal analysis; research areas; most cited articles; analysis by institutions; and author analysis.

3.1 Documents Retrieved for Each Keyword

This indicator allows knowing the ranking of the keywords, the most used decision models will be revealed, and the percentage of total papers with the specific keywords-combination. The results indicate that Analytic Hierarchical Process (AHP) leads the ranking with 8,801 papers, representing 40.4% of total documents with the specific keywords-combination of the documents, followed by Branch and Bound with 5,695 articles, which represents 26.1%, and Hamming Distance with 1,919 papers, which represents 8.8%. Branch and Bound was in the first position in the decades (1990–1999; 2000–2009) and that AHP assumed the first position in the decade 2010–2018.

The consolidated result indicates that AHP's dominance is mainly due to the results of the last decade. There are several reasons for AHP's power, the main ones being: firstly, its simplicity, and the results achieved in various complex applications. Secondly, AHP has been adopted as a decision support tool by companies in multiple countries, and thirdly, because it is the basis of many software packages designed for complex decision-making processes [14].

3.2 Papers by year

This indicator reveal the evolution and trends of research on a topic in the period from 1990 to 2018 (see Fig. 1). It also provides details on the most commonly used keywords in various countries. The positive trend over time in the number of articles published is evident. The reduction in the number of citations is visually perceived from 2015 onwards. Another interesting fact concerns the gradual growth of the average number of authors per document, which goes from 1.8 in 1990 to 3.5 in 2018. In general, the results are quite consistent and show strength at the academic level of the 17 decision models in uncertainty. AHP leads in 15 countries, and "Branch and Bound" in five countries.

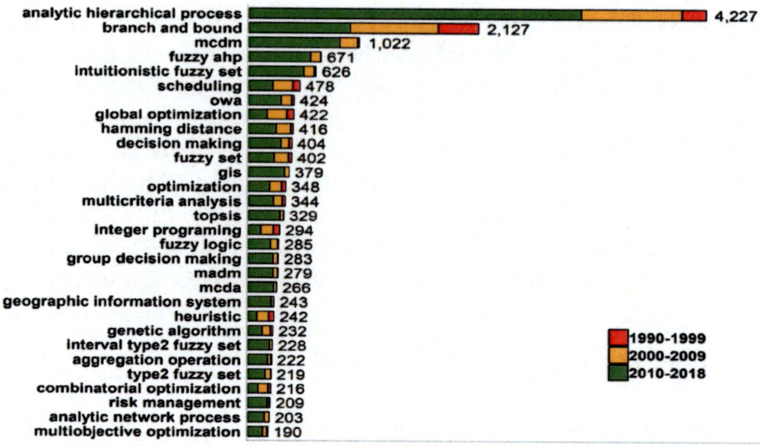

Fig. 1. Papers by year. Source: Own elaboration based on WoS 2019.

Secondly, China leads the ranking among the nations, followed by the United States, Taiwan, Iran, and the United Kingdom.

In summary, the results show that this topic is relevant to academia and that the frontier of knowledge is increasingly being extended. The main reason for the result would be in the application and usefulness of the models to solve complex situations.

3.3 Keywords Analysis

This sub-section presents a keyword analysis in which it shows the ranking and evolution of the top ten keywords from1990 to 2018 (see Fig. 2). This graph visually shows the growth of AHP compared to other keywords over the last decade.

It can be seen as relevant that AHP leads the ranking with 4,227 papers, followed by "Branch and Bound" with 2,127 articles, and Multiple Criteria Decision Making (MCDM) with 1,022 papers. The graph displays the development of all the keywords over the period, but also visually reinforces AHP's leadership and its growth from 2008 onwards.

3.4 Countries Collaboration

This sub-section shows the result of the collaboration between the countries from 2010 to 2018 (see Fig. 3). The size of each sphere represents the number of publications per country; for example, China leads the number of publications, followed by the United States and Iran. The color of each sphere indicates the clusters of collaboration between countries; for example, China has stronger collaborative links with Taiwan, Australia, and Vietnam. On the other hand, the United States has more cooperative relations with India, Singapore, Denmark, and Israel.

In 2018, China led the ranking with 781 papers, followed by the United States with 262 articles and India with 219 papers. The results also indicate China's leadership as the country with the most publications as of 2010 and the rather rapid increase in recent years.

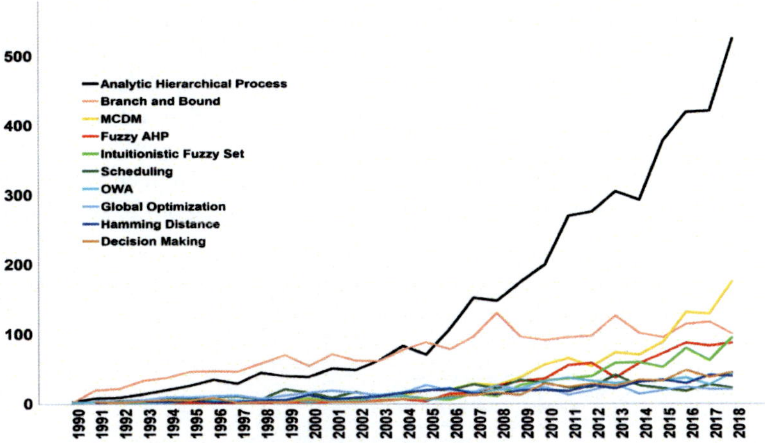

Fig. 2. Evolution of the top ten keywords Source: Own elaboration based on WoS 2019.

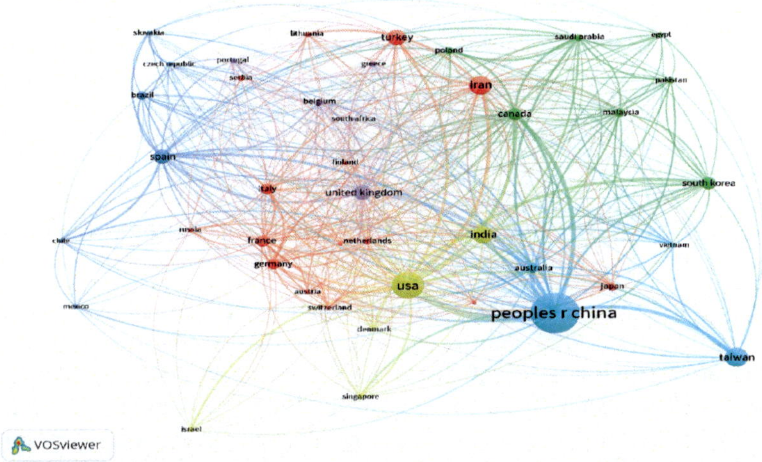

Fig. 3. Countries collaboration (2010–2018). Source: VOSviewer based on WoS 2019.

In terms of country productivity, the United States would have 137,345 citations and 32.3 citations per document (CPD), and China already had 107,090 citations and 22.8 CPD. Another result shows China's leadership with 4,705 published articles, representing 17.1%, followed by the United States with 4,256 published articles (15.4%), and Taiwan with 1,608 published articles (5.8%).

3.5 Journal Analysis

This subsection presents an analysis of the journals between 1990 and 2018 (see Fig. 4). The results indicate that the European Journal of Operational Research leads the ranking with 734 publications, followed by Expert Systems with Applications with 446 publications, and Computers & Operations Research with 375 papers. It can be observed that the three journals are related to research in operations and application systems, which shows the importance of the research fields. The results also indicate the emergence of new journals in recent decades, such as Applied Soft Computing, Soft Computing; Environmental Earth Sciences; Mathematical problems in Engineering; Int. Journal of Inf. Technology Decision Making; and International Journal of Fuzzy Systems.

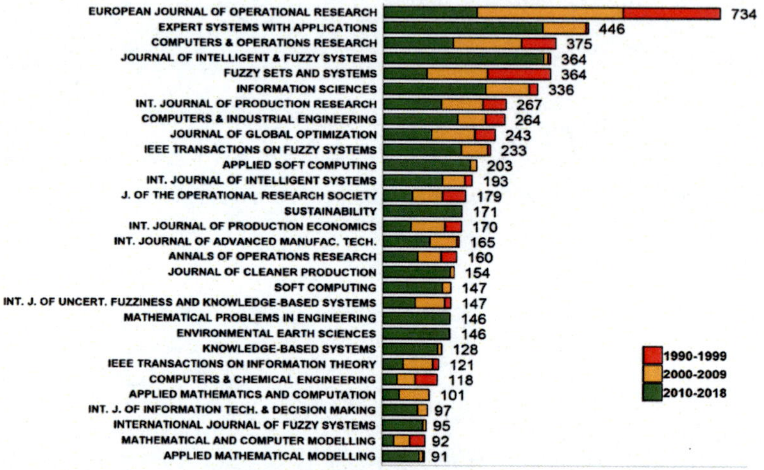

Fig. 4. Journals analysis. Source: Own elaboration based on WoS 2019.

The consolidated results (1990–2018) have presented, highlighting the journals with the most publications (TPJ) on the seventeen decision models in uncertainty, in the following order: Int. Journal of Inf. Technology Decision Making (13.5%), International Journal of Uncertainty Fuzziness and Knowledge-Based Systems (12.2%), International Journal of Intelligent Systems (11.5%), and International Journal of Fuzzy Systems (11.5%). A comparison with other scientific journals has shown a greater interest in these four journals for the seventeen decision models in uncertainty.

3.6 Research Areas

This subsection presents the results of the leading research areas related to the seventeen decision models in uncertainty, covering the period between 1990 and 2018 (see Fig. 5). The top three positions are Operations Research & Management Science with 4,560 papers, Computer Science, Artificial Intelligence with 3,511 articles, and Engineering, Electrical & Electronic with 2,427 documents. Also, the results have shown the occurrence of new areas of research linked to the seventeen models of the decision in uncertainty in recent decades, for example, Green & Sustainable Science & Technology, Energy & Fuels; Environmental Studies; and Geosciences. This result reinforces the interest and usefulness of the models in issues related to sustainable development.

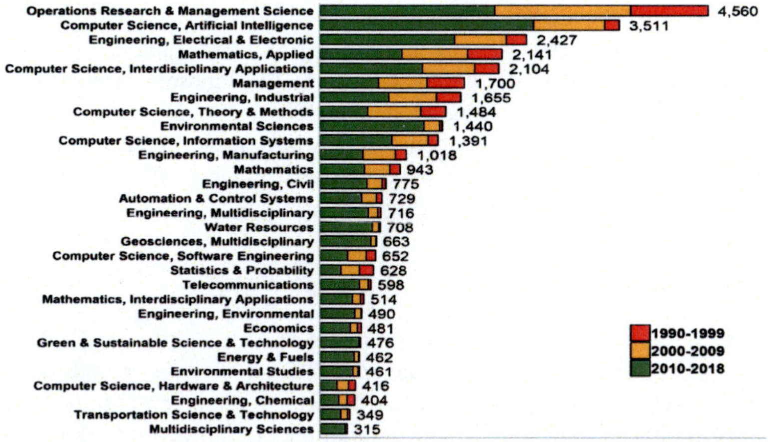

Fig. 5. Research area analysis. Source: Own elaboration based on WoS 2019.

AHP leads the areas as Computer Science, Artificial Intelligence with 394 papers, and Engineering, Electrical & Electronic with 278 articles. On the other hand, "Branch and Bound" leads the areas as Operations Research & Management Science with 1,220 papers, and Mathematics, Applied with 518 articles.

3.7 Most Cited Articles

This sub-section is dedicated to presenting the most frequently cited articles. It is interesting to note that the first three articles are from the journal "Diabetes Research and Clinical Practice" in the area of Endocrinology & Metabolism. This result indicates that decision models in uncertainty are applied in various fields and can be very useful in medicine.

In "Cites per year" (CPY), the article "How to make a decision: The analytical hierarchy process" by Saaty (1990) presents the best result with 0.61 CPY, followed by Whiting, David R., et al. (2011) 's article with 0.42 CPY, and by Guariguata, L. et al. (2014) with 0.39 CPY. The result indicates Saaty's paper as a reference for AHP and corroborates its strength among the seventeen models analyzed in this study.

3.8 Analysis by Institutions

This subsection presents an analysis of the institutions, in which it detailed a ranking of the institutions considering the total of papers (TP), the total percentage of papers in each period (PP), and the participation of the leading institution collaborator in the institution's publications (PPI).

The result indicates that the Chinese Academy of Sciences leads the ranking with 339 papers, followed by the Indian Institute of Technology with 327 articles, and the Islamic Azad University with 300 papers.

The Chinese Academy of Sciences has achieved impressive growth in the last decade, with seven papers in 1990–1999, 61 articles in 2000–2009, and 271 papers in 2010–2018. The results have revealed that the University of Chinese Academy of Sciences is its foremost collaborating institution with 25% PPI.

The results have shown the emergence of new institutions in this ranking since 2000, such as the Islamic Azad University; the Southeast University China; the Central South University; and the Yildiz Technical University.

3.9 Author Analysis

This subsection presents the analysis of authors considering the total of papers (TP), a total of cites (TC), and cites per document (CPD). This type of analysis is essential to know the ranking of the most published authors in the area of study. It also allows identifying through the number of cites the importance of the research, and the emergence of new researchers.

Zeshui Xu leads the ranking with 172 TP, 15,264 TC, and 89 CPD, followed by Ronald R. Yager with 112 TP, 9,284 TC, and 83 CPD, and Jose M. Merigo with 92 TP, 4120 TC, and 45 CPD. Also, the results have indicated the emergence of researchers in recent years, such as Merigo, Kahraman, Zavadskas, Chen, Garg, Liu, Zhou, and Pourghasemi. These results have indicated a renewal of researchers, and at the same time, it has shown the continuity of the lines of research on decision models in uncertainty.

Finally, the consolidated results (1990–2018) have indicated that the authors with the most CPD are: Herrera-Viedma in first place with 181, followed by Mendel with 171, and Chiclana with 104. It has shown the relevance of the research, and also, it has indicated that the author is a reference to other researchers.

4 Conclusions

In recent years scientific studies on decision models in uncertainty have gained relevance. Fuzzy Logic has a significant role to play in knowledge by reducing uncertainties and facilitating decision making. In the same vein, bibliometric testing has proven its usefulness through a combined approach with scientific mapping and performance analysis, allowing for comprehensive reporting on a specific topic. In this context, the article has presented a bibliometric examination of seventeen uncertainty decision models validated by specialists.

The results revealed a positive trend in the models' publications, with AHP being the most widely used, and China being the country with the most scientific collaborations. The growth of the lines of research linked to sustainable development and the appearance of specialized journals on uncertainty decision models were verified. Finally, the study showed the emergence of new institutions dedicated to the topic and new researchers' presence. This result has shown the strengthening and continuity of these lines of research.

As contributions, the manuscript has advanced in the frontier of knowledge by deepening the studies on seventeen decision models in uncertainty through a bibliometric analysis. Likewise, the article has contributed to understanding the evolution and trends of research on Fuzzy Logic. Finally, the study allows for the orientation of new researchers interested in these topics. The most significant limitation concerns the number of decision models in uncertainty included in the analysis. Future research lines are open to topics such as the management of pandemics and socio-economic crises, and sustainable development.

Acknowledgments. The authors wish to thank the Royal Academy of Economic and Financial Sciences, Spain, CENTRUM Católica Graduate Business School, Peru, and the University of Barcelona, Spain.

References

1. Zadeh, L.A.: Fuzzy sets. Inf. Control **8**, 338–353 (1965)
2. Gil-Aluja, J.: Elements for a Theory of Decision in Uncertainty; Applied Optimization. Springer, Boston, vol. 32 (1999). https://doi.org/10.1007/978-1-4757-3011-1, ISBN 978–1–4419–4817–5
3. de Paula, L.B., Lafuente, A.G.: Una contribución al desarrollo sostenible de las empresas a partir de Lógica Borrosa. Cuadernos del Cimbage **1**, 51–83 (2018)
4. Bellman, R.E., Zadeh, L.A.: Decision-making in a fuzzy environment. Manage. Sci. **17**, B-141-B-164 (1970)
5. Blanco-Mesa, F., León-Castro, E., Merigó, J.M.: A bibliometric analysis of aggregation operators. Appl. Soft Comput. J. (2019). https://doi.org/10.1016/j.asoc.2019.105488
6. Merigó, J.M., Gil-Lafuente, A.M., Yager, R.R.: An overview of fuzzy research with bibliometric indicators. Appl. Soft Comput. J. (2015). https://doi.org/10.1016/j.asoc.2014.10.035
7. Blanco-Mesa, F., Merigó, J.M., Gil-Lafuente, A.M.: Fuzzy decision making: a bibliometric-based review. J. Intell. Fuzzy Syst. **32**, 2033–2050 (2017)
8. Noyons, E.C.M., Moed, H.F., Luwel, M.: Combining mapping and citation analysis for evaluative bibliometric purposes: a bibliometric study. J. Am. Soc. Inform. Sci. **50**, 115–131 (1999)
9. Noyons, E.C.M., Moed, H.F., Van Raan, A.F.J.: Integrating research performance analysis and science mapping. Scientometrics **46**, 591–604 (1999)
10. Cobo, M.J., Lopez-Herrera, A.G., Herrera-Viedma, E., Herrera, F.: An approach for detecting, quantifying, and visualizing the evolution of a research field: a practical application to the fuzzy sets theory field. J. Informet. **5**, 146–166 (2011)
11. Sanner, M.F.: Python: a programming language for software integration and development. J. Mol. Graph. Model. **17**, 57–61 (1999)
12. Aria, M., Cuccurullo, C.: Bibliometrix : an r-tool for comprehensive science mapping analysis. J. Informet. **11**, 959–975 (2017)
13. Van Eck, N.J., Waltman, L.: Software survey: VOSviewer, a computer program for bibliometric mapping. Scientometrics **84**, 523–538 (2010). https://doi.org/10.1007/s11192-009-0146-3
14. Osorio Gómez, J., Orejuela Cabrera, J.: El proceso de análisis jerárquico (AHP) y la toma de decisiones multicriterio. Ejemplo de aplicación. Scientia et Technica **2**, 247–252 (2008)

Data Driven Business Model: A Strategy for Transparency in the Sharing Economy. A Bibliometric Overview

Carolina Luis-Bassa(✉) and Thibisay González-Rodríguez

UPF Barcelona School of Management, Barcelona, Spain
{carolina.luis,thibisay.gonzalez}@bsm.upf.edu

Abstract. Three concepts are generating special interest in research: business models based on data, sharing economic and transparency in business.

The present work raises, through a research literature analysis, the link that could exist between these three concepts that, together, can be an element to consider in the competitive advantage of companies.

Working with the Web of Science databases, and focusing the search in the business, management, finance and economy sector more specifically with the terms "Shared economy", "Transparency in business" and "Data Driven Business Model". The analysis presents a bibliometric overview that considers a wide range of indicators that include the number of citations and publications, the *h*-index, leading journals and authors, evolution of terms over time and the cites per paper.

The search was carried out term by term and subsequently combining these terms between them. The study also searched for related words with the terms used in order to analyze the evolution of these terms over time.

The results reveal that, to date, no paper that combines the three concepts proposed in this work has been published in the Business, Management, Economics, or Finance areas of the Web of Science Core Collection. In consequence, this opens a new line of research to be developed in the future.

Keywords: Data Driven Business Model · Sharing Economy · Transparency

1 Introduction

The boom in the sharing economy in recent years has allowed the development of new businesses based on data management through digital platforms. The relative ease and speed to develop products and services offer companies and individuals the possibility of earning money or generating value by sharing their assets and knowledge. According to the *Forbes Los Angeles Business Council*[1], a sharing economy growth is projected from $ 15 billion in 2014 to $ 335 billion in 2025.

[1] Forbes Los Angeles Business Council. https://www.forbes.com/sites/forbeslacouncil/2019/03/04/the-sharing-economy-is-still-growing-and-businesses-should-take-note/#bfe97ab4c339 last accessed 2019/12/20.

A. M. Gil-Lafuente et al. (Eds.): REDCID 2019, LNNS 388, pp. 226–260, 2022.
https://doi.org/10.1007/978-3-030-93787-4_14

However, the sharing business model is generating requirements in terms of security and transparency by assigning responsibilities to each of the parts that comprise it: platform, service provider and consumer. [1], in his book entitled The Trusted Advisor, argues that "... the sharing economy is not only a business model with a structure that protects people who interact, but helps people trust each other ".

Data-driven business models (DDBM), can become an interesting response for companies, to maintain visibility, transparency, and traceability of the information generated through transactions between the parties. To the extent that companies manage increasingly large and complex data, they will have to invest in more and better resources not only for the understanding of their business but to put their data open for consultation and willingness of their partners.

This paper presents the link between the correct management of data and transparency in sharing economy business models, as an element to be considered in the competitive advantage of this type of company.

This paper is organized as follows. Section 2 presents the theoretical framework by putting the reader in the context of the concepts of sharing economy, transparency in business and a business model based on data. Section 3 presents the methodology used to perform the bibliometric, where the indicators used to perform the search and analyze the bibliographic data of a set of articles related to these three concepts, are presented. Section 4 presents the results obtained from the Web of Science Core Collection database. For each search, a structured analysis is developed in the number of publications, the number of citations, authors and most-cited journals and evolution of the concepts throughout the last years. In Sect. 5, graphic visualization of the results obtained in the searches carried out is developed, using the similarity visualization software (VOS). Finally, Sect. 6 presents a summary of the findings, as well as the conclusions of the article and the next proposed lines of research.

2 Theoretical Framework

2.1 Sharing Economy

There has been widespread ambiguity and even confusion about the term "sharing economy" among academics and the public alike. One reason is due to a common misconception about the sharing economy: its [2].

Many terms have been used in recent years to define the same phenomena: "sharing economy", "collaborative economy", "gig economy", "on-demand economy", "collaborative commons", "peer-to-peer economy", "access economy", "the mesh" [2].

The European Commission defines collaborative economy models as "*business models where collaborative platforms facilitate activities creating an open marketplace for the temporary usage of goods or services often provided by private individuals*". These business models involve service providers either sharing assets, resources, time and / or skills and who may be either private individuals offering services on an occasional basis ("peers") or acting on a professional basis ("professional service"), users of those services and the intermediaries which connect providers with users and facilitate the transactions between them through an online platform ("collaborative platforms") [3].

The term "collaborative consumption" was first coined by Ray Algar in the article 'Collaborative consumption' published in the Leisure Report newsletter of April 2007. Then it began to become popular in 2010 with the publication of the book 'What's Mine Is Yours: The Rise of Collaborative Consumption" by [4]. In the Oxford English Dictionary, it is defined as "an economic system in which goods or services are shared between individuals, either for free or for a fee, usually via the Internet".

Sharing economic means taking advantage of existing infrastructure and promoting the exchange between consumers [5]. This concept has evolved and now sharing consumption allows the exchange also between legal persons.

PWC UK developed a report to assess the size and presence of the collaborative economy in the European Union [6]. The results obtained reflect that the five key sectors (peer-to-peer accommodation, peer-to-peer transportation, on-demand household services, on-demand professional services,) of the collaborative economy generated revenues of almost 4,000 million euros and facilitated transactions of 28,000 million euros in Europe.

The world has witnessed a steep rise and infiltration of the sharing economy facilitated by the growing digital platform and willingness of consumers to try mobile apps that facilitate peer-to-peer business models, shared entrepreneurial enterprises among others. We are moving from the 20th-century model where the corporation accumulates resources and produces goods and services toward the 21st-century model where we can avail certain platforms. These platforms are large companies, but they have drawn resources from a distributed crowd with digital spaces on the rise. Sharing economies allow individuals and groups to make money from underutilized assets [7].

Sharing economy firms are disrupting traditional industries across the globe. As an example, look no further than Airbnb which, at $10 billion, can boast a higher valuation than the Hyatt hotel chain. Uber is currently valued at $18.2 billion relative to Hertz at $12.5 billion and Avis at $5.2 billion. Beyond individual firms, there are now more than 1,000 cities across four continents where people can share cars. The global sharing economy market was valued at $26 billion in 2013 and some predict it will grow to become a $110 billion revenue market in the coming years, making it larger than the U.S. chain restaurant industry. The revenue flowing through the sharing economy directly into people's wallets will surpass $3.5 billion this year, with growth exceeding 25%, according to Forbes [8].

2.2 Transparency in Business

The report prepared in 2017 by [9], argues that "A culture based on data is not an end, it is not even a means, it is a consequence. Generating or informing about a large amount of data does not mean that you have a data-based culture. When data allows us to understand why things happen in our organization, we are on the right track. When that data helps us predict future events, we will find ourselves even closer to a data-based culture". The Good Rebels report considers that a company is data driven when it uses data as an integral part of the value chain and its daily reality and defines a methodology based on 4 key principles: agility, transparency, team integration and integration from various disciplines.

But what is meant by transparency in the field of business? The Business Dictionary provides a formal definition of transparency in the context of companies and institutions. The second entry for transparency in this dictionary is: "Lack of hidden agendas and conditions, accompanied by the availability of full information required for collaboration, cooperation, and collective decision making". The fourth entry defines transparency as the "Essential condition for a free and open exchange whereby the rules and reasons behind regulatory measures are fair and clear to all participants".

Expanding on these definitions of transparency in the context of economic exchange, in this work, there is interest in the following dimensions:

The amount and type of information that each of the parties taking part in the exchange share.
The degree of completion of the terms and conditions that each party accepts.
The degree of clarity of the terms and conditions for both parties.
The regulatory framework that the parties are facing.

It is worth noting that when writing establishing the conditions of an economic exchange, more completion comes at the expense of the clarity of these conditions (i.e. if it was feasible to write a perfect contract that accounted for every possible future circumstance, this contract would be unreadable).

In the context of this article, the focus will be on the principle of transparency that must be present in an organization with a DDBM.

Transparency is no longer an option but a requirement. 20 years ago, transparency was a value-added feature that businesses could implement to appeal to customers but nowadays has become a key element to succeed [10]. In a competitive market, building a transparent brand is more important than ever to engage loyal customers and big companies are doing efforts to improve the transparency mechanism every day either if it has been forced or active transparency. There are some reasons to consider why people demand more transparent businesses, but the internet has a big stake in it.

[10] wrote, "consumers are no longer isolated, in the age of digitalization consumers are self-organized and inter-connected". The amount of information available on the Internet and the connectivity across the world provide enough tools and inputs when making a purchase decision. Especially, the feedback and reviewing system between peers in social networks has forced companies to improve their customer relationship management.

The growth of the internet has driven to greater transparency in business-to-customer but not only to counter the peer-to-peer bidirectional flow. Big data management is also a reason for consumers to demand transparency. People are willing to provide private information, but they want to know the purpose and the real use of these powerful data. As companies become bigger, they have more responsibility for their consumers as they require more transparency in exchange [11].

Transparency is the seed of trust. No matter the value a company can offer to its customers, trust is the key[11]. In recent years we have experienced a transition from product dominance to service business models, earlier was easier to prove quality as most attributes were tangible, however, service must focus on customer experience to differentiate. Consumers face a trade-off in service transactions, which is the optimal point to deliver data in exchange to value [11].

2.3 Data Driven Business Model (DDBM)

In the academic literature, a precise definition of data-driven business model has not been found, however, the term is referenced in the book Big Data Analytics for Entrepreneurial Success [12], noting that "It has become in an increasingly important area of study and application and places the data at the center of value creation".

When referring to the term DDBM, there is generally a business approach. It is about making strategic decisions based on the analysis of data and its interpretation, allowing it to examine and organize the data of the organization to promote actions focused on its customers. This results in marketing strategies based on data analysis and can then talk about data driven marketing.

Data driven marketing then represents one of the most significant changes with the greatest impact in making decisions to design strategic marketing actions, thanks to the increasing quality and quantity of data available, together with the technological advances that facilitate its automation. This has allowed us to customize and optimize the marketing actions carried out by companies and brands. [13] provides this definition for data driven marketing: "It is the process of collecting complex data through online and offline channels, analyzing them to understand the psyche and consumer buying patterns, thus helping the marketing team to develop a highly personalized strategy to connect with the target audience ". Thanks to the proliferation of information provided by consumers through online media, companies increasingly have greater access to relevant data about their customers and their purchasing behavior. This new scenario requires companies to allocate appropriate resources to maximize the potential of this data. To the extent that companies manage increasingly large and complex data, known as big data, they will have to invest in more and better resources to take advantage of them. [14], state that the resource-based theory (RBT) that has been used by numerous marketing experts in recent years offers a valuable explanation of the impact of big data on Marketing actions. This resource-based theory suggests that resources that are valuable, difficult to imitate and non-substitutable position a company better for long-term success. These strategic resources can provide the basis for developing strong capabilities that lead to superior performance over time. Capabilities are needed to group, manage, and exploit resources in ways that provide added value to customers and generate advantages over the competition.

3 Bibliometrics Methods

To analyze the state of the art concerning the concepts presented in this work, "Shared economy", "Transparency in business" and "Data Driven Business Model", a search of scientific works that contemplate these concepts, specifically in the area of business and economy, has been carried out. The documentary analysis intends to find links between the three concepts to subsequently propose an empirical study to look for evidence that can relate these three concepts together, with the competitive advantage of companies. The result of the analysis is a bibliometrics that shows the studies found that relate these concepts. Bibliometrics analyzes bibliographic material and allows classifying information according to variables that include journals, institutions, and countries [15–17].

This work uses the Web of Science Core Collection database which is usually considered as one of the leading databases for classifying academic research. Currently, the Web of Science includes more than fifteen thousand journals and fifty million documents. Also, some additional documents of great prestige are collected, such as some proceedings of the congresses with the greatest international impact. The search in the databases was also expanded: Current Contents Connect, Derwent Innovations Index, KCI-Korean Journal Database, MEDLINE®, Russian Science Citation Index and SciELO Citation Index.

The search was performed by TOPIC, which shows the results of the keyword search in title, abstract, author keywords, and some other important aspects. The search was performed by TOPIC, which shows the results of the keyword search in title, abstract, author keywords, and some other important aspects.

The result was also leaked only in the BUSINESS. MANAGEMENT, FINANCE AND ECONOMICS areas of the Web of Science, and only for articles and / or congress proceedings.

The search focused on all three concepts studied: Sharing Economy, Transparency in business and Data Driven Business Model. As these are relatively new concepts, similar terms were also considered, used with the same connotation during the last years.

It was also sought by combining the three concepts. For example, articles with themes related to DDBM and Sharing Economy, DDBM and Transparency, Sharing Economy and Transparency and, finally, DDBM, Sharing Economy and transparency, all together, were searched.

The sharing economy is also known as the access economy, peer-to-peer (P2P) economy, gig economy, collaborative consumption or collaborative economy, so those keywords were also searched. There was a risk of superimposing some articles that use, in their content, different ways to refer to the Sharing Economy, but it was considered important to review all the options currently used.

To perform the search, the bibliometric indicators representing the results selected in the analysis were selected. This work focused on the total number of documents and citations because these two indicators are very useful for measuring productivity and influence [18–20] To combine both concepts, the study also uses citations by article and the h index [21–23].

On the other hand, the analysis focused on the search in academic journals, specifically in Business, Management, Finance and Economics areas. An analysis was also made of the most cited authors as well as an analysis of the evolution of the topics discussed.

4 Results and Discussion

This section shows the results of the bibliometric analysis performed on the three concepts presented: Data Driven Business Models, Sharing Economy (and their synonyms) and Transparency in Business. All these concepts were searched through the TOPIC field of the Web of Science.

For each concept, the analysis focuses on the evolution of publications from 1900 to the date of consultation, the most cited journals, the most cited papers, the most cited authors and the most cited works. It should be noted that the exposed data have been collected on the Web of Science, in November and December 2019. It should also be noted that in the search using keywords, some errors can occur since there are works with the word key requested but without the desired scientific content. As for the Transparency keyword, some of these cases have occurred.

The searches were carried out by refining by the Web of Science Categories, Management, Business, Business Finance and Economics. The Document Types consulted were Articles and Proceedings Paper.

The Databases consulted were Current Contents Connect, Derwent Innovations Index, KCI-Korean Journal Database, MEDLINE®, Russian Science Citation Index and SciELO Citation Index.

In some cases, because they are relatively new concepts, the number of papers found was small, so the results are presented directly in tables. For cases with numerous results, the analysis is complemented with a graph that simplifies visualization.

4.1 Sharing Economy

The term Sharing Economy has been used in several research sectors, being BUSINESS ECONOMICS and SOCIAL SCIENCES OTHER TOPICS, the sectors in which most papers have been published with this term. Figure 1 shows the Tree map generated by Web of Science to visualize the different sectors in which the term Sharing Economy is being used.

Fig. 1. Sectors in which papers with the term Sharing Economy have been published. Source: Web of Science 2019

The Sharing Economy concept appears for the first time, in the categories studied, in 2005 with the article "The sharing economy: Solidarity networks transforming globalization", written by Williams CC for the European Urban and Regional Studies magazine.

After more than a few years of publications, as of 2013, publications with the term "Sharing Economy" have grown to a total of 187 publications in 2019, with a total of 550 papers. Figure 2 shows the evolution of papers published on Sharing Economy since 2005.

Fig. 2. Annual number of publications using "Sharing Economy" in the Web of Science field TOPIC. Source: Own using Web of Science research results.

The journal with the highest number of citations related to the concept of Sharing Economy was the Journal of Business Research with 831 citations. The Journal of Business Research applies theory developed from business research to actual business situations. Recognizing the intricate relationships between the many areas of business activity and is published by Elsevier[2]. Figure 3 shows the 30 journals with the highest citations of papers related to the Sharing Economy.

[2] Journal of Business Research: https://www.journals.elsevier.com/journal-of-business-research retrieved on December 2019.

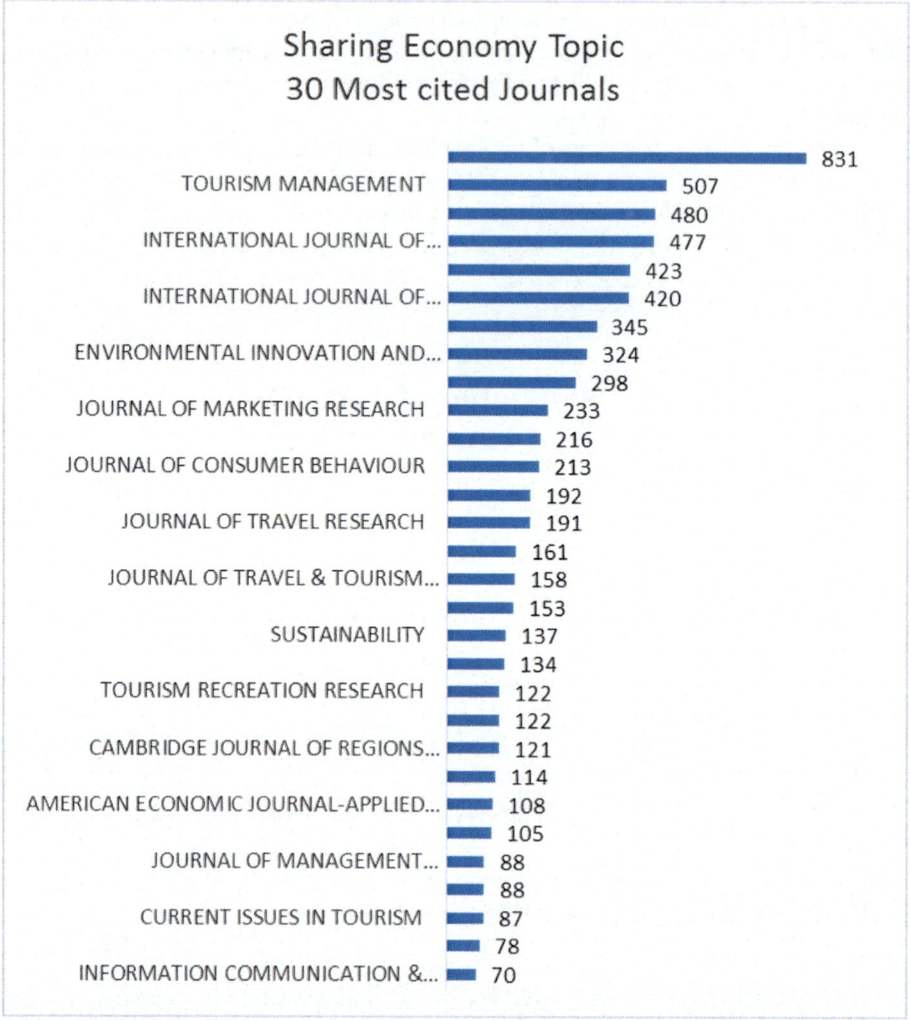

Fig. 3. "Sharing Economy" topic 30 most cited journals. Source: Own using Web of Science research results.

It is interesting to note that the term "Sharing Economy" has been published in journals of various sectors, including the Hospital and Tourism sector. There is also a link between this term and Sustainability, Ecology, Environment and Society journals. There is also a relationship between Marketing and Management magazines. Table 1 shows the various sectors in which papers related to Sharing Economy are being published.

Regarding the most cited author, Russell Belk has a total of 629 citations for his article "You are what you can access: Sharing and collaborative consumption online", published in the Journal of Business Research. Figure 4 shows the 20 most cited authors for papers related to Sharing Economy.

Table 1. 30 most cited journals with Sharing Economy in the TOPIC field Source: Web of Science, December 2019.

R	Sharing Economy topic journal	30 Most cited journals
1	Journal of Business Research	831
2	Tourism Management	507
3	Journal of The Association for Information Science and Technology	480
4	International Journal of Hospitality Management	477
5	Technological Forecasting and Social Change	423
6	International Journal of Contemporary Hospitality Management	420
7	Ecological Economics	345
8	Environmental Innovation and Societal Transitions	324
9	Journal of Cleaner Production	298
10	Journal of Marketing Research	233
11	Organization & Environment	216
12	Journal of Consumer Behaviour	213
13	Computers In Human Behavior	192
14	Journal of Travel Research	191
15	Geoforum	161
16	Journal of Travel & Tourism Marketing	158
17	Gaia-Ecological Perspectives for Science and Society	153
18	Sustainability	137
19	Business Horizons	134
20	Annals of Tourism Research	122
21	Tourism Recreation Research	122
22	Cambridge Journal of Regions Economy and Society	121
23	Journal of Tourism Futures	114
24	American Economic Journal-Applied Economics	108
25	International Journal of Communication	105
26	Applied Energy	88
27	Journal of Management Information Systems	88
28	Current Issues In Tourism	87
29	Communications of The Acm	78
30	Information Communication & Society	70

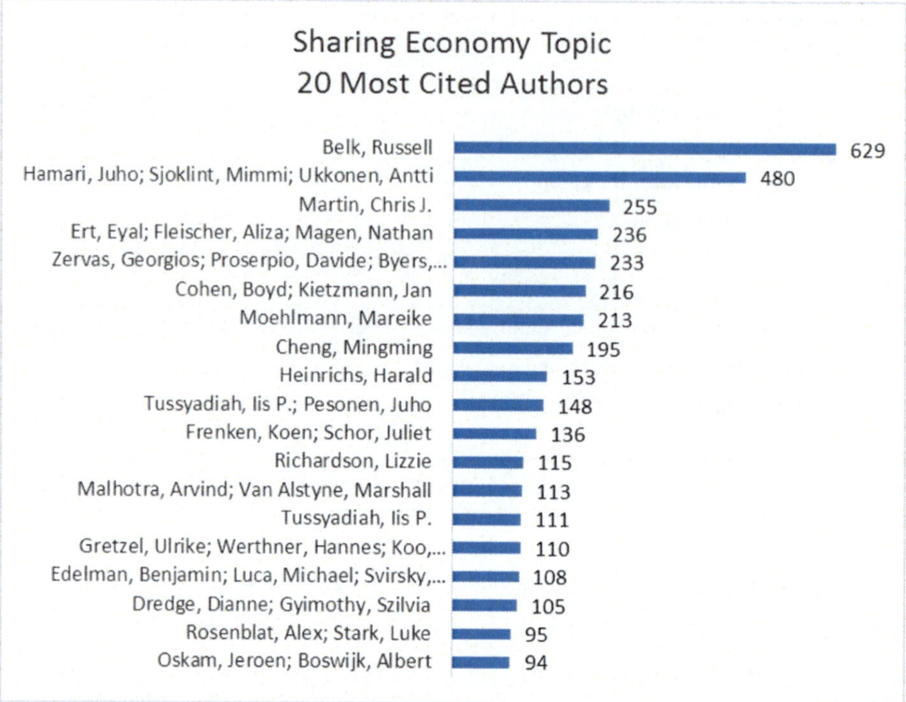

Fig. 4. Sharing Economy topic. 20 most cited authors. Source: Own using Web of Science research results.

Two articles lead the number of citations of papers that include the concept of Sharing Economy in their content: "You are what you can access: Sharing and collaborative consumption online" by [24] with 629 citations, and "The sharing economy: Why people participate in collaborative consumption "by [5] with 480 citations. Table 2 shows the 5 most cited articles related to Sharing Economy.

The citation report of the Web of Science summarizes the bibliometric information of the "Sharing Economy" concept in Topic in 550 articles, with an *h*-index of 32, an average of citations of 9.79, a total of 5,384 citations and 2,492 cited articles. Figure 5 shows the bibliometric indicators of the Web of Science related to publications on the Sharing Economy.

Other Terms for Sharing Economy

As mentioned in previous sessions, the sharing economy is also known as the access economy, peer-to-peer (P2P) economy, gig economy, collaborative consumption or collaborative economy, therefore results using these keywords will be displayed.

It is important to highlight that the search carried out for each of these terms resulted in articles that coincided since, several of them, used these terms interchangeably within the same text of the paper.

Figure 6 shows that the term "Collaborative Consumption" begins to be used in 2012 and is the term that has more mentions, reaching 50 publications in 2017 and

Table 2. 5 most cited papers related to Sharing Economy Source: Web of Science, December 2019.

R	Title	Authors	Source title	Year	Citations
1	You are what you can access: Sharing and collaborative consumption online	Belk, Russell	Journal of Business Research	2014	629
2	The sharing economy: Why people participate in collaborative consumption	Hamari, Juho; Sjoklint, Mimmi; Ukkonen, Antti	Journal of The Association for Information Science and Technology	2016	480
3	The sharing economy: A pathway to sustainability or a nightmarish form of neoliberal capitalism?	Martin, Chris J	Ecological Economics	2016	255
4	Trust and reputation in the sharing economy: The role of personal photos in Airbnb	Ert, Eyal; Fleischer, Aliza; Magen, Nathan	Tourism Management	2016	236
5	The Rise of the Sharing Economy: Estimating the Impact of Airbnb on the Hotel Industry	Zervas, Georgios; Proserpio, Davide; Byers, John W	Journal of Marketing Research	2017	233

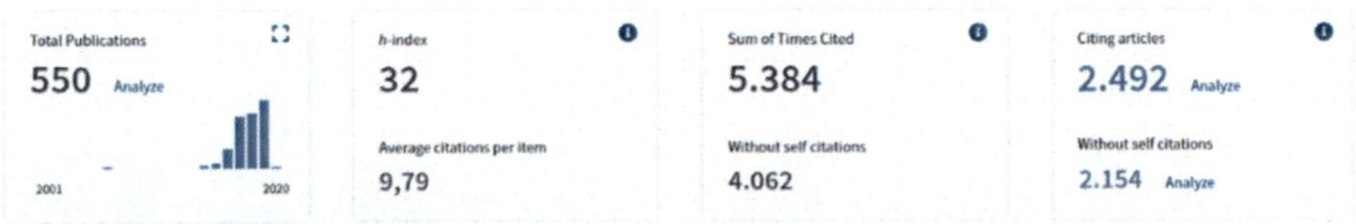

Fig. 5. Sharing economy topic. bibliometric analysis. Source: Web of Science, December 2019

maintaining a growing evolution until 2019. The next term that presents an evolution increasing in mentions, it is the term "Gig Economy" that begins to be used timidly in 2014 and evolves increasingly until reaching 43 mentions in 2019. The term "Collaborative Economy" begins to be used in 2016, reaching its maximum of mentions in 2017, presenting a decreasing evolution since then with 12 mentions in 2019. The least used terms, according to the analysis, are those of "Peer-to-peer Economy" and "Access Economy". The first has been used in the years 2016 and 2017 with 2 and 3 mentions respectively, while the second began to be used in 2017 presented few mentions (4, 2 and 3) until 2019. In general, the terms that show a growing evolution for 2019 are "Collaborative Consumption" and "Gig Economy".

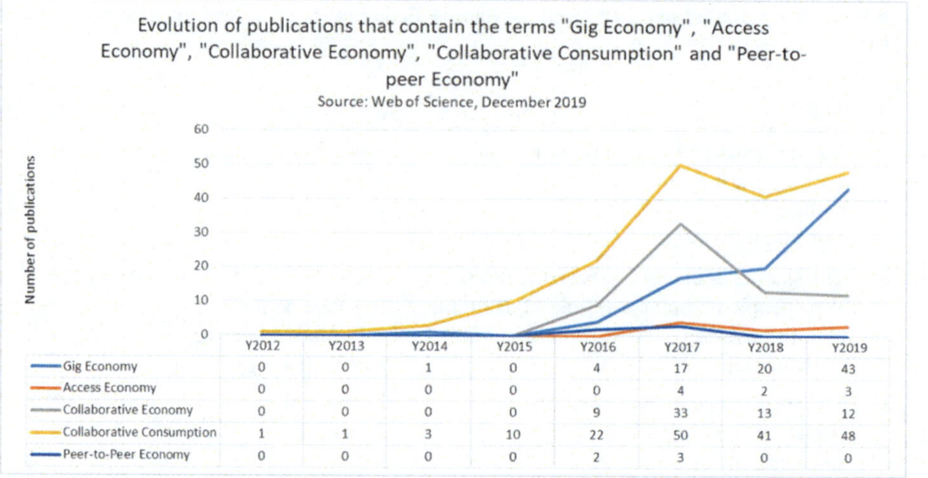

	Y2012	Y2013	Y2014	Y2015	Y2016	Y2017	Y2018	Y2019
Gig Economy	0	0	1	0	4	17	20	43
Access Economy	0	0	0	0	0	4	2	3
Collaborative Economy	0	0	0	0	9	33	13	12
Collaborative Consumption	1	1	3	10	22	50	41	48
Peer-to-Peer Economy	0	0	0	0	2	3	0	0

Fig. 6. Evolution of publications that contain the terms "Gig Economy", "Access Economy", "Collaborative Economy", "Collaborative Consumption" and "Peer-to-peer Economy". Source: Own using Web of Science research results.

Looking at Fig. 7, a comparison of the number of publications that use the term "Sharing Economy" versus the number of publications with terms that are synonymous with "Sharing Economy", can be seen. The analysis confirms that the "Sharing Economy" is the term that has been most used in scientific publications and that it shows a growing evolution in mentions for 2019.

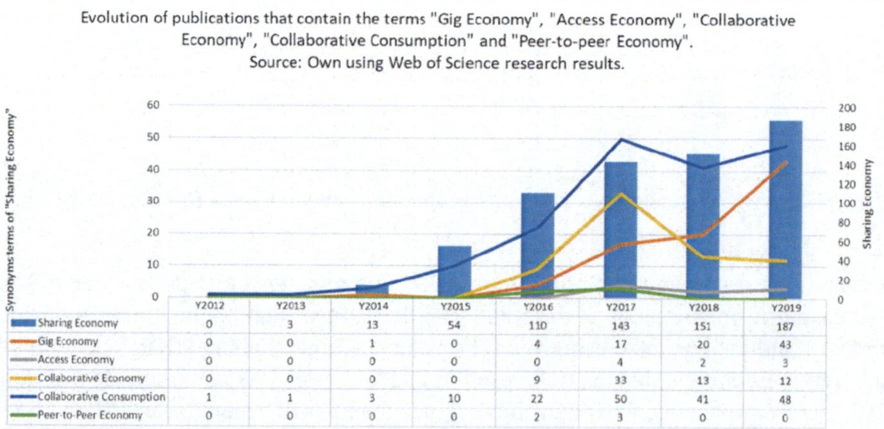

	Y2012	Y2013	Y2014	Y2015	Y2016	Y2017	Y2018	Y2019
Sharing Economy	0	3	13	54	110	143	151	187
Gig Economy	0	0	1	0	4	17	20	43
Access Economy	0	0	0	0	0	4	2	3
Collaborative Economy	0	0	0	0	9	33	13	12
Collaborative Consumption	1	1	3	10	22	50	41	48
Peer-to-Peer Economy	0	0	0	0	2	3	0	0

Fig. 7. Comparison of publications that use the term "Sharing Economy" versus terms that are synonyms. Source: Own using Web of Science research results.

Appendix A shows a table with the papers that use synonyms terms for the "Sharing Economy" together.

4.2 Transparency in Business

The term "Transparency in Business" begins to be used in 2009 closely related to the reputation of companies as an indicator of credibility. The author who mentions the term is [25] in his paper "Measuring risks to reputation" as proceedings of the conference Reputation Capital: Building and Maintaining Trust in the 21st Century.

The paper with the highest number of citations is Information Transparency in Business-to-Consumer Markets: Concepts, Framework, and Research Agenda, written by [26] in the magazine INFORMATION SYSTEMS RESEARCH and which reached 54 citations.

Table 3 shows the result of publications with the term Transparency in Business, specifying authors, journals, and number of citations.

Table 3. Articles published with using the term "Transparency in Business" Source: Web of Science, December 2019.

R	Title	Authors	Source title	Year	Citations
1	Measuring risks to reputation	Herkenhoff, Frank	Reputation Capital: Building and Maintaining Trust In The 21st Century	2009	0
2	Information Transparency in Business-to-Consumer Markets: Concepts, Framework, and Research Agenda	Granados, Nelson; Gupta, Alok; Kauffman, Robert J	Information Systems Research	2010	54
3	The Influence of Internal and External Codes on CSR Practice: The Case of Companies Operating in Serbia	Mijatovic, Ivana S.; Stokic, Dusan	Journal of Business Ethics	2010	27
4	China's Basic Standard for Enterprise Internal Control: An Opportunity for eBusiness	Carasik, Bob; Zhang, Yuxin	Ninth Wuhan International Conference On E-Business, Vols I-Iii	2010	0
5	Visionary or criminal: From profit through morality to socially sustainable entrepreneurship	Bechervaise, Neil E.; Benjamin, Colin G	Proceedings of 9th International Strategic Management Conference	2013	3

(*continued*)

Table 3. (*continued*)

R	Title	Authors	Source title	Year	Citations
6	Market Transparency in Business-to-Business e-Commerce: A Simulation Analysis	Ozcelik, Yasin; Ozdemir, Zafer D	Mobile Applications and Knowledge Advancements In E-Business	2013	0
7	Foreign corporations and the culture of transparency: Evidence from Russian administrative data	Braguinsky, Serguey; Mityakov, Sergey	Journal of Financial Economics	2015	14
8	Rise and Fall of Multinational Enterprises in Vietnam: Survival Analysis Using Census Data during 2000–2011	Vu, Tien Manh; Yamada, Hiroyuki; Otsuki, Tsunehiro	Asian Economic Journal	2017	1
9	The benefits and implementation of performance transparency: The why and how of letting your customers 'see-through' your business	Merlo, Omar; Eisingerich, Andreas; Auh, Seigyoung; Levstek, Jaka	Business Horizons	2018	0
10	Information Transparency in Business-to-Business Auction Markets: The Role of Winner Identity Disclosure	Lu, Yixin; Gupta, Alok; Ketter, Wolfgang; van Heckd, Eric	Management Science	2019	0

The total number of publications using the term Transparency in Business is 10 and has been declining. In the year 2019, only one article was published. The combination of the term Transparency together with the word Business may be irrelevant.

Because it is a very general word, another search was carried out by placing only the term "Transparency" in the TITLE field and filtering the results only articles of the BUSINESS AND ECONOMICS sector.

The Tree map of Fig. 8, generated by the Web of Science, shows the result of the different sectors that contain the term "Transparency" in the title. The sector that has published the most papers with this term is the Business and Economics sector with 2,235 articles.

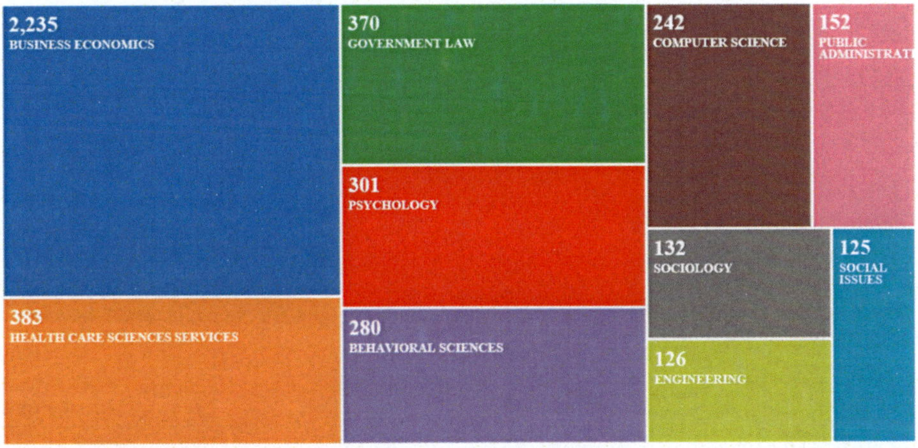

Fig. 8. Sectors that contain the term "Transparency" in the publication title. Source: Web of Science, December 2019.

Figure 9 shows the evolution of the publications with the word "Transparency" in the title. It is noted that this term has been used since 1959 and that since 2007 the publications have increased progressively until reaching 211 in 2018. The slight decrease to 186 papers in 2019 maybe because, as of the search date, there are still papers to publish.

Table 4 shows the 10 papers with the highest number of citations. The most cited article is "What determines corporate transparency?" written by [27] for the Journal of Accounting Research. Most Journals are related to Finance, Accounting and Economics, with some also being observed in the areas of Health, Engineering and Biology. The fact that the word transparency is associated with financial and accounting aspects seems to indicate that the aspect related to money is one of the most sensitive to rigor and clarity to generate trust. It also seems to be an indicator that it has already been considered a competitive advantage for companies that apply transparency in business.

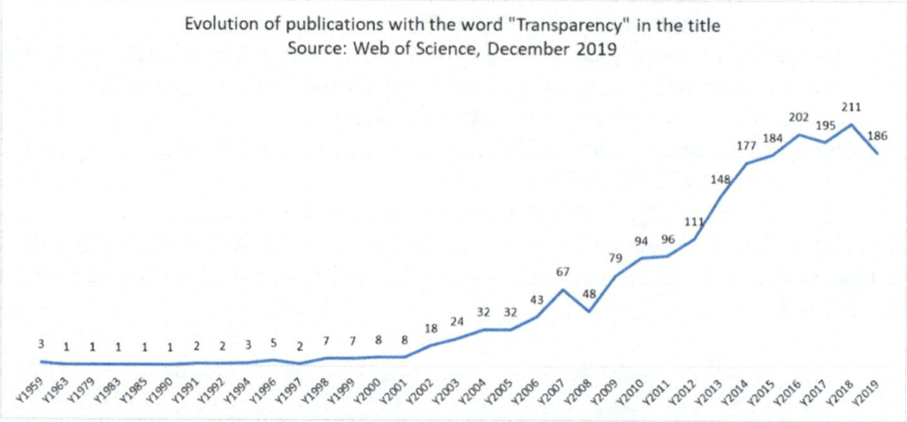

Fig. 9. Evolution of publications with the word "Transparency" in the title. Source: Own using Web of Science research results.

Table 4. 10 papers with the highest number of citations with the word Transparency in title Source: Web of Science, December 2019.

R	Title	Authors	Source title	Year	Total citations
1	What determines corporate transparency?	Bushman, RM; Piotroski, JD; Smith, AJ	Journal of Accounting Research	2004	642
2	The personalization privacy paradox: An empirical evaluation of information transparency and the willingness to be profiled online for personalization	Awad, NF; Krishnan, MS	Mis Quarterly	2006	298
3	Political relationships, global financing, and corporate transparency: Evidence from Indonesia	Leuz, Christian; Oberholzer-Gee, Felix	Journal of Financial Economics	2006	275

(*continued*)

Table 4. (*continued*)

R	Title	Authors	Source title	Year	Total citations
4	Accounting and Environmentalism - An Exploration of the Challenge of Gently Accounting For Accountability, Transparency And Sustainability	Gray, R	Accounting Organizations and Society	1992	210
5	Corporate bond market transaction costs and transparency	Edwards, Amy K.; Harris, Lawrence E.; Piwowar, Michael S	Journal of Finance	2007	209
6	Is Transparency the Key to Reducing Corruption in Resource-Rich Countries?	Kolstad, Ivar; Wiig, Arne	World Development	2009	206
7	Does search-facilitating technology improve the transparency of financial reporting?	Hodge, FD; Kennedy, JJ; Maines, LA	Accounting Review	2004	201
8	Central Bank transparency	Geraats, PM	Economic Journal	2002	183
9	The wrong kind of transparency	Prat, A	American Economic Review	2005	181
10	No one is perfect: The limits of transparency and an ethic for 'intelligent' accountability	Roberts, John	Accounting Organizations and Society	2009	176

4.3 Data Driven Business Model

The term Data Driven Business Model has been used in areas related to Business, and engineering. Figure 10 shows the Tree map generated by Web of Science.

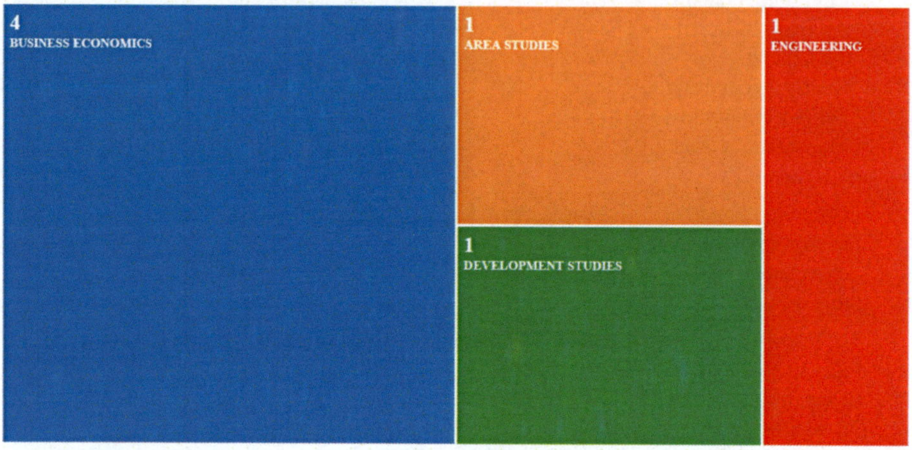

Fig. 10. Sectors that contain the term "Data Driven Business Model" in the TOPIC field. Source: Web of Science, December 2019.

This term is published for the first time, within the area of Business, Management, Economics, and Finance, in 2016. The article mentioned is "Capturing value from big data - a taxonomy of data-driven business models used by start-up corporations "written by [28] for the INTERNATIONAL JOURNAL OF OPERATIONS & PRODUCTION MANAGEMENT. This article has 30 citations.

Table 5 shows the four publications made to date of this work. During 2018 and 2019, no article with this term was published in TOPIC.

4.4 Combination of Terms

To find a relationship between the three concepts presented in this work, "Sharing Economy", "Transparence in Business" and "Data Driven Business Model", a combination of the words was used to search the Web of Science papers that have already related these concepts to each other. All possible combinations were made while maintaining the original terms. The results are analyzed below.

Combination 1: "Sharing Economy" and "Transparency in Business" and "Data Driven Business Model"
The result of the search with this combination of terms was ZERO published papers, therefore, to date there is no publication on the Web of Science with these concepts together in a single article.

Combination 2: "Sharing Economy" and "Transparency in Business"
The result of the search with this combination of terms was ZERO published papers,

Table 5. Papers with the terms Data Driven Business Model Source: Web of Science, December 2019.

R	Title	Authors	Source title	Year	Citations
1	Capturing value from big data - a taxonomy of data-driven business models used by start-up firms	Hartmann, Philipp Max; Zaki, Mohamed; Feldmann, Niels; Neely, Andy	International Journal of Operations & Production Management	2016	30
2	Data-Driven Business Model Innovation	Sorescu, Alina	Journal of Product Innovation Management	2017	9
3	Big data-driven business model innovation by traditional industries in the Chinese economy	Cheah, Sarah; Wang, Shenghui	Journal of Chinese Economic and Foreign Trade Studies	2017	6
4	The Influence of Big Data on Transformation of Traditional Manufacturing Enterprises from Business Model Perspective	Chen Jie	Proceedings of The Fourth International Symposium - Management, Innovation & Development, Bk One & Two	2017	0

therefore, to date there is no publication on the Web of Science with these concepts together in a single article.

Combination 3: "Sharing Economy" and "Data Driven Business Model"
The result of the search with this combination of terms was ZERO published papers, therefore to date there is no publication on the Web of Science with these concepts together in a single article.

Combination 4: "Transparency in Business" and "Data Driven Business Model".
The result of the search with this combination of terms was ZERO published papers, therefore, to date there is no publication on the Web of Science with these concepts together in a single article.

The results show that the three concepts presented in this work have not been published together in any paper of the Web of Science. This may be an indicator that it may be a new line of research to address.

To broaden the range, the search was carried out again using only the word "Transparency" and combining it with the terms "Sharing Economy" and "Data Driven Business Model", all in the TOPIC field. The results are shown below.

Combination 5: "Transparency" and "Data Driven Business Model"
The result of the search with this combination of terms was ZERO published papers, therefore, to date there is no publication on the Web of Science with these concepts together in a single article.

Combination 6: "Transparency" and "Sharing Economy"
The result was 20 articles in various areas that can be seen in Fig. 11, which shows the Tree map of the Web of Science.

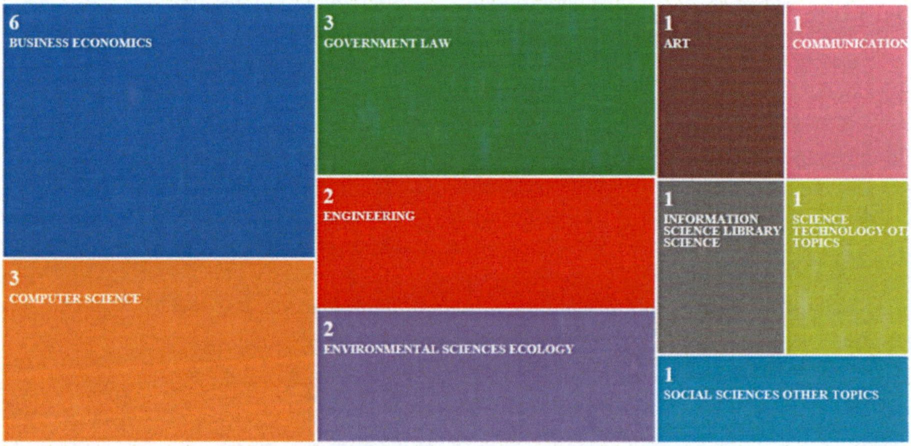

Fig. 11. Sectors that contain the combination of the terms "Transparency" and "Sharing Economy". Source: Web of Science, December 2019.

When filtering by the Business, Management, Economics, and Finance areas, the result is 6 papers with the combination of Transparency and Sharing Economy. Only one article presents 13 citations: "Identify Innovative Business Models: Can Innovative Business Models Enable Players to React to Ongoing or Unpredictable Trends?", Written by [29] for the ENTREPRENEURSHIP RESEARCH JOURNAL.

The titles of the papers can be seen in Table 6 and suggest a relationship of the terms sought with issues such as sustainability, crowdfunding, and accounting methods.

5 Graphical Analysis

The results shown in the previous section have been analyzed based on specific criteria and using visualization tools such as graphs and tables. However, an interesting aspect to analyze is how the bibliographic data obtained are connected, to identify relationships between different variables. One way to do this is to perform a graphic analysis

Table 6. Papers with the terms Transparency and Sharing Economy Source: Web of Science, December 2019.

R	Title	Authors	Source title	Year	Citations
1	Identify Innovative Business Models: Can Innovative Business Models Enable Players to React to Ongoing or Unpredictable Trends?	Pisano, Paola; Pironti, Marco; Rieple, Alison	Entrepreneurship Research Journal	2015	13
2	Business model development for sustainable apparel consumption The case of Houdini Sportswear	Holtstrom, Johan; Bjellerup, Charlotte; Eriksson, Johanna	Journal of Strategy and Management	2019	0
3	Technology environment and crowdfunding platforms in Brazil	Lima, Afonso; Araujo, Francisco Fabiano Mapurunga	Rege-Revista De Gestao	2019	0
4	Close encounters and the illusion of accountability in the sharing economy	McDaid, Emma; Boedker, Christina; Free, Clinton	Accounting Auditing & Accountability Journal	2019	0
5	An Assessement of Algorithmic Accountability Methods	Voinea, Cristina; Uszkai, Radu	Proceedings of the 12th International Management Conference: Management Perspectives In The Digital Era (Imc 2018)	2018	0
6	Digitalization Fostering Business Cooperation - Reality or Fiction?	Jankowska, Barbara; Polowczyk, Jan	Innovation, Entrepreneurship and Digital Ecosystems	2016	0

[30]. The VOS viewer software developed by [31], allows visualizing the bibliographic information showing the connections between the elements.

For the graphic visualization using the VOS software, a single content file was created of all the articles and proceedings found in the Web of Science, in the areas of Business, Management, Economics, and Finance, referring to the concepts "Sharing Economy", "Transparency in Business "and" Data Driven Business Model ". A total of 12,480 papers were analyzed.

First, let us look at the co-citations between the journals analyzed. In this case, the VOS software found 245 relationships that met a minimum of 20 citations between them. Figure 12 shows the result for publications that present co-citations between journals.

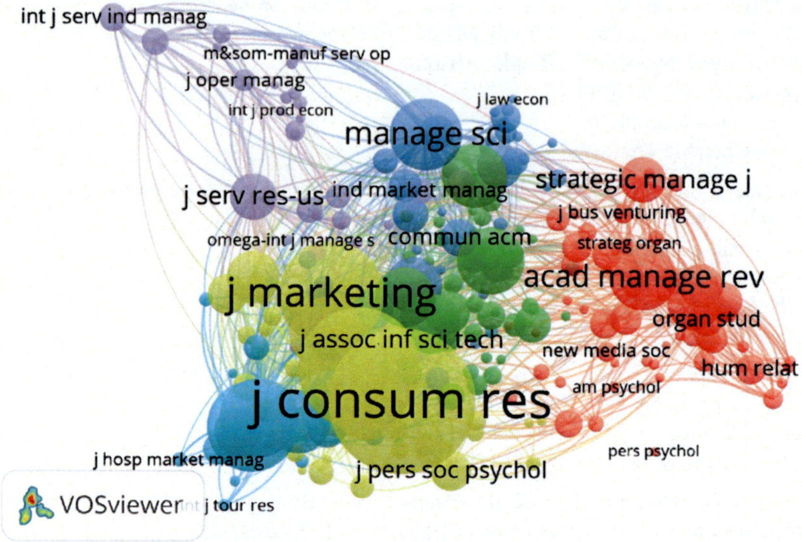

Fig. 12. Co-citation of journals of papers that contain the terms "Sharing Economy", "Transparency in Business" and "Data Driven Business Model". Source: VOS viewer software, December 2019.

Figure 13 shows the relationships between the keywords of the papers analyzed. The software identifies seven clusters with 210 relationships between the elements. The most prominent cluster is the one led by the word Sharing Economy.

Finally, Fig. 14 shows the map of co-citations by country. In this case, the software recognizes 6 clusters made up of 69 countries. The leading country is the United States, followed by China and England.

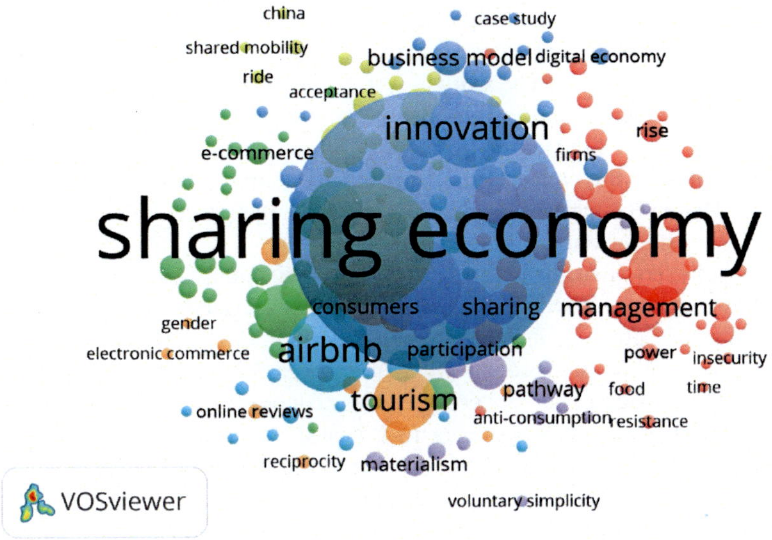

Fig. 13. Keywords visualization. Source: VOS viewer software, December 2019.

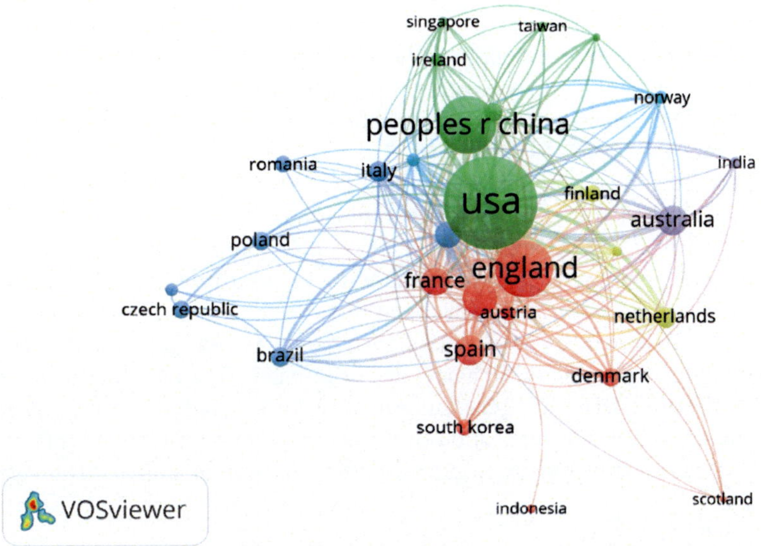

Fig. 14. Countries co-citation. Source: VOS viewer software, December 2019.

6 Conclusions

This study provides a bibliometric overview of the publications registered in the Web of Science Core Collection, which use the concepts of "Sharing Economy", "Transparency in Business" and "Data Driven Business Model". The analysis collected information until December 2019, limiting the search to articles and proceedings in the Business,

Management, Economics, and Finance areas of the Web of Science. Several rankings were developed for different criteria that include papers, authors, journals and the evolution of concepts over time. The work also uses the VOS viewer software to graphically map the bibliographic data using co-citation between journals, keywords, and countries.

The objective of the study is to identify the relationship existing in the scientific literature, between the concepts of "Sharing Economy", "Transparency in Business" and "Data Driven Business Model" to identify if, together, can be an element to consider in the competitive advantage of companies.

The results show that, in the case of the "Sharing Economy" concept, several scientific papers have been developed with a total of 550 papers. It has also been seen that interest in the subject is growing in recent years. The evolution of the term and the use of synonyms such as "access economy", "peer-to-peer (P2P) economy", "gig economy", "collaborative consumption" or "collaborative economy" could also be seen.

In the case of the "Transparency in Business" concept, the result found was 10 papers that used this term. Therefore, the search was extended by placing only the term "Transparency" to obtain a total of 2,235 papers with that term in the title. The analysis of the term "Transparency" shows that it is associated with financial and accounting aspects which seems to indicate that the aspect related to money is one of the aspects that is most considered to generate trust. It could also be an indicator that transparency is being considered as a competitive advantage of companies.

The search for the concept of "Data Driven Business Model" yielded a total of 6 articles. The term is very new, and conclusions based on the results cannot yet be drawn. As well, the term has been used in other areas such as engineering and health, in addition to business. This opens a branch of scientific interest due to the growth of data-based business' models.

The combination of the terms to search for papers showed that there are no registered works on the Web of Science with these combinations. The only combination that showed results was that of "Transparency" and "Sharing Economy" with 20 published papers. It is worth delving into the analysis of these works to ratify if it is a direct relationship.

The fact that no paper has been found that combines the three concepts, "Sharing Economy", "Transparency in Business" and "Data Driven Business Model", can be interpreted as an opportunity for a new line of research. The bibliometric analysis indicates that the three terms are generating interest for themselves for the companies, so, to advance in a deeper investigation of the union of these three concepts, it can give results that allow modeling good practices to achieve competitive advantages for the companies.

Appendix A

Table Appendix A. Papers that use synonyms terms for "Sharing Economy" together Source: Web of Science, December 2019.

R	Title	Authors	Terms used in the same paper
1	A comparison of GMVN and SE model based on the 4C framework	Luo Yi-ning; Rong Ke; Shi Yong-jiang	Collaborative Consumption, Peer-to-peer Economy
2	A triadic framework for collaborative consumption (CC): Motives, activities and resources & capabilities of actors	Benoit, Sabine; Baker, Thomas L.; Bolton, Ruth N.; Gruber, Thorsten; Kandampully, Jay	Collaborative Consumption, Sharing Economy
3	Benefits and Threats of Collaborative Consumption	Silwanowicz, Tomasz	Collaborative Consumption, Collaborative Economy
4	Business model innovation and value-creation: the triadic way	Andreassen, Tor Wallin; Lervik-Olsen, Line; Snyder, Hannah; Van Riel, Allard C. R.; Sweeney, Jillian C.; Van Vaerenbergh, Yves	Collaborative Consumption, Sharing Economy
5	Can You Gig It? An Empirical Examination of the Gig Economy and Entrepreneurial Activity	Burtch, Gordon; Carnahan, Seth; Greenwood, Brad N	Gig EconomySharing Economy
6	Chinese travelers' behavioral intentions toward room-sharing platforms. The influence of motivations, perceived trust, and past experience	Wu, Jiang; Zeng, Minne; Xie, Karen L	Collaborative Consumption, Sharing Economy
7	Colaborative economy: a new market for the social economy	Alfonso Sanchez, Rosalia	Collaborative Consumption, Collaborative Economy
8	Collaborative apparel consumption in the digital sharing economy: An agenda for academic inquiry	Park, Hyejune; Armstrong, Cosette Marie Joyner	Collaborative Consumption, Sharing Economy
9	COLLABORATIVE BRANDS - DIGITAL VALUE CREATION IN THE COLLABORATIVE ECONOMY	Harmaala, Minna-Maari	Collaborative Consumption, Collaborative EconomyPeer-to-peer Economy

(*continued*)

Table Appendix A. (*continued*)

R	Title	Authors	Terms used in the same paper
10	Collaborative Consumption: Strategic and Economic Implications of Product Sharing	Jiang, Baojun; Tian, Lin	Collaborative Consumption, Sharing Economy
11	Collaborative consumption: the role of familiarity and trust among Millennials	Mittendorf, Christoph	Collaborative Consumption, Sharing Economy
12	Collaborative Economy and Tourism	Dredge, Dianne; Gyimothy, Szilvia	Collaborative Consumption, Collaborative Economy
13	Collaborative Economy As An Indication of Sustainable Consumption In Results of Examined People	Tluczak, Agnieszka; Kauf, Sabina	Collaborative Consumption, Collaborative Economy
14	Collaborative Economy: Analysis of Scientific Production In Academic Magazines	Duran-Sanchez, Amador; Alvarez-Garcia, Jose; de la Cruz del Rio-Rama, Maria; Patricia Maldonado-Erazo, Claudia	Collaborative Consumption, Collaborative Economy
15	Collaborative fashion consumption and its environmental effects	Iran, Samira; Schrader, Ulf	Collaborative Consumption, Sharing Economy
16	Collaborative practices and consumerist habitus: An analysis of the transformative mechanisms of collaborative consumption	Herbert, Maud; Collin-Lachaud, Isabelle	Collaborative Consumption, Collaborative Economy
17	Commercial orientation in grassroots social innovation: Insights from the sharing economy	Martin, Chris J.; Upham, Paul; Budd, Leslie	Collaborative Consumption, Sharing Economy
18	Community and Connection: Exploring Non-monetary Aspects of the Collaborative Economy Through Recreation Vehicle Use	Hardy, Anne	Collaborative Consumption,Collaborative Economy

(*continued*)

Table Appendix A. (*continued*)

R	Title	Authors	Terms used in the same paper
19	Definitions and Mapping the Landscape in the Collaborative Economy	Gyimothy, Szilvia; Dredge, Dianne	Collaborative Consumption,Sharing Economy
20	De-ownership orientation and collaborative consumption during turbulent economic times	Lindblom, Arto; Lindblom, Taru	Collaborative Consumption,Sharing Economy
21	Describing The Collaborative Economy: Forms of Food Sharing Initiatives	Privitera, Donatella	Collaborative Consumption,Collaborative Economy
22	Design and the creation of meaningful consumption practices in access-based consumption	Gruen, Adele	Access Economy Collaborative Consumption,
23	Diversifying and de-growing the circular economy: Radical social transformation in a resource-scarce world	Hobson, Kersty; Lynch, Nicholas	Collaborative Consumption,Sharing Economy
24	Does shared consumption affect consumers' values, attitudes, and norms? A panel study	Roos, Daniel; Hahn, Ruediger	Collaborative Consumption,Sharing Economy
25	Dynamic Pricing in a Labor Market: Surge Pricing and Flexible Work on the Uber Platform	Chen, M. Keith	Gig Economy Sharing Economy
26	Entrepreneurship for Long-term Care in Sharing Economy	Huarng, Kun-Huang	Collaborative Consumption,Sharing Economy
27	From Ownership to Access: How The Sharing Economy Is Changing The Consumer Behavior	Barbu, Catalin Mihail; Laurentiu Florea, Dorian; Ogarca, Radu Florin; Barbu, Mihai Constantin Razvan	Collaborative Consumption,Collaborative Economy Sharing Economy

(*continued*)

Table Appendix A. (*continued*)

R	Title	Authors	Terms used in the same paper
28	Future scenarios of the collaborative economy: Centrally orchestrated, social bubbles or decentralized autonomous?	Fehrer, Julia A.; Benoit, Sabine; Aksoy, Lerzan; Baker, Thomas L.; Bell, Simon J.; Brodie, Roderick J.; Marimuthu, Malliga	Collaborative Consumption,Collaborative Economy
29	I do not own a car anymore: An analysis of possessions' disposal and changes in consumers' identities	Sandes, Fabio Shimabukuro; Leandro, Julio; Boaventura, Patricia; da Silva Junior, Adelson Ferreira	Access Economy Collaborative Consumption,Collaborative Economy
30	Improvising Economy: Everyday Encounters and Tourism Consumption	Johannesson, Gunnar Thor; Lund, Katrin Anna	Collaborative Consumption,Collaborative Economy
31	In sharing economy, we trust: the effects of host attributes on short-term rental purchases	Wu, Jiang; Ma, Panhao; Xie, Karen L	Peer-to-peer Economy Sharing Economy
32	Innovations in The Institutional Modelling Of The Sharing Economy	Popov, Evgeniy; Hercegova, Katariprimena; Semyachkov, Konstantin	Gig Economy Sharing Economy
33	Intermediation in a Sharing Economy: Insurance, Moral Hazard, and Rent Extraction	Weber, Thomas A	Collaborative Consumption,Sharing Economy
34	Is the Sharing Economy Unfair Competition And Should It Be Regulated?	Svecova, Lenka; Veber, Jaromir	Access Economy Collaborative Consumption,
35	Lateral Exchange Markets: How Social Platforms Operate in a Networked Economy	Perren, Rebeca; Kozinets, Robert V	Collaborative Consumption,Sharing Economy
36	Mapping the sharing economy for sustainability research	Plewnia, Frederik; Guenther, Edeltraud	Collaborative Consumption,Collaborative Economy Sharing Economy
37	Mapping themes in the study of new work practices	Aroles, Jeremy; Mitev, Nathalie; de Vaujany, Francois-Xavier	Collaborative Economy Gig Economy

(*continued*)

Table Appendix A. (*continued*)

R	Title	Authors	Terms used in the same paper
38	Mine is yours? Using sentiment analysis to explore the degree of risk in the sharing economy	Chang, Wei-Lun; Wang, Jia-Yin	Collaborative Consumption,Sharing Economy
39	Negotiating labour standards in the gig economy: Airtasker and Unions New South Wales	Minter, Kate	Gig Economy Sharing Economy
40	Peer-To-Peer Accommodation: Drivers and User Profiles	Pesonen, Juho; Tussyadiah, Iis	Collaborative Consumption,Collaborative Economy Sharing Economy
41	Peer-to-peer collaborative consumption for fashion products in the sharing economy: Platform operations	Choi, Tsan-Ming; He, Yanyan	Collaborative Consumption,Sharing Economy
42	Peer-to-Peer Product Sharing: Implications for Ownership, Usage, and Social Welfare in the Sharing Economy	Benjaafar, Saif; Kong, Guangwen; Li, Xiang; Courcoubetis, Costas	Collaborative Consumption,Sharing Economy
43	Policy and Regulatory Challenges in the Tourism Collaborative Economy	Dredge, Dianne	Collaborative Economy Sharing Economy
44	Product Pricing in a Peer-to-Peer Economy	Weber, Thomas A	Peer-to-peer Economy Sharing Economy
45	Product Pricing in a Peer-to-Peer Economy	Weber, Thomas A	Collaborative Consumption,Collaborative Economy
46	Promises and paradoxes of the sharing economy: An organizing framework	Acquier, Aurelien; Daudigeos, Thibault; Pinkse, Jonatan	Access Economy Collaborative Consumption,Sharing Economy
47	Seismic Shifts in the Sharing Economy: Shaking Up Marketing Channels and Supply Chains	Ferrell, O. C.; Ferrell, Linda; Huggins, Kyle	Access Economy Gig Economy Sharing Economy

(*continued*)

Table Appendix A. (*continued*)

R	Title	Authors	Terms used in the same paper
48	Share more, drive less: Millennials value perception and behavioral intent in using collaborative consumption services	Hwang, Jiyoung; Griffiths, Merlyn A	Collaborative Consumption,Collaborative Economy Sharing Economy
49	Sharing economies: moving beyond binaries in a digital age	Davies, Anna R.; Donald, Betsy; Gray, Mia; Knox-Hayes, Janelle	Gig Economy Sharing Economy
50	Sharing economy workers: selling, not sharing	Ravenelle, Alexandrea J	Gig Economy Sharing Economy
51	Sharing Economy: Challenges and Opportunities In Tourism	Onete, Cristian Bogdan; Plesea, Doru; Budz, Sonia	Collaborative Consumption,Collaborative Economy
52	Sharing intangibles: Uncovering individual motives for engagement in a sharing service setting	Milanova, Veselina; Maas, Peter	Collaborative Consumption,Sharing Economy
53	Social Initiatives in Food Consumption and Distribution as Part of Sustainable Consumption and Sharing Economy	Bachnik, Katarzyna; Szumniak-Samolej, Justyna	Collaborative Consumption,Collaborative Economy
54	Sticky Norms, Endogenous Preferences, and Shareable Goods	Fremstad, Anders	Collaborative Consumption,Sharing Economy
55	Supplier Resources in the Sharing Economy: Three Regulatory Concerns	Thorne, Debbie M.; Quinn, Floyd F	Gig Economy Sharing Economy
56	Tackling Bogus Self-Employment: Some Lessons from Romania	Williams, Colin C.; Horodnic, Ioana Alexandra	Collaborative Economy Gig Economy
57	The Collaborative Economy Based Analysis of Demand: Study of Airbnb Case in Spain and Portugal	Palos-Sanchez, Pedro R.; Correia, Marisol B	Collaborative Consumption,Collaborative Economy Sharing Economy

(*continued*)

Table Appendix A. (*continued*)

R	Title	Authors	Terms used in the same paper
58	The Collaborative Economy: a new Economic Model	Manuel Sastre-Centeno, Jose; Elena Inglada-Galiana, Maria	Collaborative Consumption,Collaborative Economy
59	The Conceptualization of the Sharing Economy as a Business Model	Siuskaite, Donata; Pilinkiene, Vaida; Zvirdauskas, Dainius	Collaborative Consumption,Collaborative Economy
60	The eruption of Airbnb in tourist cities: Comparing spatial patterns of hotels and peer-to-peer accommodation in Barcelona	Gutierrez, Javier; Carlos Garcia-Palomares, Juan; Romanillos, Gustavo; Henar Salas-Olmedo, Maria	Collaborative Consumption,Sharing Economy
61	The Global Platform Economy: A New Offshoring Institution Enabling Emerging-Economy Microproviders	Lehdonvirta, Vili; Kassi, Otto; Hjorth, Isis; Barnard, Helena; Graham, Mark	Gig Economy Sharing Economy
62	The integration of item-sharing and crowd shipping: Can collaborative consumption be pushed by delivering through the crowd?	Behrend, Moritz; Meisel, Frank	Collaborative Consumption,Sharing Economy
63	The Reflection of The Shared Economy In National Accounting - The Czech Republic Case	Makovsky, Petr	Collaborative Economy Gig Economy
64	The role of technology in collaborative consumer communities	Guillemot, Samuel; Privat, Helene	Collaborative Consumption,Collaborative Economy
65	The Sharing Economy and the Future of Personal Mobility: New Models Based on Car Sharing	Novikova, Olga	Collaborative Consumption,Sharing Economy
66	The sharing economy as the commons of the 21st century	Bradley, Karin; Pargman, Daniel	Collaborative Economy Sharing Economy

(*continued*)

Table Appendix A. (*continued*)

R	Title	Authors	Terms used in the same paper
67	The sharing economy in social media: Analyzing tensions between market and non-market logics	Laurell, Christofer; Sandstrom, Christian	Collaborative Consumption,Sharing Economy
68	The sharing economy: A pathway to sustainability or a nightmarish form of neoliberal capitalism?	Martin, Chris J	Collaborative Consumption,Collaborative Economy Sharing Economy
69	The sharing economy: Your business model's friend or foe?	Kathan, Wolfgang; Matzler, Kurt; Veider, Viktoria	Collaborative Consumption,Sharing Economy
70	To earn is not enough: A means-end analysis to uncover peer-providers' participation motives in peer-to-peer carsharing	Wilhelms, Mark -Philipp; Henkel, Sven; Falk, Tomas	Collaborative Consumption,Sharing Economy
71	Understanding collaborative consumption: Test of a theoretical model	Barnes, Stuart J.; Mattsson, Jan	Collaborative Consumption,Collaborative Consumption,Sharing Economy
72	Upsides and downsides of the sharing economy: Collaborative consumption business models' stakeholder value impacts and their relationship to context	Dreyer, Betine; Ludeke-Freund, Florian; Hamann, Ralph; Faccer, Kristy	Collaborative Consumption,Sharing Economy
73	Value co-creation and co-destruction in the Airbnb sharing economy	Camilleri, Jeannette; Neuhofer, Barbara	Collaborative Consumption,Sharing Economy
74	What makes an Airbnb host a superhost? Empirical evidence from San Francisco and the Bay Area	Gunter, Ulrich	Collaborative Consumption,Sharing Economy
75	What managers should know about the sharing economy	Habibi, Mohammad Reza; Davidson, Alexander; Laroche, Michel	Collaborative Consumption,Peer-to-peer Economy Sharing Economy

(*continued*)

Table Appendix A. (*continued*)

R	Title	Authors	Terms used in the same paper
76	When the sharing economy becomes neoliberalism on steroids: Unravelling the controversies	Murillo, David; Buckland, Heloise; Val, Esther	Collaborative Consumption,Sharing Economy
77	Why people participate in the sharing economy: an empirical investigation of Uber	Lee, Zach W. Y.; Chan, Tommy K. H.; Balaji, M. S.; Chong, Alain Yee-Loong	Access Economy Collaborative Consumption,Sharing Economy
78	You are what you can access: Sharing and collaborative consumption online	Belk, Russell	Collaborative Consumption,Sharing Economy

References

1. Matzler, K., Veider, V., Kathan, W.: Adapting to the sharing economy. MIT Sloan Manag. Rev. **56**(2), 71–77 (2015)
2. Carballa Smichowski, B.: Data as a common in the sharing economy: a general policy proposal. HAL, (2016)
3. Alfonso Sánchez, R.: Economía colaborativa: un nuevo mercado para la economía social. Revista de Economía Pública, Social y Cooperativa **88**, 231–258 (2017)
4. Botsman, R., Rogers, R.: What's Mine is Yours: How Collaborative Consumption is Changing the way we Live. Collins, London (2010)
5. Hamari, J., Sjöklint, M., Ukkonen, A.: The sharing economy: Why people participate in collaborative consumption. J. Am. Soc. Inf. Sci. **67**(9), 2047–2059 (2016)
6. Vaughan, R., Daverio, R.: Assessing the size and presence of the collaborative economy in Europe. Publications Office of the European Union (2016)
7. Yaraghi, N., Ravi, S.: The current and future state of the sharing economy. Brookings India IMPACT 032017 (2017)
8. Cannon, S., Summers, L.H.: How uber and the sharing economy can win over regulators. Harv. Bus. Rev. **13**(10), 24–28 (2014)
9. Good Rebels. Data Driven Marketing the DNA of customer oriented companies. https://www.goodrebels.com/wp-content/uploads/2017/10/data_driven.pdf. Accessed 15 Nov 2019
10. Tapscot, D., Ticoll, D.: The Naked Corporation: How the Age of Transparency will Revolutionize Business. Simon and Schuster, New York (2013)
11. Morey, T., Forbath, T., Schoop, A.: Customer data: designing for transparency and trust. Harv. Bus. Rev. **93**(5), 96–105 (2015)
12. Sedkaoui, S.: Data Analytics and Big Data. Wiley, Hoboken
13. Grandhi, B., Patwa, N., Saleem, K.: Data driven marketing for growth and profitability. In: 10th Annual Conference of the EuroMed Academy of Business, pp. 675–694. Euromed Press, Rome (2018)
14. Erevelles, S., Fukawa, N., Swayne, L.: Big data consumer analytics and the transformation of marketing. J. Bus. Res. **69**(2), 897–904 (2016)

15. Merigó, J.M., Gil-Lafuente, A.M., Yager, R.R.: An overview of fuzzy research with bibliometric indicators. Appl. Soft Comput. **27**, 420–433 (2015)
16. Blanco-Mesa, F., Merigó, J.M., Gil-Lafuente, A.M.: Fuzzy decision making: a bibliometric-based review. J. Intell. Fuzzy Syst. **32**(3), 2033–2050 (2017)
17. Merigó, J.M., Mas-Tur, A., Roig-Tierno, N., Ribeiro-Soriano, D.: A bibliometric overview of the journal of business research between 1973 and 2014. J. Bus. Res. **68**(12), 2645–2653 (2015)
18. Merigó, J.M., Yang, J.B.: A bibliometric analysis of operations research and management science. **73**, 37–48 (2017)
19. Valenzuela, L.M., Merigó, J.M., Johnston, W.J., Nicolas, C., Jaramillo, J.F.: Thirty years of the journal of business and industrial marketing: a bibliometric analysis. J. Bus. Ind. Mark. (2017)
20. Laengle, S., Modak, N.M., Merigo, J.M., Zurita, G.: Twenty-five years of group decision and negotiation: a bibliometric overview. Group Decis. Negot. **27**(4), 505–542 (2018)
21. Hirsch, J.E.: An index to quantify an individual's scientific research output. Proc. Natl. Acad. Sci. **102**(46), 16569–16572 (2005)
22. Gaviria-Marin, M., Merigó, J.M., Baier-Fuentes, H.: Knowledge management: a global examination based on bibliometric analysis. Technol. Forecast. Soc. Chang. **140**, 194–220 (2019)
23. Merigó, J.M., Pedrycz, W., Weber, R., de la Sotta, C.: Fifty years of information sciences: a bibliometric overview. Inf. Sci. **432**, 245–268 (2018)
24. Belk, R.: You are what you can access: sharing and collaborative consumption online. J. Bus. Res. **67**(8), 1595–1600 (2014)
25. Herkenhoff, F.: Measuring risks to reputation. In: Klewes, J., Wreschniok, R. (eds) Reputation Capital. Springer, Berlin, Heidelberg (2009). https://doi.org/10.1007/978-3-642-01630-1_13
26. Granados, N., Gupta, A., Kauffman, R.J.: Research commentary—information transparency in business-to-consumer markets: concepts, framework, and research agenda. Inf. Syst. Res. **21**(2), 207–226 (2010)
27. Bushman, R.M., Piotroski, J.D., Smith, A.J.: What determines corporate transparency? J. Account. Res. **42**(2), 207–252 (2004)
28. Hartmann, P.M., Zaki, M., Feldmann, N., Neely, A.: Capturing value from big data–a taxonomy of data-driven business models used by start-up firms. Int. J. Oper. Prod. Manage. (2016)
29. Pisano, P., Pironti, M., Rieple, A.: Identify innovative business models: can innovative business models enable players to react to ongoing or unpredictable trends? Entrep. Res. J. **5**(3), 181–199 (2015)
30. Cobo, M.J., López-Herrera, A.G., Herrera-Viedma, E., Herrera, F.: Science mapping software tools: review, analysis, and cooperative study among tools. J. Am. Soc. Inform. Sci. Technol. **62**(7), 1382–1402 (2011)
31. Waltman, L., Van Eck, N.J., Noyons, E.C.: A unified approach to mapping and clustering of bibliometric networks. J. Informet. **4**(4), 629–635 (2010)

Application of the Forgotten Effects Theory to the Qualitative Analysis of the Operational Risk Events

Agustín Torres Martínez[1(✉)], Anna Maria Gil-Lafuente[1], Aras Keropyan[2], and José M. MerigóLindahl[3,4]

[1] Department of Business Administration, University of Barcelona, Av. Diagonal 690, 08034 Barcelona, Spain
amgil@ub.edu
[2] GRIT-EAE Research Group, EAE Business School, Barcelona, Spain
aras.keropyan@campus.eae.es
[3] Department of Management Control and Information Systems, University of Chile, Santiago, Chile
jmerigo@fen.uchile.cl
[4] School of Information, Systems and Modelling, University of Technology Sydney, Sydney, Australia
Jose.Merigo@uts.edu.au

Abstract. The second agreement of the Committee on Banking Supervision (Basel II) represents a challenge for financial institutions, risk managers and researchers in terms of operational risk management, due to the complexity of developing effective systems to identify, measure, monitor, control and mitigate this risk. This requires performing a lost scenario analysis with expert judgment to decide the priority actions in order to manage properly the different events of loss associated to the operational risk, as well as taking measures derived from loss distributions. In this paper, we have used the forgotten effects theory, which allows us to determine the causal direct and indirect relationships between the causes and effects of the operational losses of a financial institution, throughout the opinions of a group of experts.

Keywords: Operational risk · Forgotten effects theory · Fuzzy logic

1 Introduction

In the last years, the way of management the risk has changed a lot, not only because of the guidelines of the Basel agreements and the regulation of the economic entities in each country, but also by other factors such as globalization and technological sophistication of the banking operations and financial markets. On the other hand, several publicly known problems such as; corporate failures, bankruptcies of entities and corruption have created a great concern about the exposure of the risks in general and particularly the operational risk.

A. M. Gil-Lafuente et al. (Eds.): REDCID 2019, LNNS 388, pp. 261–270, 2022.
https://doi.org/10.1007/978-3-030-93787-4_15

Basel II[1] defined a framework for quantifying the operational risk, suggesting that the calculation of capital requirements related to operational risk are carried out by: Basic Indicator Approach, Standardized Approach, Advanced Measurement Approach (AMA). The first two methods are calculated in a simplified way applying to the average gross annual income the capital requirement. In the third method, the calculation is made on the actual exposure of the institution to the operational risk. In the AMA method, both statistical models and scenario analysis are taken into account, in the opinion of the experts plays an important role to set the risk and define the improvement actions.

For the analysis of qualitative information, there are different tools such as self-assessment questionnaires, risk maps, KRI risk indicators, scorecards and scenario analysis which are mainly adequate to identify potential exposure risks. Other tools for the analysis of the information in uncertain environments can be realized by fuzzy logic techniques.

The forgotten effects theory is a very useful tool for the analysis of complex problems, which requires additional analysis a part from determining the incidence relations between causes and effects to see the interrelations between the causes with the causes and the effects with the effects in order to obtain the indirect relationships that cannot be detected in the first analysis. This technique has been used by several researchers in different fields such as: hospitality industry [1], entrepreneurship [2], social responsibility and environment [3], economic growth and social welfare [4], among others.

The objective of this study is to apply the forgotten effects theory in the qualitative analysis of the operational risk with the opinion of the experts, and seeking the indirect relation impacts between different types of events and operational losses. This approach will enable to improve strategies to mitigate the impact of this type of risks in financial institutions.

2 Operational Risk

The Basel II Committee defines the operational risk as the:

"Risk of loss resulting from inadequate or failed internal processes, people and systems or from external events."

This definition includes the legal risk[2], but excludes the strategic[3] and reputational risks[4].

On the other hand, Basel II defines the event type classification and loss effect types through the loss related to the operational risk. Following, in Tables 1 and 2 these definitions are given:

[1] Basel II is the second agreement of the Basel Committee on Banking Supervision published in 2004 with the aim of establishing necessary capital requirements to protect the institutions against financial and operational risks.

[2] Legal Risk: the possibility of being punished, fined or forced to pay punitive damages resulting from supervisory actions or private agreements between the parties.

[3] Strategic Risk: the possibility of choosing a wrong strategy that may affect the results.

[4] Reputational risk: the possibility of causing loss when the entity's name is affected by the decrease of the confidence of the whole institution.

Table 1. Loss event type classification

Category	Definition
Internal fraud	Losses due to acts of a type intended to defraud, misappropriate property or circumvent regulations, the law or company policy, excluding diversity/ discrimination events, which involves at least one internal party
External fraud	Losses due to acts of a type intended to defraud, misappropriate property or circumvent the law, by a third party
Employment Practices and Workplace Safety	Losses arising from acts inconsistent with employment, health or safety laws or agreements, from payment of personal injury claims, or from diversity / discrimination events
Clients, Products & Business Practices	Losses arising from an unintentional or negligent failure to meet a professional obligation to specific clients (including fiduciary and suitability requirements), or from the nature or design of a product
Damage to Physical Assets	Losses arising from loss or damage to physical assets from natural disaster or other events
Business Disruption and System Failures	Losses arising from disruption of business or system failures
Execution, Delivery and Process Management	Losses from failed transaction processing or process management, from relations with trade counterparties and vendors

Source: QIS 2 - Operational Risk Loss Data. Bank for International Settlements (BIS) (2001) [5]

The losses associated with this type of risk can arise from failures of processes, in the technology, in the performance of people, and also due to the occurrence of extreme external events. For its evaluation, normally following two variables are used:

- The frequency or the probability of occurrence of a risk event, which consists of the detected number of causal risk events.
- The severity or importance of the impact of risk events on the results or assets of the company.

These variables are fundamental to quantify the losses using statistical models. However, the identification of operational risk involves a self-evaluation process and diagnosis that may be contributed by internal audits, being very useful qualitative tools to identify and monitor risks.

Table 2. Loss effect types

Category	Definition
Legal Liability	Judgments, settlements and other legal costs
Regulatory Action	Fines or the direct payment of any other penalties, such as license revocations
Loss or Damage to Assets	Direct reduction in value of physical assets due to some kind of accident (e.g., neglect, accident, fire, earthquake)
Restitution	Payments to third parties on account of operational losses for which the bank is legally responsible
Loss of Recourse	Losses experienced when a third party does not meet its obligations to the bank, and which are attributable to an operational mistake or event (i.e., which could have been avoided even though the counterparty refused or was unable to pay)
Write-down	Direct reduction in the value of assets due to theft, fraud, unauthorized activity or market or credit losses arising as a result of operational events

Source: QIS 2 - Operational Risk Loss Data. Bank for International Settlements (BIS) (2001) [5]

3 Proposed Model

From extensive and deep studies on the incidence or causal relationships that take place at all levels of the natural or social structures, the Forgotten Effects Theory [6] is developed. From this theory some models have been built which allow obtaining all the direct and indirect causality relations, recovering those elements which initially could have been forgotten, totally or partially. Despite the existence of a good control system, there is always the possibility of not considering or forgetting, either voluntarily or involuntarily some incidence relationships which are not always explicit, evident or viewable, and usually they are not perceived directly. It is common that these incidence relationships remain existing and hidden over effects, and provoked by the accumulation of causes.

The incidence concept [7] can be associated to the function idea. Precisely, in all processes of sequential nature, where incidents are transmitted in chains, it is usual to ignore, voluntarily or involuntarily, some stages. Each forgetfulness behaves as a side effect and rebounds on the whole network of incidence relationships, being a kind of combinatorial process.

We start with our approach [8] with the existence of indirect relationship, which means, a cause-effect matrix defined by two sets of elements:

$A = \{a_i/i = 1, 2, \ldots, n\}$, that behave as causes.
$B = \{b_j/j = 1, 2, \ldots, m\}$, that behave as effects.

and a causality relation matrix $[M]$ defined by:

$$[M] = \{\mu_{a_i b_j} \in [0, 1]/i = 1, 2, \ldots, n; j = 1, 2, \ldots m\} \quad (1)$$

being $\mu_{a_ib_j}$ membership characteristic functions of each of the array elements $[M]$ (formed by the corresponding rows of the matrix elements A-causes- and the corresponding columns of the matrix elements B-effects-).

By this way, we can say that the $[M]$ matrix is composed of the estimations in relation to all the effects that the elements of the set A prosecute on the elements of the set B.

The desired objective is to obtain a new matrix that reflects both direct and indirect causal relationships, called as the second-generation relationships.

To achieve this objective, it is necessary to establish the mechanism that will make possible to represent the fact that different causes may have effects on themselves and, at the same time, take into account that some certain effects may also have incidents on themselves. For this reason, it will be necessary to build two additional causal relationships, which will gather the possible effects arising from causes related to each other, on one hand, and effects between each other. These two auxiliary matrices are defined as following:

$$[A] = \left\{\mu_{a_ia_j} \in [0, 1]/i, j = 1, 2, \ldots, n\right\} \tag{2}$$

$$[B] = \left\{\mu_{b_ib_j} \in [0, 1]/i, j = 1, 2, \ldots, m\right\} \tag{3}$$

Both $[A]$ and $[B]$ matrixes are square and reflexive but they are not symmetric.

The obtained result by integrating $[M]$, $[A]$ and $[B]$ matrices will be a new matrix shown by $\left[M^*\right]$ which collects incidents between causes and effects of the first and second generation, i.e., the initial causal relationships to which the potential impact of any other cause or / and any effect is added.

On the other hand, it is acceptable that the difference between the effect matrix of the first and second generation and the initial direct impact matrix will allow understanding the extent of some causal relationships that have been forgotten or avoided:

$$[O] = \left[M^*\right] - [M] \tag{4}$$

It is also possible to know, from the degree of ignoring any incident, the element (either cause or effect) that makes liaison.

This means that the secondary implications of some incidents that have not been considered or taken into account in the right intensity, can lead to some erroneous actions or at least poorly estimated.

4 Application of the Model

The model of operational risk qualitative analysis is applied in the Retail Banking business line, to a financial institution taking the following variables explained in Sect. 2:

Operational risk events variables:

C_1: Internal fraud.
C_2: External fraud.
C_3: Employment Practices and Workplace Safety.
C_4: Clients, Products & Business Practices.

C_5: Damage to Physical Assets.
C_6: Business Disruption and System Failures.
C_7: Execution, Delivery and Process Management.

Types of operational risk losses variables:

E_1: Legal Liability.
E_2: Regulatory Action.
E_3: Loss or Damage to Assets.
E_4: Restitution.
E_5: Loss of Recourse.
E_6: Write-down.

To analyze the relationships between these variables, the following semantic scale has been used:

0: Without incidence.
0.1: Virtually without incidence.
0.2: Almost without incidence.
0.3: Very weak incidence.
0.4: Weak incidence.
0.5: Average incidence.
0.6: Substantial incidence.
0.7: Sufficient incidence.
0.8: Strong incidence.
0.9: Very Strong incidence.
1: The highest incidence.

A group of five experts have estimated the impacts between these variables. The opinion of the experts has been taken into account with equal weight with the following results (Table 3, 4, 5 and 6):

After obtaining the above matrices, we proceed to calculate the second order matrix with cumulative effects, through the composition:

$$\text{MAX-MIN}\left(M^* = [A] \circ [B] \circ [M]\right) \tag{5}$$

Finally, the forgotten effect matrix is calculated $\left(O = \left[M^*\right] - [M]\right)$, by la the difference between the matrix of indirect effects $\left[M^*\right]$ and the matrix of direct effects $[M]$:

Table 3. Estimated incidences between causes and effects [M]

	E_1	E_2	E_3	E_4	E_5	E_6
C_1	0.8	0.9	0.3	0.9	0.7	0.9
C_2	0.7	0.5	0.8	0.5	0.7	0.9
C_3	0.7	0.6	0.9	0.6	0.1	0
C_4	0.9	0.7	0.2	0.8	0.6	0.5
C_5	0.3	0.1	1	0.2	0.2	0
C_6	0.3	0.4	0.8	0.5	0.9	0.6
C_7	0.8	0.7	0.4	0.6	0.8	0.9

Table 4. Estimated incidences between causes [A]

	C_1	C_2	C_3	C_4	C_5	C_6	C_7
C_1	1	0.9	0.6	0.5	0.4	0.7	0.6
C_2	0.8	1	0.3	0.6	0.5	0.7	0.4
C_3	0.4	0.6	1	0.4	0.8	0.8	0.5
C_4	0.5	0.5	0.4	1	0.3	0.6	0.8
C_5	0.2	0.4	0.6	0.3	1	0.5	0.4
C_6	0.9	0.7	0.5	0.5	0.6	1	0.4
C_7	0.7	0.6	0.4	0.5	0.5	0.6	1

Table 5. Estimated incidences between effects [*B*]

↱	E_1	E_2	E_3	E_4	E_5	E_6
E_1	1	0.9	0.4	0.7	0.8	0.7
E_2	0.8	1	0	0.9	0.6	0.5
E_3	0.5	0.3	1	0.6	0.2	0.2
E_4	0.8	0.9	0.3	1	0.4	0.3
E_5	0.5	0.5	0.6	0.7	1	0.8
E_6	0.9	0.9	0.1	0.5	0.6	1

Table 6. Convolution among matrices MAX-MIN$[M^*]$

↱	E_1	E_2	E_3	E_4	E_5	E_6
C_1	0.9	0.9	0.8	0.9	0.8	0.9
C_2	0.9	0.9	0.8	0.8	0.8	0.9
C_3	0.7	0.7	0.9	0.7	0.8	0.8
C_4	0.9	0.9	0.6	0.8	0.8	0.8
C_5	0.6	0.6	1	0.6	0.6	0.6
C_6	0.9	0.9	0.8	0.9	0.9	0.9
C_7	0.9	0.9	0.6	0.7	0.8	0.9

5 Analysis of Results

The Table 7 with the forgotten effects matrix shows us the causalities relations that have not been qualified by the experts. In this case, two most incidences have been taken, which are 0.7 (sufficient incidence) and 0.8 (strong incidence).

Table 7. Forgotten effects matrix [O]

↱	E_1	E_2	E_3	E_4	E_5	E_6
C_1	0.1	0	0.5	0	0.1	0
C_2	0.2	0.4	0	0.3	0.1	0
C_3	0	0.1	0	0.1	0.7	0.8
C_4	0	0.2	0.4	0	0.2	0.3
C_5	0.3	0.5	0	0.4	0.4	0.6
C_6	0.6	0.5	0	0.4	0	0.3
C_7	0.1	0.2	0.2	0.1	0	0

The first case shows that, the experts had qualified the impact between C3 (Employment Practices and Workplace Safety) and E5 (Loss of Recourse) with the value 0.1 (virtually without incidence), but throughout the forgotten effects theory, the impact is found as 0.7 when the relationship between these two variables are analyzed with C6 (Business Disruption and System Failures). This can be shown graphically as following (Fig. 1):

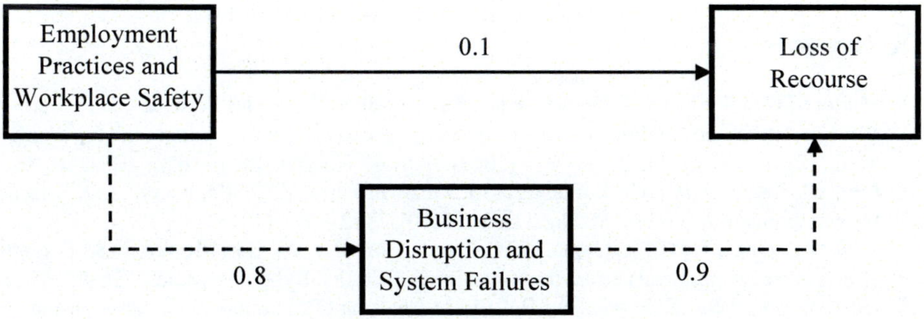

Fig. 1. Forgotten effect between variables C_3 and E_5

In the second case, the relation between C3 (Employment Practices and Workplace Safety) and E6 (Write-down) the experts qualified it initially with the 0 value (Without incidence). Nevertheless, the impact of those two variables is 0.8 (strong incidence) taking into account the forgotten effect impact in which is involved C6 (Business Disruption and System Failures) and E5 (Loss of Recourse). It is important to underline that the first case is included in the second case. This can be shown graphically as following (Fig. 2):

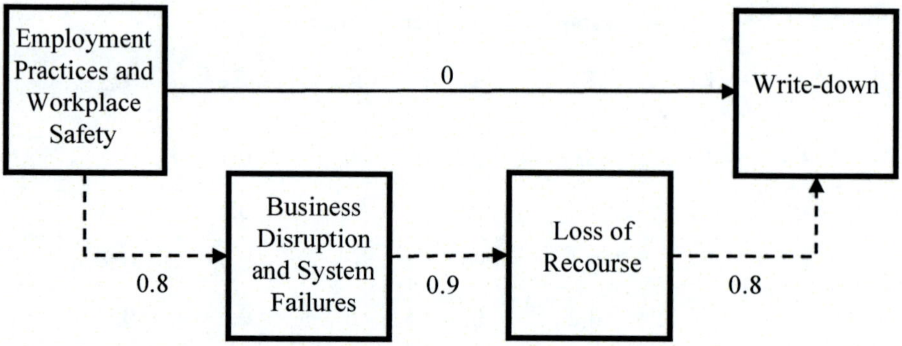

Fig. 2. Forgotten effect between variables C_3 and E_6

6 Conclusion

The great complexity in the operational risk management requires the usage of multiple quantitative and qualitative tools. This will allow taking the best decisions mitigating its economic impact as well as in other impacts such as the quality of service, reputation and social responsibility among others. In qualitative analysis, this study provides a new insight into the causal relationships between the events types and the impact of operational losses.

Final results show that the cause and effect that had no impact initially may indirectly get to have a strong impact by analyzing the intermediate relationships with other causes and effects. This analysis enables to improve the decision-making with a significant impact on the measures taken to mitigate the impact of operational risk in any business.

References

1. Martorell-Cunill, O., Gil-Lafuente, A.M., Socias Salvà, A., Mulet Forteza, C.: The growth strategies in the hospitality industry from the perspective of the forgotten effects. Comput. Math. Organ. Theory **20**(2), 195–210 (2013). https://doi.org/10.1007/s10588-013-9167-9
2. Maqueda Lafuente, F.J., Gil-Lafuente, A.M., Guzman-Parra, V.F., Gil Lafuente, J.: Key factors for entrepreneurial success. Manag. Decis. **51**(10), 1932–1944 (2013)
3. Vizuete Luciano, E., Gil-Lafuente, A.M., García González, A., Boria-Reverter, S.: Forgotten effects of corporate social and environmental responsibility. Kybernetes **42**(5), 736–753 (2013)
4. Gil-Lafuente, A.M., Keropyan, A., Balvey, J.: Analysis of the relationship between the economic activity of regions and welfare of its inhabitants. Econom. Comput. Econom. Cybernet. Stud. Res. **49**(3), 215–232 (2015)
5. Bank for International Settlements (BIS): QIS 2 - Operational Risk Loss Data (2001). http://www.bis.org/bcbs/qisoprisknote.pdf. Accessed 05 Sept 2019
6. Kaufmann, A., Gil Aluja, J.: Modelos para la investigación de efectos olvidados. Milladoiro, Vigo (1988)
7. Gil-Lafuente, A.M.: Elements teòrics i Tècnics de la incertesa, Incertesa i Bio-enginyeria, Reial Academia de Doctors, Barcelona (2008)
8. Gil-Lafuente, A.M.: Fuzzy Logic in Financial Analysis. Springer, Heidelberg (2005)
9. Gil-Aluja, J., Gil-Lafuente, A.M.: Towards an Advanced Modelling of Complex Economic Phenomena. Springer, Heidelberg (2012)

Business Informality: An Application of the Forgotten Effects Model

Marlene Rocío Moscoso-Quiceno(✉) and Sonia Elena Godoy-Hortua

Corporación de Estudios Tecnológicos del Norte del Valle, Cartago-Valle, Colombia
investigacion@cotecnova.edu.co

Abstract. Informal Business as a multidimensional and heterogeneous phenomenon, whose definition and scope can be observed, analyzed and intervened from different perspectives, has become one of the greatest obstacles for the growimg and consolidation of companies, considered fundamental pillars for the progress and development of the socioeconomic situation of nations regarding this type of business; not only for their contribution to productive expansion, but also as generators of employment and wealth. From this perspective, the theory of forgotten effects proposed by Kaufmann and Gil Aluja (1988), applied to the practical case of business informality in the municipality of Cartago, Valle del Cauca, Colombia, this factor allows revealing relationships of causality and effect, based on all those forgotten elements, which under conditions of uncertainty and at first glance are not easy to consider in the business formalization process, but, if not taken into account, it would ignore important variables for decision-making and the design of strategies for growing and social welfare.

Keywords: Forgotten effects · Business informality · Economic growth

1 Introduction

The business community is one of the determinants of prosperity and economic development for the regions. In this context, competitiveness and innovation has become a great challenge for all the nations in the world. However, one of the greatest obstacles for business development[1] is the so-called informality, which for the purposes of this study has been approached from a business perspective, which involves all those productive activities of licit goods and services that have not been registered with the authorities .

[1] "Business development articulates different elements with which the entrepreneur can lead an organization towards the achievement of its objectives. Elements such as economic growth, business culture, leadership, knowledge management and innovation". (Delfín and Acosta 2016, p. 187).

A. M. Gil-Lafuente et al. (Eds.): REDCID 2019, LNNS 388, pp. 271–287, 2022.
https://doi.org/10.1007/978-3-030-93787-4_16

And whose definition has been transformed over the time, giving rise to a combination of approaches that highlight the interrelationship between labor informality characterized by informal employment and business informality supported in the informal sector[2].

From a business perspective, the structuralist approach is supported by the postulate that the informal economy is made up of economic units (workers and micro-enterprises) that serve capitalist companies as a means to increase competitiveness, Using the reduction of inputs and cheap labor (Portes et al. 1989). From the institutionalist approach, informality is a "voluntary decision" and free of individuals to carry out informal activities, which occurs as a result of legal and institutional barriers imposed by the State (Perry et al. 2007). On the subject, the ILO (2013) considers that, to understand the compendium of informality, a broad, interdisciplinary and coherent approach is necessary; oriented to an adequate transition to formality.

On the other hand, Kaufmann and Gil Aluja (1988) argue that when trying to explain a reality or solve a problem, natural errors or forgetfulness are usually incurred voluntarily or involuntarily, given that there is a natural interdependence of causes and effects, which even under conditions of strict control are not appreciated directly. In the same way, "Given the process of globalization in the economic situation of all the countries it is mandatory to establish a series of mechanisms that allow to work in of the information either objective as subjective" Gil-Lafuente et al. (2020).

In this argumentative framework, the objective of the study is to present a mathematical model on the causes and effects that influence business informality in the municipality of Cartago-Valle-Colombia, approached from the forgotten effects model of Kaufmann and Gil Aluja, come close to with the Support from a team of ten experts in the business area, who through the valuation of previously defined and contrasted variables, allowed to identify some forgotten effects, which if not taken into account, would affect the incidence relationships necessary for adequate decision-making of public and private policy of the municipality.

2 Informality as a Heterogeneous Phenomenon

The business structure is considered an engine of economic growth and a determining factor of wealth and employment. According to the categorization by size, worldwide, companies are classified into micro, small, medium and large companies; whose contributions to the business fabric are marked by the characteristics of each typology.

One of the most relevant characteristics of –mipymes[3]- is that they serve as suppliers to large companies, in addition to generating added value in terms of generating

[2] The term informal sector or business informality was born in the early 1970s, as an economic-social phenomenon that brings together a set of activities to survive the economic crisis. The ILO (1993) describes it as "a set of units dedicated to the production of goods or the provision of services with the primary purpose of creating jobs and generating income for the people who participate in this activity" (p. 2).

[3] Mipyme is the acronym for micro, small and medium enterprises. It corresponds to smaller independent units, whose definition varies according to the country, but which mostly involves variables such as the number of employees, the sales value and the economic sector, among others.

employment and income. Similarly, this category of company is characterized by the fact that the vast majority of them work in informality[4], which according to De Soto (1989) "occurs when the law imposes rules that exceed the socially accepted normative framework, not it protects the expectations, choices and preferences of those who cannot comply with such rules and the State does not have sufficient coercive capacity "(p. 12).

According to the United Nations Organization (2017), MSMEs are considered "the backbone of most of the world's economies", given that "they represent more than 90% of all companies, generate between 60% and 70% of employment and they are responsible for 50% of the Gross Domestic Product (GDP) worldwide" (Agencia EFE 2018). However, despite the significant contribution to the growth and development of the regions, a large percentage of microenterprises and SMEs around the world (74%) work in informality (Agencia EFE 2018), negatively compromising productivity, competitiveness, innovation, growth and economic and social development of the regions.

According to Porter (1991) productivity is one of the variables with the greatest incidence on competitiveness. From this premise, the limitations inherent to informal companies, in terms of personnel, technological management, organizational capacity, strategic competence, among others (Sáez de Viteri 2000), generate barriers that make it difficult to penetrate new markets, creating difficulties in the access to development credit, as well as wasted training opportunities, minimal use of production and information systems. In this way, the low levels of productivity of informal companies limit the opportunities to grow and compete in environments of formalization, corroborating the negative relationship between informality and the productivity of companies (Perry et al. 2007).

On the other hand, innovation considered one of the engines of the development of the economy and one of the greatest limitations of companies that develop in informality, represents one of the fundamental elements for the expansion of new markets, which translates into greater profitability and income, development and social welfare, hat according to Bueno Campos (1988) intangible[5] assets are increasingly relevant in the creation of value and the growth possibilities of companies.

In line with the above, proposing the causal relationship of business informality in the municipality of Cartago-Valle-Colombia, from the theoretical perspective of the forgotten effects of Kaufmann and Gil Aluja (1988), allows establishing the incidence of business informality in the economic growth and development of the municipality of Cartago-Valle-Colombia, through the cause-effect relationship; involving those elements that define informality and that, if forgotten or not taken into account, would affect the design of policies and strategies in favor of business formalization.

---

4 For Schneider and Enste (2000) and Mincer (1976) informality is conceived as a phenomenon that brings together those productive activities of legal goods and services that have not been registered with the authorities or that fail to comply with fiscal and tax policies. In addition, it covers aspects such as not having a commercial registry, not having accounting, not making contributions to social security, not having paid taxes and not making any type of expense in licenses or commercial registry (Cárdenas and Mejía 2007, p. 8).

5 Among intangible assets, technological intangibles stand out, guarantors of expenses in innovation and development, in which companies invest.

3 Theory of Forgotten Effects

The models for the investigation of forgotten effects arise as a result of the vague subsets theory[6], initially worked on Lotfi Zadeh (1965) and supported on the principles of uncertainty and the subjective of Kaufmann and Gil Aluja (1986) as elements applicable to the multiple fields of action in uncertain environments.

Kaufmann and Gil Aluja (1988) point out that when trying to examine or solve a problem, "there is always a certain risk of negligence, or recklessness, which should be avoided. Risks are not always explicit, visible or perceived immediately: they happen sometimes, and the most terrible is that they are hidden, that they are nothing more than effects on effects, of an accumulation of causes" (p. 7).

In this way, incidence relationships are generated, which "propagate in a network of chains in which many stages are omitted and conclusions are forgotten, more or less voluntarily separated" (Kaufmann and Gil Aluja 1988).

3.1 Description of the Model

According to Kauffman and Gil Aluja (1988), the forgotten effects model arises from causal relationships and effects of the elements of two sets:

$$A = \{a_i/i = 1, 2, 3, \ldots, n\} \text{ act as causes} \tag{1}$$

$$B = \left\{b_j/j = 1, 2, 3, \ldots, m\right\} \text{ act as effects} \tag{2}$$

Supported in the causal relationship, it is considered that there is an incidence of (a_i) on (b_j), if the value of the characteristic function of membership of the pair (a_i, b_j) is valued[7] at [0, 1]. In this way, the degree of incidence of each (a_i) on (b_j), is defined by the function:

$$\mu : AxB \rightarrow [0, 1] \tag{3}$$

in which,

$$\forall\left(a_i, b_j\right) \in AxB; \mu\left(a_i, b_j\right) \in [0, 1] \tag{4}$$

The matrix of direct or first-order incidents then arises, which shows the causal relationships between the elements of set A, which act as causes, and the elements of set B, as effects, expressing the relationship through the matrix $[\underset{\sim}{M}]$, shown in Fig. 1.

In the search to recover the incidence levels that were forgotten or were not manifest in the previous stage, a new set of elements (C) is determined, which operate as effects of the set (B):

$$C = \{c_k|k = 1, 2, \ldots, \mathrm{p}\}. \tag{5}$$

6 For Kaufmann and Gil Aluja (1986) "The theory of fuzzy subsets is a part of mathematics that is perfectly adapted to the treatment of both the subjective and the uncertain" (p. 18).

7 For Gil-Lafuente et al. (2017), valuation is a subjective concept similar to a numerical assignment, expressed in truth levels and confidence intervals [0,1] that allows obtaining correspondences.

$[\underset{\sim}{M}] =$

	b_1	b_2	....	b_m
a_1	$\mu_{a_1b_1}$	$\mu_{a_1b_2}$	....	$\mu_{a_1b_m}$
a_2	$\mu_{a_2b_1}$	$\mu_{a_2b_2}$	....	$\mu_{a_2b_m}$
$\vdots$	$\vdots$	$\vdots$	$\vdots$	....
a_n	$\mu_{a_nb_1}$	$\mu_{a_nb_2}$	....	$\mu_{a_nb_m}$

Fig. 1. Direct incidence matrix. Source: adapted from Kauffman and Gil Aluja (1988)

$[\underset{\sim}{N}] =$

	c_1	c_2	....	c_p
b_1	$\mu_{b_1c_1}$	$\mu_{b_1c_2}$	....	$\mu_{b_1c_p}$
b_2	$\mu_{b_2c_1}$	$\mu_{b_2c_2}$	....	$\mu_{b_2c_p}$
$\vdots$	$\vdots$	$\vdots$	$\vdots$	$\vdots$
b_m	$\mu_{b_mc_1}$	$\mu_{b_mc_2}$	....	$\mu_{b_mc_p}$

Fig. 2. Direct incidence matrix. Source: adapted from Kaufmann and Gil Aluja (1988)

In Fig. 2 the incidence matrix $[\underset{\sim}{N}]$ is represented.

With the purpose of outlining a new relation of incidences of (A) on (C), and using the operator of maximum and minimum [Max. - Min.], A new incidence relation $\underset{\sim}{P}$ is established between the elements $[A]$ and $[C]$, expressed as:

$$\underset{\sim}{P} = [\underset{\sim}{M}] \circ [\underset{\sim}{N}]; \tag{6}$$

in which,

$\underset{\sim}{P}$: establishes causal relationships between the elements of the set (A) and the elements of the set (C), also considering the elements of the set (B).

[o]: corresponds to the symbol that represents the composition [Max. - Min.] Of two uncertain relationships, where:

$$\forall (a_i, c_p) \in AXB,\ (a_i, c_p)_{M \circ N} = \bigvee_{b_j} \left(\mu_M (a_i, b_j) \bigwedge \mu_N (b_j, c_p) \right) \tag{7}$$

3.2 Direct and Indirect Causal Relationship

In order to consider the cause-effect relationships that may be voluntarily or involuntarily hidden, progress is made in the process, taking into account the incidence relationships defined in the previous phase.

To get a new matrix that reveals the indirect incidences, relationships of causes over causes and effects on effects are established, generating a square matrix $[\underset{\sim}{A}]$ and a square matrix $[\underset{\sim}{B}]$, of this shape:

$$[\underset{\sim}{A}] = \{\mu_{a_i a_j} \in [0, 1]/i, \mathrm{j} = 1, 2, \ldots.., n\}, \text{ that relates causes over causes} \tag{8}$$

$$[\underset{\sim}{B}] = \{\mu_{b_i b_j} \in [0, 1]/i, \mathrm{j} = 1, 2, \ldots.., \mathrm{m}\}, \text{ that relates effects over effects} \tag{9}$$

It should be noted that these matrices are reflective, that is, any of their elements (cause-effect), affects itself.

$$\mu_{a_i a_j} = 1; \forall i, \ j = 1, 2, \ldots, \mathrm{n} \tag{10}$$

$$\mu_{b_i b_j} = 1; \forall i, \ j = 1, 2, \ldots, \mathrm{m} \tag{11}$$

Giving continuity to the process, the accumulated effects matrix $[\underset{\sim}{M}^*]$ is generated. This matrix is elaborated from the convulation of maxima and minima and the direct incidence matrices $[\underset{\sim}{M}]$ and indirect $[\underset{\sim}{A}, \underset{\sim}{B}]$, generating incidents between second generation causes and effects, like this:

$$[\underset{\sim}{M}^*] = [\underset{\sim}{A}] \circ [\underset{\sim}{M}] \tag{12}$$

$$[\underset{\sim}{M}^*] = [\underset{\sim}{A}] \circ [\underset{\sim}{M}] \circ [\underset{\sim}{B}] \tag{13}$$

Starting from the difference between the accumulated effects matrix $[\underset{\sim}{M}^*]$ and the direct incidence matrix $[\underset{\sim}{M}]$, the forgotten effects matrix $[\underset{\sim}{O}]$ (Fig. 3), which allows evidence of forgetfulness in causal relationships, expressed as:

$$[\underset{\sim}{O}] = [\underset{\sim}{M}^* - \underset{\sim}{M}] \tag{14}$$

Finally, the analysis of the forgotten effects matrix is carried out taking into account that "as the value obtained between an element of (a_i) and another of (b_j), within the matrix $[\underset{\sim}{O}]$ is higher, the forgotten effect between the two is higher; which implies a greater possibility of error for the subject who makes decisions"(Gil Lafuente et al. 2017, p. 44).

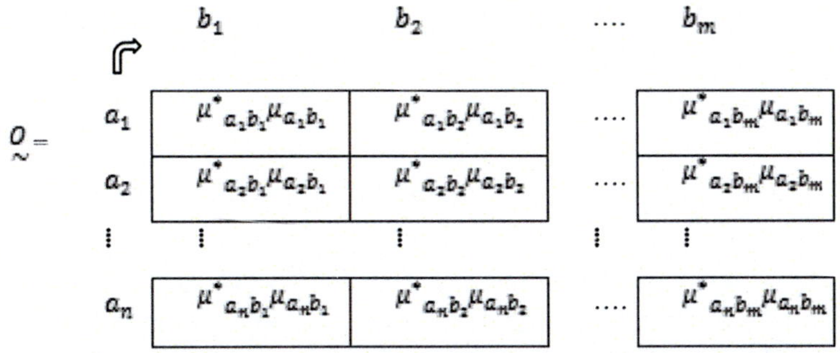

Fig. 3. Forgotten effects matrix. Source: adapted from Kaufmann and Gil Aluja (1988)

4 Case Analysis

The case of analysis corresponds to the application of the forgotten effects model of Kaufmann and Gil Aluja in (1988) in the causal relationship of business informality in the municipality of Cartago Valle del Cauca Colombia, a city with an approximate area of 260 km^2 and a total population of 135,729 people, of which 97% are concentrated in the municipal seat[8] (Urban) and only 3% in the rural sector (DANE 2018). The municipality is characterized by particularities in its socioeconomic dynamics, among which an unemployment rate of 31.2% and a marked concentration of micro and small businesses stand out, most of which coexist in informality and are oriented to commercial activity (54.8%) and the provision of services (13.14%). Only 6.93% of the establishments are dedicated to Industry and 25.13% to various activities such as agriculture, construction and other services (Cartago Chamber of Commerce 2017).

In this context, the incidence of informality in the economic growth of the municipality of Cartago, from the perspective of the forgotten effects of Kaufmann and Gil Aluja (1988), is relevant given that "in real life there is always the possibility of stopping consider or forget voluntarily or not some causal relationships that are not always clear, so they are not perceived in the analysis and problem solving processes" (Gil Lafuente et al. 2017. p. 53).

For the application of the model, a group of 10 experts in the business area was formed, who, using the Fuzzy-Delphi[9] method, evaluated the variables defined for the study, considering those that could be relegated or discarded due to omission or forgetfulness.

Two groups of elements were then taken into consideration (Table 1):

A: criteria that define business informality and act as causes.

[8] Capital or main population of a territory.

[9] For Gil-Lafuente et al. (2017) "Delphi is a scientific forecasting method based on the opinions of experts, created by the Rand Corporation in California in 1964." (p. 31). For the authors, the method is based on three characteristics: "the anonymity" of the experts, "the iteration and controlled feedback" and "the response of the group is presented in statistical form, showing not only the point of view of the majority, but all opinions indicating the degree of agreement obtained". (p. 31–32).

B: elements of the economic environment, which determine economic growth and development and act as effects.

Table 1. Defined criteria for the analysis. Source: Authors' elaboration

Causes		Effects	
a_1	Excessive bureaucratic procedures	b_1	Productivity
a_2	Lack of knowledge about the formalization procedures and requirements	b_2	Competitiveness
a_3	Formalization costs	b_3	Employment generation
a_4	Business size	b_4	Internationalization
a_5	Lack of ethics and values	b_5	Market expansion
a_6	Affiliation to subsidized health system	b_6	Financial sector
a_7	Ignorance of tax benefits	b_7	Social security system
a_8	Orientation towards an economic sector	b_8	Quality of products and services
a_9	Oppressive legal and regulatory framework	b_9	Innovation
a_{10}	Weak State Supervision and Enforcement Capacity	b_{10}	Nation's revenue
a_{11}	Weak productive structure	b_{11}	Investment
a_{12}	Unemployment		
a_{13}	Poverty in the agricultural sector and a forced displacement		
a_{14}	Education level		
a_{15}	Ease of access to the informal sector		

Using the endecadarian scale proposed by Kaufmann and Gil Aluja (1988), the group of experts continues the valuation process, based on the causal relationship that occurs between the elements of set A and the elements of set B, using semantic scales that define levels of truth in confidence intervals valued between $\{0, 1\}$ (Table 2).

From the relationship of causes and effects of the first generation, between the criteria that define business informality and act as causes and the elements that require the growth and economic development of the municipality and act as effects, the first incidence matrix is established direct $[\underset{\sim}{M}]$, shown in the table (Table 3).

Consistent with the approaches of the model, and in order to establish the incidents that directly or indirectly affect the elements that define business informality and economic growth, square matrices are generated, which relate causes to causes $[\underset{\sim}{A}]$ and effects on effects and $[\underset{\sim}{B}]$, expressed in Tables 4 and 5.

Table 2. Semántic Scal. Source: adapted from Kaufmann and Gil Aluja (1988)

Level	Label
0	Without incidence
0,1	Virtually no incident
0,2	Almost without incident
0,3	Very weak incidence
0,4	Weak incidence
0,5	Medium incidence
0,6	Considerable incidence
0,7	Quite incidence
0,8	Strong incidence
0,9	Very strong incidence

Table 3. Estimated incidents between causes and effects. Source: prepared by the authors, based on expert valuation

↱	E_1	E_2	E_3	E_4	E_5	E_6	E_7	E_8	E_9	E_{10}	E_{11}
C_1	0,6	0,7	0,8	0,9	0,7	0,6	0,8	0,3	0,1	0,8	0,7
C_2	0,7	0,8	0,6	0,8	0,6	0,8	0,8	0,6	0,2	0,7	0,6
C_3	0,9	0,8	0,9	0,8	0,5	0,5	0,8	0,5	0,1	0,8	0,7
C_4	0,8	0,9	0,8	0,8	0,8	0,9	0,8	0,7	0,8	0,8	0,8
C_5	0,5	0,5	0,6	0,6	0,5	0,3	0,5	0,8	0,2	0,7	0,7
C_6	0,1	0,1	0,2	0,1	0,1	0,1	0,8	0,1	0,1	0,7	0,2
C_7	0,7	0,7	0,8	0,7	0,6	0,5	0,6	0,4	0,6	0,6	0,5
C_8	0,7	0,8	0,8	0,7	0,7	0,5	0,5	0,6	0,7	0,7	0,8
C_9	0,8	0,8	0,8	0,7	0,7	0,7	0,8	0,7	0,7	0,7	0,7
C_{10}	0,6	0,6	0,7	0,5	0,4	0,7	0,9	0,6	0,2	0,8	0,6
C_{11}	0,9	0,8	0,9	0,8	0,8	0,8	0,8	0,8	0,8	0,8	0,8
C_{12}	0,8	0,8	1	0,7	0,6	0,6	0,8	0,7	0,6	0,8	0,8
C_{13}	0,7	0,7	0,5	0,4	0,5	0,4	0,5	0,5	0,1	0,7	0,6
C_{14}	0,7	0,7	0,8	0,7	0,7	0,5	0,2	0,7	0,7	0,7	0,6
C_{15}	0,6	0,6	0,8	0,5	0,5	0,7	0,8	0,5	0,5	0,7	0,7

Table 4. Incidence between the different causes. Source: prepared by the authors, based on expert valuation

↱	C1	C2	C3	C4	C5	C6	C7	C8	C9	C10	C11	C12	C13	C14	C15
C1	1	0,6	0,7	0,6	0,4	0,4	0,5	0,2	0,6	0,4	0,6	0,6	0,4	0,1	0,8
C2	0,1	1	0,2	0,1	0,1	0,8	0,9	0,2	0,1	0,1	0,6	0,6	0,5	0,1	0,8
C3	0,1	0,2	1	0,8	0,4	0,9	0,1	0,4	0,3	0,3	0,8	0,5	0,4	0,1	0,8
C4	0,7	0,8	0,9	1	0,1	0,9	0,8	0,7	0,1	0,1	0,6	0,7	0,2	0,6	0,7
C5	0,3	0,5	0,1	0,4	1	0,5	0,1	0,1	0,8	0,8	0,8	0,5	0,7	0,3	0,8
C6	0,2	0,1	0	0	0	1	0	0	0,1	0,1	0,6	0,1	0	0	0
C7	0,1	0,2	0,1	0,5	0,2	0,4	1	0,1	0,3	0,1	0,8	0,7	0,6	0,1	0,2
C8	0,5	0,2	0,7	0,8	0	0,1	0,5	1	0,1	0,1	0,7	0,6	0,8	0,5	0,6
C9	0,9	0,7	0,8	0,7	0,8	0,7	0,4	0,5	1	0,5	0,8	0,8	0,9	0,4	0,8
C10	0,5	0,7	0,7	0,2	0,8	0,7	0,7	0,1	0,9	1	0,8	0,7	0,8	0,7	0,9
C11	0,7	0,6	0,7	0,8	0,7	0,8	0,5	0,8	0,6	0,5	1	0,9	0,8	0,7	0,8
C12	0,3	0,4	0,3	0,9	0,7,	0,9	0,7	0,8	0,3	0,1	0,9	1	0,9	0,8	0,8
C13	0,1	0,1	0,1	0,1	0,5	0,9	0,1	0,7	0,1	0,1	0,6	0,8	1	0,7	0,8
C14	0,2	0,7	0,2	0,7	0,5	0,7	0,8	0,7	0,6	0,6	0,8	0,8	0,7	1	0,7
C15	0,5	0,6	0,4	0,8	0,5	0,7	0,7	0,8	0,3	0,1	0,8	0,7	0,7	0,6	1

Table 5. Incidence between the different effects. Source: prepared by the authors, based on expert valuation

↱	E1	E2	E3	E4	E5	E6	E7	E8	E9	E10	E11
E1	1	0,8	0,8	0,7	0,8	0,7	0,6	0,8	0,8	0,8	0,7
E2	0,8	1	0,7	0,8	0,8	0,6	0,5	0,8	0,8	0,8	0,7
E3	0,6	0,6	1	0,5	0,6	0,5	0,6	0,5	0,4	0,7	0,5
E4	0,8	0,8	0,7	1	0,9	0,6	0,6	0,8	0,8	0,8	0,7
E5	0,7	0,8	0,8	0,8	1	0,5	0,5	0,8	0,8	0,8	0,7
E6	0,6	0,7	0,5	0,6	0,6	1	0,3	0,3	0,5	0,6	0,6
E7	0,6	0,6	0,7	0,4	0,5	0,2	1	0,1	0,1	0,6	0,3
E8	0,7	0,8	0,7	0,8	0,8	0,6	0,4	1	0,7	0,6	0,7
E9	0,9	0,9	0,8	0,8	0,7	0,6	0,3	0,9	1	0,6	0,6
E10	0,7	0,7	0,8	0,7	0,7	0,7	0,8	0,6	0,7	1	0,8
E11	0,8	0,7	0,8	0,7	0,8	0,7	0,6	0,7	0,7	0,7	1

Using the Fuzzy Log[10] program and based on the composition of maximums and minimums, in addition to the matrices $[\underset{\sim}{M}]$, $[\underset{\sim}{A}]$, $[\underset{\sim}{B}]$, it is created the maximum and minimum convulation matrix $[\underset{\sim}{M}^{*}]$, which relates second generation causes and effects (Table 6).

$$[\underset{\sim}{M}^{*}] = [\underset{\sim}{A}] \circ [\underset{\sim}{M}] \tag{15}$$

$$[\underset{\sim}{M}^{*}] = [\underset{\sim}{A}] \circ [\underset{\sim}{M}] \circ [\underset{\sim}{B}] \tag{16}$$

Table 6. Maximum - minimum convolution between matrices. Source: Prepared by the authors, using the Fuzzy Log program

↱	E_1	E_2	E_3	E_4	E_5	E_6	E_7	E_8	E_9	E_{10}	E_{11}
C_1	0,7	0,7	0,8	0,9	0,7	0,7	0,8	0,6	0,6	0,8	0,7
C_2	0,7	0,8	0,8	0,8	0,6	0,8	0,8	0,6	0,6	0,7	0,7
C_3	0,9	0,8	0,9	0,8	0,8	0,8	0,8	0,8	0,8	0,8	0,8
C_4	0,9	0,9	0,9	0,8	0,8	0,9	0,8	0,7	0,8	0,8	0,8
C_5	0,8	0,8	0,8	0,8	0,8	0,8	0,8	0,8	0,8	0,8	0,8
C_6	0,6	0,6	0,6	0,6	0,6	0,6	0,8	0,6	0,6	0,7	0,6
C_7	0,8	0,8	0,8	0,8	0,8	0,8	0,8	0,8	0,8	0,8	0,8
C_8	0,8	0,8	0,8	0,8	0,8	0,8	0,8	0,7	0,8	0,8	0,8
C_9	0,8	0,8	0,8	0,9	0,8	0,8	0,8	0,8	0,8	0,8	0,8
C_{10}	0,8	0,8	0,8	0,8	0,8	0,8	0,9	0,8	0,8	0,8	0,8
C_{11}	0,9	0,8	0,9	0,8	0,8	0,8	0,8	0,8	0,8	0,8	0,8
C_{12}	0,9	0,9	1	0,8	0,8	0,9	0,8	0,8	0,8	0,8	0,8
C_{13}	0,8	0,8	0,8	0,7	0,7	0,7	0,8	0,7	0,7	0,8	0,8
C_{14}	0,8	0,8	0,8	0,8	0,8	0,8	0,8	0,8	0,8	0,8	0,8
C_{15}	0,8	0,8	0,8	0,8	0,8	0,8	0,8	0,8	0,8	0,8	0,8

[10] Fuzzy Log is a fuzzy logic tool that allows the development and processing of models based on the mathematics of uncertainty to recover forgotten effects in causal relationships, providing reliable results in graphical and numerical form.

In the same way, with the accumulated effects matrix is generated, which includes intervening incidents between causes and second-generation effects (Table 7).

Table 7. Cumulative effects. Source: prepared by the authors, using the Fuzzy Log program

↱	E_1	E_2	E_3	E_4	E_5	E_6	E_7	E_8	E_9	E_{10}	E_{11}
C_1	0,8	0,8	0,8	0,9	0,9	0,7	0,8	0,8	0,8	0,8	0,8
C_2	0,8	0,8	0,8	0,8	0,8	0,8	0,8	0,8	0,8	0,8	0,7
C_3	0,9	0,8	0,9	0,8	0,8	0,8	0,8	0,8	0,8	0,8	0,8
C_4	0,9	0,9	0,9	0,8	0,8	0,9	0,8	0,8	0,8	0,8	0,8
C_5	0,8	0,8	0,8	0,8	0,8	0,8	0,8	0,8	0,8	0,8	0,8
C_6	0,7	0,7	0,7	0,7	0,7	0,7	0,8	0,6	0,7	0,7	0,7
C_7	0,8	0,8	0,8	0,8	0,8	0,8	0,8	0,8	0,8	0,8	0,8
C_8	0,8	0,8	0,8	0,8	0,8	0,8	0,8	0,8	0,8	0,8	0,8
C_9	0,8	0,8	0,8	0,9	0,9	0,8	0,8	0,8	0,8	0,8	0,8
C_{10}	0,8	0,8	0,8	0,8	0,8	0,8	0,9	0,8	0,8	0,8	0,8
C_{11}	0,9	0,8	0,9	0,8	0,8	0,8	0,8	0,8	0,8	0,8	0,8
C_{12}	0,9	0,9	1	0,8	0,8	0,9	0,8	0,8	0,8	0,8	0,8
C_{13}	0,8	0,8	0,8	0,8	0,8	0,7	0,8	0,8	0,8	0,8	0,8
C_{14}	0,8	0,8	0,8	0,8	0,8	0,8	0,8	0,8	0,8	0,8	0,8
C_{15}	0,8	0,8	0,8	0,8	0,8	0,8	0,8	0,8	0,8	0,8	0,8

Finally, to reflect oversights in relations of coincidence, matrix effects forgotten is generated as a result of the difference between the matrix cumulative effects $[\underset{\sim}{M}^*]$ and matrix directly affect $[\underset{\sim}{M}]$ (Table 8).

$$[\underset{\sim}{O}] = [\underset{\sim}{M}^* - \underset{\sim}{M}] \tag{17}$$

Table 8. Forgotten effects. Source: own elaboration using the Fuzzy Log program

	E_1	E_2	E_3	E_4	E_5	E_6	E_7	E_8	E_9	E_{10}	E_{11}
C_1	0,2	0,1	0	0	0,2	0,1	0	0,5	0,7	0	0,1
C_2	0,1	0	0,2	0	0,2	0	0	0,2	0,6	0,1	0,1
C_3	0	0	0	0	0,3	0,3	0	0,3	0,7	0	0,1
C_4	0,1	0	0,1	0	0	0	0	0,1	0	0	0
C_5	0,3	0,3	0,2	0,2	0,3	0,5	0,3	0	0,6	0,1	0,1
C_6	0,6	0,6	0,5	0,6	0,6	0,6	0	0,5	0,6	0	0,5
C_7	0,1	0,1	0	0,1	0,2	0,3	0,2	0,4	0,2	0,2	0,3
C_8	0,1	0	0	0,1	0,1	0,3	0,3	0,2	0,1	0,1	0
C_9	0	0	0	0,2	0,2	0,1	0	0,1	0,1	0,1	0,1
C_{10}	0,2	0,2	0,1	0,3	0,4	0,1	0	0,2	0,6	0	0,2
C_{11}	0	0	0	0	0	0	0	0	0	0	0
C_{12}	0,1	0,1	0	0,1	0,2	0,3	0	0,1	0,2	0	0
C_{13}	0,1	0,1	0,3	0,4	0,3	0,3	0,3	0,3	0,7	0,1	0,2
C_{14}	0,1	0,1	0	0,1	0,1	0,3	0,6	0,1	0,1	0,1	0,2
C_{15}	0,2	0,2	0	0,3	0,3	0,1	0	0,3	0,3	0,1	0,1

Table 9. Cause-effect relationships associated with forgotten effects with intervening variables

Variables with forgetting effect	interposed key variables		Variables with degree of involvement
Excessive bureaucratic procedures	Weak productive structure	Internationalization	
Formalization costs	Business size	Productivity	
	Unemployment	Competitiveness	Innovation
Poverty in the agricultural sector and forced displacement		Market expansion	

Following Gil-Lafuente et al. (2015) "It is common for these incidence relationships to remain hidden because they are effects on effects, there is an accumulation of causes that provoke them" (p. 107).

Next, a summary of the cause-effect relationships is presented, associated with the forgotten effects and the most outstanding intervening variables, since they involve degrees of incidence equal to or greater than 0.7, which is the value closest to unity (Table 9).

Given that the results provide the elements that have contributed to the indirect effects, the cause-effect relationship associated with the forgotten effect costs of formalization is presented below:

4.1 Cause-Effect Relationship Associated with Forgotten Effect Costs of Formalization

The incidence relationship (a3–b9) indicates that in the first instance the team of experts established a valuation of (0.1) in the incidence of formalization costs on innovation; However, the results reveal that this relationship increases to 0.8 given the intervention of some variables such as business size, productivity and internationalization as intervening elements, which in turn affect innovation, accumulating effects in the causal relationship. Thus, the high costs of formalization become a barrier for the expansion of the market, which in a context of interaction hinders the innovative capacity (Fig. 4).

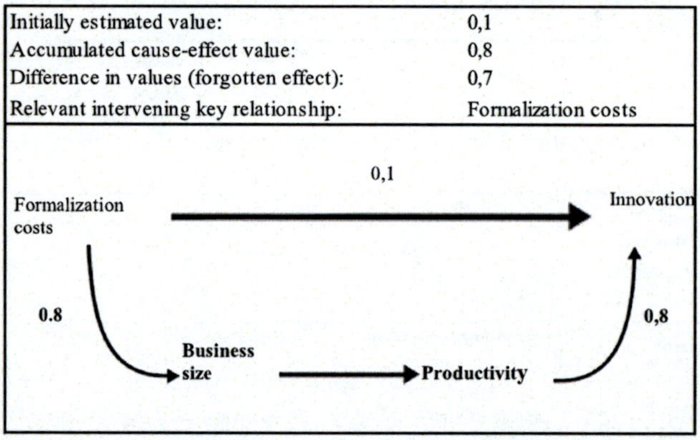

Fig. 4. Relationship of incidence costs of formalization – innovation. Source: prepared by the authors, using the Fuzzy Log program.

Figure 5. Represents the total direct for this relation of causalaty incidences.

The results coincide with some approaches of the ILO (2017) when it maintains that the non-compliance of the formalization requirements by the companies is due to some extent to the inability to cover the formalization costs. In the same way, they show a close relationship with the Colombian and North Valley of the Cauca socioeconomic reality, characterized by a weak business structure, with a high concentration of micro, small and medium-sized enterprises, of which a high percentage is developed in informality, and

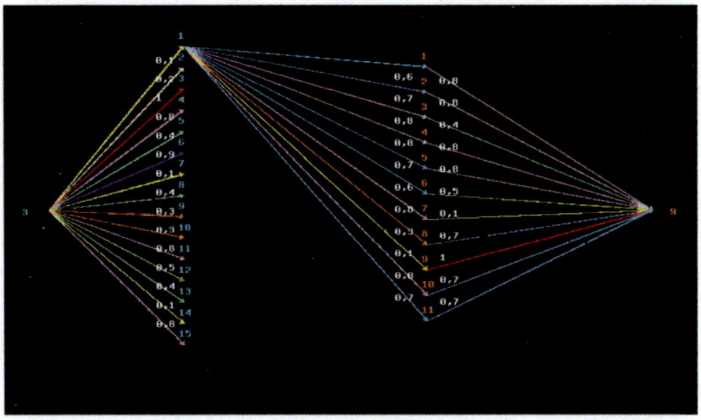

Fig. 5. Total incidences of formalization costs on innovation

they identify due to low levels of productivity, competitiveness and innovation, avoiding local economic growth and development.

5 Conclusions

The application of the theory of forgotten effects to the study of business informality in the municipality of Cartago, made it possible to show the relationships of causality and effect, based on some forgotten variables such as excessive bureaucratic procedures, formalization costs and poverty in the sector. agriculture-forced displacement, and the incidence of innovation, as an inherent element of economic growth, which involves relationships and interrelationships between companies and the institutional regulatory framework, regulating and conditioning business formalization activities, which with the application of other methodologies are not easily discernible, despite being of great importance for the global understanding of the phenomenon in question, given that they are relevant for the design and intervention of public and private policies.

Based on the results and the conclusions raised, new methodologies will be used for future research, which will open the field of knowledge to new researching alternatives of the business fabric, in favor of local economic growth and development.

References

Alfaro, G., Zaragoza, A., Alfaro, V., Gil Lafuente, A.M.: Innovation capabilities and innovation systems: a forgotten effects analysis of their components. Adv. Intell. Syst. Comput. (894), 51–62 (2020)

Bueno, C.: El capital intangible como clave estratégica en la competencia actual. Bol. Estud. Econ. **53**(164), 205–229 (1998)

Cámara de Comercio Cartago: Informe económico de los municipios perteneciente a la jurisdicción de la Cámara de Comercio de Cartago (2017). https://issuu.com/camaradecomerciocartago/docs/informe_econ__mico_2016_o.e. Accessed 09 Mar 2020

Cárdenas, M., Mejía, C.: Informalidad en Colombia: nueva evidencia. Coyuntura Econ. **72**(2), 27–54 (2007)

DANE: Resultados Censo Nacional de Población y Vivienda (2018). https://www.dane.gov.co/index.php/estadisticas-por-tema/demografia-y-poblacion/censo-nacional-de-poblacion-y-vivienda-2018. Accessed 25 Nov 2019

Delfin, F., Acosta, M.; Importancia y análisis del desarrollo empresarial. Pensamiento gestión (40), 184–202 (2016)

De Soto, H.: The Other Path: The Invisible Revolution in the Third World. Harper and Row publishers, New York (1989)

Gil Lafuente, A.M., González, F., Flores, B.: Teoría de los efectos olvidados en la incidencia de la actividad económica en la calidad de vida de los habitantes y cuantificación de los efectos para un reequilibrio territorial. Inceptum **10**(19), 105–124 (2015)

Gil Lafuente, A.M., Keropyan, A., Balvey, J.: Analysis of the relationship between the economic activity of regions and welfare of its inhabitants. Econ. Comput. Econ. Cybern. Stud. Res. **49**(3), 215–232 (2015)

Gil Lafuente, A.M., García, I., Souto, L., Blanco, B., Ortiz, M., Zamora, T.; Management and decision-making in the Cuban business system. Royal Academy of Economic and Financial Sciences, Barcelona (2017)

Gil Lafuente, A.M., Amiguet, L., Torres, A.: Modelo de efectos olvidados en el análisis estratégico de medios de comunicación. Inquietud Empresarial **20**(1), 73–85 (2020)

González, F., Flores, B., Gil Lafuente, A.M., Amiguet, J.: Teoría de los efectos olvidados y su aplicación en el desarrollo de la empresa. Cuadernos Cimbage **19**(2), 51–77 (2017)

Kaufmann, A., Gil Aluja, J.: Introducción de la teoría de los subconjuntos borrosos a la gestión de las empresas. Ed. Milladoiro, Santiago de Compostela (1986)

Kaufmann, A., Gil Aluja, J.: Modelos para la investigación de efectos olvidados. Ed. Milladoiro, Vigo (1988)

Mincer, J.: Unemployment effects of minimum wages. J. Polit. Econ. **84**(4), S87-104 (1976)

Merigó, J.M., Gil Lafuente, A.M., Yager, R.R.: An overview of fuzzy research with bibliometric indicators. Appl. Soft Comput. **27**, 420–433 (2015)

ILO. Resolution concerning statistics of employment in the informal sector, adopted by the Fifteenth International Conference of Labour Statisticians (1993). https://www.ilo.org/wcmsp5/groups/public/---dgreports/---stat/documents/normativeinstrument/wcms_087484.pdf. Accessed 07 May 2020

ILO. Statistical definition of informal employment: Guidelines on a statistical definition of informal employment, adopted by the Seventeenth International Conference of Labor Statisticians (2003). https://ilo.org/public/english/bureau/stat/download/papers/def.pdf. Accessed 02 Oct 2019

ILO. The Informal Economy and Decent Work: A Policy Resource Guide Supporting the Transition to Formality (2013). Recovered from https://www.ilo.org/wcmsp5/groups/public/---ed_emp/---emp_policy/docume. Accessed 22 May 2020

ILO. Enterprise Formalization (2017). https://www.ilo.org/wcmsp5/groups/public/---ed_emp/---emp_ent/---ifp_seed/documents/publication/wcms_549523.pdf. Accessed 17 Feb 2020

Porter, M.: The competitive advantage of the nations. Plaza y Janes, Buenos Aires (1991)

Perry, G., Maloney, W., Arias, O., Fajnzylber, P., Mason, A., Saavedra, J.: Informality: Exit and Exclusion. World Bank, Washington (2007)

Portes, A., Castells, M., Benton, L.: The Informal Economy: Studies in Advanced and Less Advanced Developed Countries. John Hopkins University Press, Baltimore (1989)

Quemba, J.: Labor informality in the colombian rural sector. Económgr. Escuela Econ. **124**, 1–39 (2018) de Viteri, D.S.: The competitive potential of the company. Invest. Eur. Dirección Econ. Empresa **6**(3), 71–86 (2000)

Schneider, F., Enste, D.: Shadow economies: size, causes and consequences. J. Econ. Lit. **38**(1), 77–114 (2000)

Zadeh, L.A.: Fuzzy sets. Inf. Control **8**, 338–353 (1965)

Adaptation of Grocery Stores to the Post-Covid-19 Environment. The Case of Barcelona

Emili Vizuete-Luciano(✉), Sefa Boria-Reverter, Maria Luisa Solé-Moro, and Anna Maria Gil-Lafuente

Faculty of Business and Economics, University of Barcelona, Avda. Diagonal 690, 08034 Barcelona, Spain
evizuetel@ub.edu

Abstract. In order to help the decision-making to the owners of grocery stores of less than 150 m^2 in Barcelona, this work presents a decision support system based in the Forgotten Effects Theory of 701 stores at the City. COVID-19 is changing consumption patterns and Grocery stores must adapt to the new reality by improving customer service, product exposure, cordiality… in order not to see their market share reduced due to the growth of online sales of food products, they also have large distributors as competitors who offer a greater variety of products to their customers since they have a much larger sales area. The focus in on the importance given in each District to the different parameters that we have determined. The experiment carried out allows us to determine the second-generation effects, which are more indirect. The results highlight the differences between the different Districts in the assessment of the different determinants of our study.

Keywords: Forgotten Effects Theory · Consumer behavior · Retail · COVID-19

1 Introduction

From the moment the first company was born, they have adapted to the different changes that have taken place in the environment in order to survive. With the passage of time, technological innovation and the fluidity with which information travels nowadays has made the changes that take place much faster in their execution as well as the possible threats that must be faced. Traditional strategies can no longer respond to these changes and therefore companies must direct their efforts towards achieving competitive advantages that allow them to differentiate themselves from their competitors and thereby achieve recognition from the market and their customers.

In addition, the financial crisis that has particularly affected the euro zone countries with higher levels of indebtedness such as Greece, Italy, Portugal and Spain; has given rise to a process where Consumers have ended up being much more demanding with

E. Vizuete-Luciano—Serra Húnter Fellow.

A. M. Gil-Lafuente et al. (Eds.): REDCID 2019, LNNS 388, pp. 288–300, 2022.
https://doi.org/10.1007/978-3-030-93787-4_17

what they buy, they value the immediate availability of the product, with long hours and all this at an adjusted price [8].

In the food distribution sector, competition is very high and in Barcelona (Spain), large distributors and international food chains have a relevant position within it. The reality that these establishments face in recent decades has evolved at breakneck speed due to the appearance of online sales; and that in May 2020 grew with respect to the previous year by 30.6% [18], being of the analyzed channels (Hypermarket, Supermarket, Discount Store, Traditional Store and other channels), which has had better behavior. If we analyze the value per channel in 2019 [19], the most important are Supermarkets with 47.8% of the total followed by Traditional Stores with 17.0%.

Buyers have at their disposal different channels to make their purchases and each of the channels offers its advantages and disadvantages in order to retain their customers.

In this context, Traditional Stores or also known as Grocery Stores, have acquired greater relevance in large Spanish cities since there has been a deregulation of hours in most of them and this has been highly valued by consumers who demand products of convenience in increasingly extended hours, including holidays.

Grocery Stores are establishments with a surface area of less than 150 m^2, with interrupted opening hours from 09:00 to 00:00 or beyond, generally run by immigrants who have come to the city with the hope of getting a better life, and They have different categories of food, personal hygiene, home cleaning, fruit and vegetables, as well as bread sales.

Citizens of Pakistan run most of these establishments in the city of Barcelona since the financial crisis that began in 2007. Pakistanis are very enterprising people who see this type of establishment as an opportunity to start their own business and thus have a better life. Another factor that can explain this growth is that Barcelona is a cosmopolitan city, a recipient of immigration and considered one of the most attractive cities in Europe to develop business.

At the end of 2019, the first news emerged about some cases of pneumonia in Asia, that rapidly spread worldwide, bringing SARS-CoV-2 to our cities, and lives, which was later, renamed COVID-19. The Spanish Authorities decide the total confinement of the population in mid-March 2020 in order to stop the high number of infections that were taking place in the country, leaving the Food, Energy and Financial Sector as essential services; As a result, online commerce grows exponentially [26]. A few months after said confinement, in September COVID-19 is still in our cities that are preparing to face the so-called second wave of the Pandemic.

Because of the above, Spaniards have changed their consumption patterns and as a sign of this, 37% of the population does not feel comfortable visiting a Shopping Center at this time [1], so their purchases have been transfer to local commerce and e-commerce. Although these changes have been, observe throughout the world [20], home delivery has also increased despite the difficulties it represents in terms of logistics [10].

The possibility that they adapt their commercial offer, improving different points such as foreign sales, distribution at the point of sale and others, can be an important boost to position themselves strongly in their neighborhoods in the next post-COVID era that we are going to experience soon. Understanding the specific factors that can have a positive impact on the satisfaction of unplanned or convenience shoppers can help these small businesses target these consumers, especially when there are economic difficulties like the current ones [30].

The objective of our work is to verify the relationship between the theory related to forgotten effects and its application in everyday Grocery stores smaller than 150 m^2 in the city of Barcelona. To do this, in Sect. 2 we carry out a thorough review of concepts about food establishments, consumer behavior and the city of Barcelona; from it, the hypotheses are empirically analyzed emerge. In Sect. 3, we develop the Theory of Forgotten Effects. In Sect. 4 we present the study carried out, we specify the methodology used and its results. Finally, in Sect. 5 we show the conclusions regarding the main findings obtained in the study.

2 Review of Concepts

2.1 Grocery

In retail, as in many disciplines, proximity has been a theme developed by several authors who have focused on proximity in the purchase of food, applied to convenience stores and the like [14].

On the other hand, the customer demands healthy products and there is generally less availability within the convenience store format due to their smaller size, which means losing sales in favor of supermarkets and hypermarkets [42].

Grocery retailers also have competition from Discount Supermarkets and Hypermarkets, which have very competitive prices [24]. In Spain, they are the market leaders [39] and, in addition, the loyalty programs that large distributors implement with their customers have been a success on different occasions, increasing both the average ticket and the frequency of purchase [13, 31].

Food stores are dynamic in nature and therefore adapt more quickly to changes in their environment. The current scenario discovers the need for an organizational framework to develop and implement strategies at the point of sale regarding the market, the company, the store and the customer [29].

How the consumer is cared for by the employee or owner is essential in most occasions and for this reason cordiality, uniformity and attention is essential to achieve high levels of satisfaction [38], is what will give place to the one determined as customer experience and that will be fundamental to increase the repetition of sales [27].

2.2 Consumer Behavior

From the Consumer point of view, it has been shown that grocery buyers rely on sight, taste, touch and smell in the act of searching, finding, sharing and navigating information during the buying process; therefore, a good organization and distribution of the point of sale is essential [33]. Therefore, an increasingly sensitive consumer with their environment has begun to demand local and sustainable products, being willing to pay a higher price for them [5, 22, 23, 37, 40]; since specialists warn of the alarming change in food purchase patterns in Spain [2].

Online shopping has grown a lot in recent months due to the effect of COVID19, although previously an attempt has been made to understand the motivational factors, social influences and perceived trust that influence consumers' intention to buy food online [12, 35]. It has also been observed that online purchases made from mobile devices such as Smartphone or Tablet, makes the number of orders placed throughout the year higher among those customers who tend to make smaller purchases [43], or if they carry out purchases of a certain size, they end up spacing them out in time [34].

Although in most cases it has costs associated with home delivery [7], the price factor is less relevant than at the point of sale [4], in addition to the associated problems (pollution, increased traffic…) at high levels of pollution [11], can pose obstacles to its development in large cities.

Being able to enjoy long opening hours makes the consumer less sensitive to price and this is due to the long working hours that citizens may have in certain countries and particularly in Spain, which make family reconciliation difficult [25].

Consumer behavior at the point of sale is highly influenced by the comfort and well-being that they can achieve while making their purchase. The use of good lighting, as well as proper cleaning and attentive staff can create an optimal environment for the customer to stay and return to the store in the near future [9]. In addition, if the point of sale has good conditions, the products are untidy, there is no information…, the consumer will be confused and this will lead them to buy what is strictly necessary, so their average ticket will be low [15].

2.3 Barcelona

It is a Spanish city on the shores of the Mediterranean and which occupies a strategic position in Southern Europe, in 2019, it has a population of more than 1.63 million people do and a surface area of 101.35 km^2, which represents a population density of 158.3 inhabitants per hectare and that we can consider as high [21]. It is considered one of the main tourist destinations worldwide and in 2019, 52.68 MM of passengers arrived by plane, while 4.63 MM did so by boat and 4.30 MM in the Barcelona-Madrid High Speed Train [32].

The city is distributed in 10 districts and 73 neighborhoods [16] (Fig. 1):

Barcelona is one of the main cities to receive investment and for years it has been a focus of investment of a real estate and business nature, being within the Top Ten in Europe [36]. Thus, with the passage of time it is also a city that receives immigration; mainly from South America with 84.461, Southern Europe with 48.949 people and from South Asia with 39.880 [3]; what has turned it into a cosmopolitan and open city where many entrepreneurs come to Barcelona to start their business; as it is a land of opportunity.

We highlight that the Barcelona Metropolitan Area has a population of around 5.00 million people, which makes it the area with the largest number of inhabitants on the Mediterranean coast and one of the most remarkable in Europe. Being specialized in construction, services, tourism, manufacturing and commerce [6].

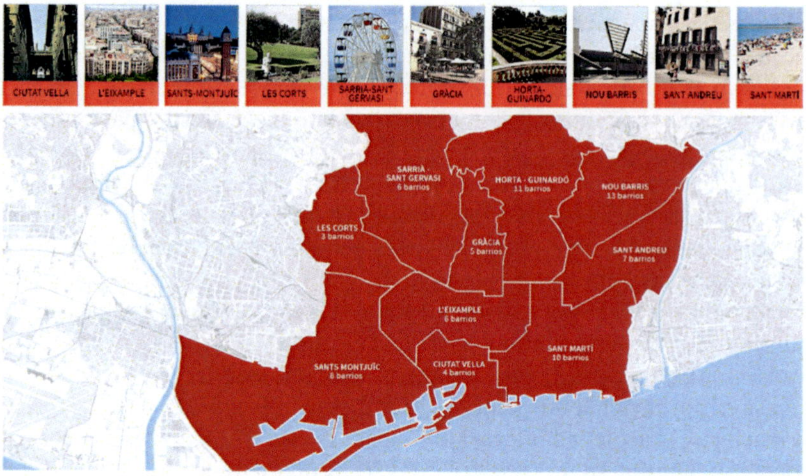

Fig. 1. Barcelona Map. https://www.barcelona.cat/es/vivir-en-bcn/vive-el-barrio

3 The Forgotten Effects Theory

When we carry out an activity be it personal, professional…, it is subject to some degree to a cause-effect relationship [28] since we are living beings and we live in a system that works in harmony.

In business life, many times the managers of the company may not consider or voluntarily omit some causal relationships that not perceived or are not directly evident. These relationships may be hidden and due to this, the result obtained in the decisions will be far from the expected. The hidden effects generated by the decisions taken in the first place must be valued, since this will lead us more easily to the achievement of the proposed objectives. For this reason, it will be advisable to use decision tools and models that allow the decision-maker to work with the information in its entirety, in order to determine the possible direct and indirect causal relationships that can be observe.

Not considering the possible incidents, they will result in secondary effects that will generate a future impact on the incidence relationships and therefore on the results expected by the Managers or Entrepreneurs [41] and that they can suppose great divergences between these and the really obtained ones.

Incidence, being a subjective concept, is difficult to parameterize [17], but if we consider it, the decision-maker will be able to achieve their objectives much faster and with it success in decision-making. In order to define the Theory of Forgotten Effects, as a starting point, we will observe two sets of elements: $A = \{a_i/i = 1, 2, \ldots, n\}$ and $B = \{b_j/j = 1, 2, \ldots, m\}$, we will define the incidence of a_i on b_j when the value of the membership function of (a_i, b_j) is between [0, 1]. Thus, the degree of incidence of a_i on b_j can be express with the function:

$$\mu : AxB \to [0, 1],\ for\ \forall(a_i, b_j) \in AxB, \mu(a_i, b_j) \in [0, 1] \tag{1}$$

The Matrix that we will call $\tilde{M}$, the "Direct Incident Matrix" determined by the set of pairs (a_i, b_j) obtained. In Table 1, it can be observe the cause-effect relationships that will occur between the elements of A (causes) and B (effects):

Table 1. Direct incident matrix

$\tilde{M} =$

	b_1	b_2	b_3	b_4	...	b_m
a_1	$\mu_{a_1b_1}$	$\mu_{a_1b_2}$	$\mu_{a_1b_3}$	$\mu_{a_1b_4}$	...	$\mu_{a_1b_m}$
a_2	$\mu_{a_2b_1}$	$\mu_{a_2b_2}$	$\mu_{a_2b_3}$	$\mu_{a_2b_4}$	...	$\mu_{a_2b_m}$
a_3	$\mu_{a_3b_1}$	$\mu_{a_3b_2}$	$\mu_{a_3b_3}$	$\mu_{a_3b_4}$	...	$\mu_{a_3b_m}$
a_4	$\mu_{a_4b_1}$	$\mu_{a_4b_2}$	$\mu_{a_4b_3}$	$\mu_{a_4b_4}$	...	$\mu_{a_4b_m}$
a_5	$\mu_{a_5b_1}$	$\mu_{a_5b_2}$	$\mu_{a_5b_3}$	$\mu_{a_5b_4}$	...	$\mu_{a_5b_m}$
a_6	$\mu_{a_6b_1}$	$\mu_{a_6b_2}$	$\mu_{a_6b_3}$	$\mu_{a_6b_4}$	...	$\mu_{a_6b_m}$
⋮	⋮	⋮	⋮	⋮	⋮	⋮
a_n	$\mu_{a_nb_1}$	$\mu_{a_nb_2}$	$\mu_{a_nb_3}$	$\mu_{a_nb_4}$	...	$\mu_{a_nb_m}$

The cause-effect relationships observed between A and B represent the matrix of direct incidence, or first order. These relationships are obtain by establishing the relationships of some elements with others and from them the forgotten effects can be analyze.

To establish the incidence relationship, we use the max-min operator on the same matrix $\tilde{M}$, where the sub-indices $i, j, i \neq j$ for $\mu_{a_i,a_j} \neq \mu_{a_j,a_i}$ and $\mu_{b_i,b_j} \neq \mu_{b_j,b_i}$. The result obtained is a new matrix $\left[\tilde{M}^*\right]$, which determines the incidents between the causes and the second-generation effects; considering the possible impact that they may have a posteriori.

From the different options that we have for calculating indirect or forgotten effects, we will choose the arithmetic difference $\left[\tilde{M}^*\right] - \left[\tilde{M}\right] = \left[\tilde{\gamma}\right]$. Where $\left[\tilde{M}^*\right]$, contains the first and second-generation effects and $\left[\tilde{M}\right]$, contains the direct effects. The subtraction between matrices will show us the forgotten effects that we have discovered.

4 The Study

In our work, we have carried out an investigation on the reality of the Grocery establishments smaller than 150 m^2 in the city of Barcelona, in order to size and understand the market of these establishments of up to 150 m^2 with free hours.

We can define them as "small supermarkets" that is; they have the categories of food, personal hygiene and cleaning for the home. With Self-service and them often sell fruit, vegetables and bread.

The daily Grocery stores of less than 150 m^2 in each of the districts of the city of Barcelona have been counted and a sample of 701 establishments has been observed, to determine the basic characteristics of each one of them.

Distributed among the different Districts (Fig. 2):

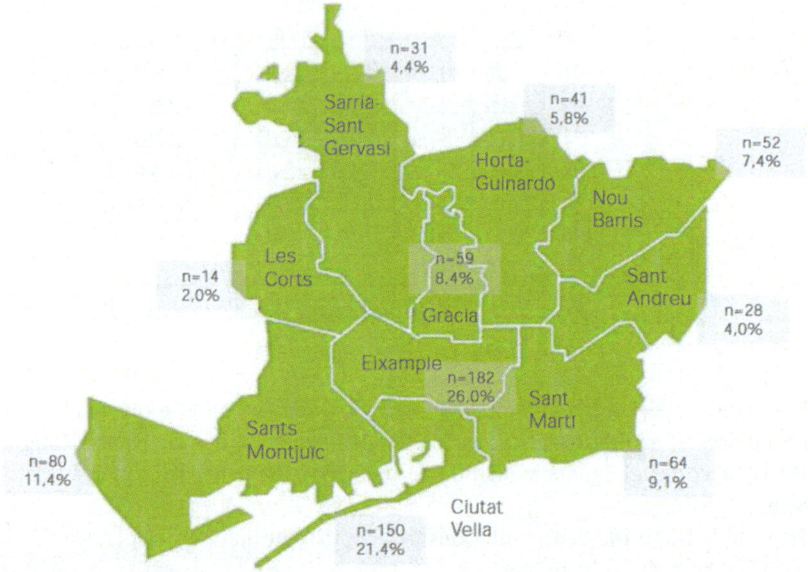

Fig. 2. Barcelona Map with number of observations in our study per districts.

Various studies have been carry out to determine the different attributes that the population in Spain takes into account when shopping for food. [2] Determined as reasons for choosing a certain type of food store among others: Food Quality, Proximity, Price, Home delivery, Freshness, Variety, Availability, Origin, Dependency, Personalized service and routines.

Three general blocks and their determinants have been consider in our study:

EXTERNAL APPEARANCE

- Opening hours
- Establishment poster
- Exterior cleaning
- Exhibition and sale of products outside the establishment

INTERNAL ORGANIZATION

- Comfort
- Good distribution and organization
- The cleanliness of the establishment

- The products correctly labelled

CUSTOMER SERVICE

- Cordiality when entering and leaving the store
- Uniformity of employees
- The care received

With the data obtained, we have determined the direct effects matrix $\left[\tilde{M}\right]$, incorporating a fictitious column with the average value of the observations made in each of the concepts that we have determined, and that we use to understand the reality of the city of Barcelona understood as a whole, see Table 2:

Table 2. Direct incident matrix observed

	Cause \ Effect	Districts										
		Ciutat Vella	Eixample	Nou Barris	Gràcia	Horta-Guinardó	Les Corts	Sant Andreu	Sant Martí	Sants - Montjuïc	Sarià Sant. Gervasi	Barcelona average
External appearance	Opening hours	0,546	0,478	0,635	0,441	0,561	0,429	0,572	0,406	0,488	0,613	0,517
	Establishment poster	0,953	0,951	0,788	0,932	0,878	0,929	0,750	0,952	0,788	0,968	0,889
	Exterior cleaning	0,847	0,786	0,731	0,864	0,610	0,786	0,714	0,578	0,788	0,968	0,767
	Exhibition outside	0,080	0,423	0,346	0,288	0,463	0,286	0,429	0,359	0,213	0,161	0,305
Internal organization	Comfort	0,364	0,368	0,356	0,407	0,305	0,500	0,286	0,266	0,463	0,419	0,373
	Good distribution	0,874	0,877	0,760	0,848	0,817	0,929	0,805	0,766	0,901	0,920	0,849
	Cleaning	0,913	0,830	0,769	0,780	0,634	0,857	0,893	0,781	0,213	0,161	0,683
	Correctly labeled	0,807	0,731	0,712	0,915	0,854	0,786	0,679	0,656	0,575	0,871	0,759
Customer service	Cordiality	0,782	0,818	0,636	0,707	0,652	0,727	0,631	0,749	0,774	0,786	0,726
	Uniformity	0,047	0,071	0,019	0,119	0,049	0,000	0,000	0,047	0,038	0,065	0,046
	Care received	0,683	0,803	0,816	0,778	0,821	1,000	0,583	0,455	0,639	0,643	0,722

Then we will proceed to calculate the max-min of the matrix $\left[\tilde{M}\right]$ on itself, obtaining the matrix $\left[\tilde{M}^*\right]$, see Table 3:

In order to determine the forgotten effects, we will apply the arithmetic difference between the matrices $\left[\tilde{M}^*\right] - \left[\tilde{M}\right] = \left[\tilde{\gamma}\right]$ and we will obtain Table 4, which is a new matrix that shows us the forgotten effects in cause-effect relationships:

In Table 5, we present the top 10 second-generation Forgotten Effects that we have determined in the study we have carried out:

The highest value of the forgotten effects that corresponds to the lack of exterior cleaning in the Sant Martí District with 0.986; we can see that directly in the Table 5 made from the valuations observed in the sample is 0.714, which shows us the increase in the importance of this factor for the residents of said district.

Table 3. Incident Matrix between causes and second-generation effects

	Effect / Cause	Districts										
		Ciutat Vella	Eixample	Nou Barris	Gràcia	Horta-Guinardó	Les Corts	Sant Andreu	Sant Martí	Sants - Montjuïc	Sarià Sant. Gervasi	Barcelona average
External appearance	Opening hours	0,635	0,635	0,635	0,635	0,610	0,635	0,635	0,578	0,635	0,635	0,635
	Establishment poster	0,951	0,951	0,816	0,932	0,878	0,929	0,805	0,951	0,901	0,951	0,889
	Exterior cleaning	0,786	0,788	0,786	0,786	0,786	0,786	0,786	0,786	0,786	0,786	0,786
	Exhibition outside	0,429	0,429	0,429	0,429	0,429	0,463	0,429	0,429	0,463	0,423	0,429
Internal organization	Comfort	0,500	0,500	0,500	0,500	0,500	0,500	0,500	0,500	0,500	0,500	0,500
	Good distribution	0,877	0,877	0,816	0,877	0,877	0,929	0,805	0,877	0,901	0,920	0,877
	Cleaning	0,893	0,857	0,788	0,848	0,830	0,857	0,893	0,830	0,857	0,857	0,849
	Correctly labeled	0,786	0,786	0,760	0,786	0,786	0,786	0,786	0,766	0,786	0,786	0,786
Customer service	Cordiality	0,818	0,818	0,788	0,818	0,818	0,818	0,750	0,818	0,788	0,818	0,818
	Uniformity	0,080	0,119	0,119	0,119	0,119	0,119	0,119	0,119	0,119	0,119	0,119
	Care received	0,874	0,877	0,788	0,848	0,817	0,929	0,805	0,803	0,901	0,920	0,849

Table 4. Forgotten effects

	Effect / Cause	Districts										
		Ciutat Vella	Eixample	Nou Barris	Gràcia	Horta-Guinardó	Les Corts	Sant Andreu	Sant Martí	Sants - Montjuïc	Sarià Sant. Gervasi	Barcelona average
External appearance	Opening hours	0,089	0,954	0,483	0,626	0,404	0,899	0,753	0,026	0,729	0,284	0,525
	Establishment poster	0,378	0,981	0,498	0,324	0,971	0,675	0,784	0,480	0,016	0,564	0,567
	Exterior cleaning	0,938	0,800	0,970	0,039	0,259	0,389	0,812	0,986	0,766	0,514	0,647
	Exhibition outside	0,387	0,594	0,009	0,651	0,689	0,319	0,368	0,922	0,581	0,269	0,479
Internal organization	Comfort	0,779	0,953	0,025	0,675	0,631	0,830	0,336	0,343	0,656	0,936	0,616
	Good distribution	0,770	0,674	0,842	0,191	0,274	0,316	0,490	0,946	0,667	0,087	0,526
	Cleaning	0,060	0,081	0,790	0,369	0,012	0,041	0,628	0,362	0,539	0,791	0,367
	Correctly labeled	0,639	0,391	0,119	0,001	0,943	0,017	0,690	0,406	0,781	0,217	0,420
Customer service	Cordiality	0,900	0,582	0,690	0,137	0,084	0,677	0,185	0,772	0,952	0,395	0,537
	Uniformity	0,264	0,430	0,065	0,176	0,203	0,835	0,683	0,515	0,748	0,969	0,489
	Care received	0,298	0,127	0,828	0,131	0,072	0,869	0,207	0,790	0,703	0,678	0,470

Table 5. Top 10 Forgotten effects

Cause	Effect	Valuation
Cleaning	Sant Martí	0,986
Establishment poster	Eixample	0,981
Establishment poster	Horta-Guinardó	0,971
Cleaning	Nou Barris	0,970
Uniformity	Sarrià Sant Gervasi	0,969
Opening hours	Eixample	0,954
Comfort	Eixample	0,953
Cordiality	Sants-Montjuïc	0,952
Good distribution	Sant Martí	0,946
Correctly labeled	Horta-Guinardó	0,943

Although the values observed in the relationships of the forgotten effects increase those that we have observed directly, we highlight the importance of Uniformity in the District of Sarrià-Sant Gervasi where the value of the forgotten effect is 0.969, while in the sample it only gave it a score of 0.065.

5 Conclusions

We have observed that the city of Barcelona has a high number of Grocery stores, shops that tend to have unregulated opening hours and that sell food products, as well as fresh products. In 2020, at the height of COVID-19, where outbreaks are occurring in many cities in Europe and which has resulted in an almost total stoppage of tourism, as well as long-distance trips, it is having devastating consequences. For cities clearly oriented to tourism and the services sector such as Barcelona. In the short and medium term, they represent an increase in unemployment, as well as a reduction in the disposable income of families.

These establishments have a very high competition and adapting to the needs of their customers, as well as considering their preferences is essential. Our study has shown notable differences between the ratings observed in the first place and the determination of the forgotten effects; it has also shown us that there are notable inequalities in the assessment given to each of the causes in the different districts of Barcelona.

For example, exterior cleaning has been a cause that we have observed in most of the city's districts, without the income level of the citizens having special relevance. Second, the cause that has generated the greatest forgotten effects in our study is the establishment's poster; we emphasize that these establishments tend to have unclear or attractive signs that attract the attention of their customers, and the largest are observed in the Eixample and Horta-Guinardó districts.

On the other hand, we highlight that in the District with a higher income, which is Sarrià Sant Gervasi, the cause that generates the greatest forgotten effect is the uniformity of the employees; we understand that it is due to the concept of a store of proximity and trust that has so much diffusion among its inhabitants.

Finally, we highlight the forgotten effects that we have determined in the Eixample District, which being one of the most cosmopolitan and open we consider it normal to consider these causes within the high quality of service that they demand in their purchases.

According to these evidences, the analyzed establishments should promote changes in the orientation of their businesses since it is observed that the quality of service is essential to increase their sales and maintain their customers since risks such as sales currently threaten them online that with COVID-19 is growing exponentially.

For this, the design of specific interventions that mitigate these negative effects among their clients is essential.

References

1. Accenture Homepage. https://newsroom.accenture.es/es/news/el-37-por-ciento-de-los-espa%C3%B1oles-es-reticente-acudir-centros-comerciales.htm. Accessed 05 Oct 2020
2. Achon, M., Serrano, M., Garcia-Gonzalez, A., Alonso-Aperte, E., Varela-Moreiras, G.: Present food shopping habits in the Spanish adult population: a cross-sectional study. Nutrients **9**(5), 508 (2017)
3. Ajuntament de Barcelona Homepage. https://www.bcn.cat/estadistica/catala/dades/inf/pobest/pobest20/part2/index.htm. Accessed 05 Oct 2020
4. Anesbury, Z., Nenycz-Thiel, M., Dawes, J., Kennedy, R.: How do shoppers behave online? An observational study of online grocery shopping. J. Consum. Behav. **15**(3), 261–270 (2016)
5. Arce-Urriza, M., Cebollada, J., Tarira, M.F.: The effect of price promotions on consumer shopping behavior across online and offline channels: differences between frequent and non-frequent shoppers. Inf. Syst. E-Bus. Manag. **15**(1), 69-87 (2017)
6. Baldasano, J.M.: COVID-19 lockdown effects on air quality by NO2 in the cities of Barcelona and Madrid (Spain). Sci. Total Environ. **741**, 140353 (2020)
7. Belavina, E., Girota, K., Kabra, A.: Online grocery retail: revenue models and environmental impact. Manag. Sci. **63**(6), 1781–1799 (2017)
8. Bellini, S., Cardinali, M.G., Grandi, B.: A structural equation model of impulse buying behavior in grocery retailing. J. Retail. Consum. Serv. **36**, 164–171 (2017)
9. Bercik, J., Horska, E., Wang, R.W.Y., Chen, Y.C.: The impact of parameters of store illumination on food shopper response. Appetite **106**, 101–109 (2016)
10. Campbell, A.M., Savelsbergh, M.V.P.: Decision support for consumer direct grocery initiatives. Transp. Sci. **39**(3), 313–327 (2005)
11. Carrabs, F., Cerulli, R., Sciomachen, A.: An exact approach for the grocery delivery problem in urban areas. Soft. Comput. **21**(9), 2439–2450 (2016). https://doi.org/10.1007/s00500-016-2406-5
12. Elms, J., de Kervenoael, R., Hallsworth, A.: Internet or store? An ethnographic study of consumers' internet and store-based grocery shopping practices. J. Retail. Consum. Serv. **32**, 234–243 (2016)
13. Filipe, S., Marques, S.H., Salgueiro, M.D.: Customers relationship with their grocery store: direct and moderating effects from store format and loyalty programs. J. Retail. Consum. Serv. **37**, 78–88 (2017)
14. Gahinet, M.C.: The dimensions of proximity applied to food purchases. Revue d'Economie Regionale et Urbaine **5–6**, 1367–1390 (2018)

15. Garaus, M., Wagner, U., Kummer, C.: Cognitive fit, retail shopper confusion, and shopping value: empirical investigation. J. Bus. Res. **68**(5), 1003–1011 (2015)
16. Garcia, X., Garcia-Sierra, M., Domene, E.: Spatial inequality and its relationship with local food environments: The case of Barcelona. Appl. Geogr. **115**, 102140 (2020)
17. Gil-Lafuente, A.M.: Fuzzy Logic in Financial Analysis. Springer Verlag, Berlin (Germany) (2005)
18. Gobierno de España Homepage. https://www.mapa.gob.es/es/prensa/ultimas-noticias/los-hogares-incrementan-sus-compras-de-alimentos-en-supermercados-y-tiendas-tradicionales-/tcm:30-538155. Accessed 21 Oct 2020
19. Gobierno de España: Informe del consumo de alimentación en España 2019 (2019)
20. Grashuis, J., Skevas, T., Segovia, M.S.: Grocery shopping preferences during the COVID-19 Pandemic. Sustainability **12**, 5369 (2020)
21. Gutiérrez, J., García-Palomares, J.C., Romanillos, G., Salas-Olmedo, M.H.: The eruption of Airbnb in tourist cities: comparing spatial patterns of hotels and peer-to-peer accommodation in Barcelona. Tour. Manag. **62**, 278–291 (2017)
22. Hoek, A.C., Pearson, D., James, S.W., Lawrence, M.A., Friel, S.: Healthy and environmentally sustainable food choices: consumer responses to point-of-purchase actions. Food Qual. Prefer. **58**, 94–109 (2017)
23. Hoek, A.C.: Healthy and environmentally sustainable food choices: consumer responses to point-of-purchase actions. Food Qual. Prefer. **58**, 94–106 (2017)
24. Hökelekli, G., Lamey, L., Verboven, F.: The battle of traditional retailers versus discounters: the role of PL tiers. J. Retail. Consum. Serv. **39**, 11–22 (2017)
25. Hunneman, A., Verhoef, P.C., Sloot, L.M.: The moderating role of shopping trip type in store satisfaction formation. J. Bus. Res. **78**, 133–142 (2017)
26. IAB Spain: Estudio eCommerce (2020)
27. Ismana-Ilisan, C.M.: The main new driver of customer experience in grocey retail – the fresh opportunity. Bull. Transilvania Univ. Brasov Ser. V – Econ. Sci. **10**(2), 133–140 (2017)
28. Kaufmann, A., Gil Aluja, J.: Modelos para la investigación de efectos olvidados. Milladoiro, Santiago de Compostela (1988)
29. Kumar, V., Anand, A., Song, H.: Future of retailer profitability: an organizing framework. J. Retail. **93**(1), 96–119 (2017)
30. Martínez-Ruiz, M.P., Blazquez-Resino, J.J., Pino, G., Capestro, M.: Store attributes and consumer satisfaction with unplanned purchases: an empirical research in Spain. In: Martínez-López, F., Gázquez-Abad, J., Ailawadi, K., Yagüe-Guillén, M. (eds.) Advances in National Brand and Private Label Marketing. Springer Proceedings in Business and Economics, pp. 19–27. Springer, Cham (2017). https://doi.org/10.1007/978-3-319-59701-0_3
31. Meyer-Waarden, L.: Effects of loyalty program rewards on store loyalty. J. Retail. Consum. Serv. **24**, 22–32 (2015)
32. Observatori del Turisme a Barcelona: Perfil i habits dels turistes a la ciutat de Barcelona 2018–2019 (2019)
33. Ocepek, M.G.: Sensible shopping: a sensory exploration of the information environment of the grocery store. Libr. Trends **66**(3), 371–394 (2018)
34. Oliveira-Castro, J., James, V.K., Foxall, G.R.: Postreinforcement pause in grocery shopping: comparing interpurchase time across products and consumers. Pshichological Rec. **57**, 483–500 (2007)
35. Pauzi, S.F.F.B., Chin, T.A., Choon, T.L., Sulaiman, Z.: Motivational factor for online grocery shopping. Adv. Sci. Lett. **23**(9), 9140–9144 (2017)
36. Price Waterhouse Coopers: Emerging Trends in Real Estate. Climate of change (2020)
37. Printezis, I., Grebitus, C.: Marketing channels for local food. Ecol. Econ. **152**, 161–171 (2018)
38. Soderlund, M.: The proactive employee on the floor of the store and the impact on customer satisfaction. J. Retail. Consum. Serv. **43**, 46–53 (2018)

39. Statista: Leading supermarkets in Spain (2019)
40. Stranieri, S., Ricci, E.C., Banterle, A.: Convenience food with environmentally-sustainable attributes: a consumer perspective. Appetite **116**, 11–20 (2017)
41. Vizuete-Luciano, E., Gil-Lafuente, A.M., García-González, A., Boria-Reverte, S.: Forgotten effects of corporate social and environmental responsibility a case study of Catalonian economy. Kybernetes **42**(5), 736–753 (2013)
42. Volpe, R., Jaenicke, E.C., Chenarides, L.: Store formats, markets structure and consumers food shopping decisions. Appl. Econ. Perspect. Policy **40**(4), 672–694 (2018)
43. Wang, R.J.H., Malthouse, E.C., Krishnamurthi, L.: On the go: How movile shopping affects customer purchase behavior. J. Retail. **91**(2), 217–234 (2015)

Author Index

A. M. Gil-Lafuente et al. (Eds.): REDCID 2019, LNNS 388, p. 301, 2022.
https://doi.org/10.1007/978-3-030-93787-4